Benchmark Papers in Genetics Series

Editor: David L. Jameson—University of Houston

(GENES, PROTEINS, AND CELLULAR AGING)

Edited by

ROBIN HOLLIDAY
National Institute for Medical Research
London, England

A Hutchinson Ross Publication

VNR VAN NOSTRAND REINHOLD COMPANY
New York

Manufactured in the United States of America.

Published by Van Nostrand Reinhold Company Inc.
115 Fifth Avenue
New York, New York 10003

Van Nostrand Reinhold Company Limited
Molly Millars Lane
Wokingham, Berkshire RG11 2PY, England

Van Nostrand Reinhold
480 Latrobe Street
Melbourne, Victoria 3000, Australia

Macmillan of Canada
Division of Gage Publishing Limited
164 Commander Boulevard
Agincourt, Ontario MIS 3C7, Canada

15   14   13   12   11   10   9   8   7   6   5   4   3   2   1

Library of Congress Cataloging-in-Publication Data
Main entry under title:
Genes, proteins, and cellular aging.
   (Benchmark papers in genetics; v. 17)
   Reprinted from various sources.
   "A Hutchinson Ross publication."
   Includes index.
   1. Cells—Aging—Addresses, essays, lectures.   2. Cytogenetics—Addresses,
essays, lectures.   3. Proteins—Addresses, essays, lectures.   I. Holliday,
R. (Robin), 1932-   II. Series: Benchmark papers in genetics; 17.   [DNLM:
1. Aging—collected works.   2. Cytogenetics—collected works.   3. Proteins—
metabolism—collected works.   W1 BE516 v.17 / WT 104 G327]
QH608.G46   1986   574.87'61   85-17823
ISBN 0-442-23246-2

CONTENTS

v

Contents

SERIES EDITOR'S FOREWORD

The study of any discipline assumes the mastery of the literature of the subject. In many branches of science, even one as new as genetics, the expansion of knowledge has been so rapid that there is little hope of learning of the development of all phases of the subject. The student has difficulty mastering the textbook; the young scholar must tend to the literature near his own research; the young instructor barely finds time to expand his horizons to meet his class-preparation requirements; the monographer copes with wider literature but usually from a specialized viewpoint; and the textbook author is forced to cover much the same material as previous and competing texts to respond to the user's needs and abilities.

Few publishers have the dedication to scholarship to serve primarily the limited market of advanced studies. The opportunity to assist professionals at all stages of their careers has been recognized by Hutchinson Ross, and by a distinguished group of editors knowledgeable in specific portions of the genetic literature. These editors have selected papers and portions of papers that demonstrate both the development of knowledge and the atmosphere in which that knowledge was developed. There is no substitute for reading great papers. Here you can learn how questions are asked, how they are approached, and how difficult and essential it is to obtain definitive answers and clear writing.

Dr. Holliday is a major contributor to the field and his ability to select, collect, and synthesize the literature will be a considerable service to science. The importance of aging increases as other sources of mortality are eliminated or reduced in importance. Dr. Holliday's study should provide an excellent benchmark to what must become an increasing number of investigators.

Hutchinson Ross has provided a valuable service to scholarship with its benchmark series. While this is the last number of the genetics volumes, I am pleased to see that they intend to maintain, as their primary target, advanced studies of scholarly excellence through the publication of research monographs.

It has indeed been a pleasure to me to work with the many authors, volume editors, and publishers.

DAVID L. JAMESON

PREFACE

In many books and monographs, the study of aging is treated as a single discipline, even though it is an extremely diverse field, using a variety of biological systems and experimental procedures. This diversity is documented by work that is mainly descriptive, in the sense that age-related changes are described but cannot easily be interpreted. For the most part, experiments have not been set up to test particular hypotheses or theories that are concerned with both the problem of the initiation of the aging process and its acceleration to the death of the organism. The many manifestations of the aging process that have been so fully documented are necessarily the result of changes at the cellular and molecular level. This volume does not cover the whole field but confines itself to studies of cells and molecules.

In studying aging organisms, the investigator must grapple with the complexities of organs and tissues that have many differentiated cell types, some of which are irreversibly postmitotic, whereas others are dividing or capable of division. In contrast, homogeneous populations of cells with finite growth potential provide many opportunities for quantitative experimental investigations and for the testing of specific theories of aging. Any reasonable definition of aging can certainly be applied to cell populations that die out after a long period of growth, although the relevance of this type of clonal aging to the aging of the whole organism remains to be unraveled.

Many of the publications that are reprinted here are concerned with the development and testing of a particular theory of aging. This is derived from the "central dogma" of molecular biology—namely, that information is encoded in DNA, which is then transferred to RNA and proteins. It proposes that errors in this pathway of information transfer may accumulate under certain circumstances and have progressively deleterious effects, and also that these may include changes in DNA itself. The *general error theory,* as it has come to be known, concerns itself with the build up of all types of errors in macromolecules and the possible interactions between different kinds of defects. Although one of its virtues is that is suggests a variety of experimental tests, these have proved to be more difficult to carry out than was originally appreciated. Error theories were formulated 20 or more years ago, but detailed experimental tests have been largely within the last 12 years or so. I feel that this is an appropriate time to take stock of our present understanding of cell aging, so that the future use of the powerful new techniques of molecular biology will not just lead to the accumulation of more empirical observations but also to the better exploration of existing concepts and theories of aging, or at least to the acquisition of information that can be meaningfully interpreted.

My interest in the field of cellular aging stems from stimulating discussions with scientific colleagues, and many of the published experiments have depended on their hard work and enthusiasm. I would therefore like to thank all of them for their innumerable contributions, and I name them here in the approximate order that they first joined in the overall program of research: Brian Harrison, Leslie Orgel, Cynthia Lewis, Gill Tarrant, Ian Buchanan, Tony Stevens, Zhores Medvedev, Rita Medvedeva, Lily Huschtscha, Tom Kirkwood, Stuart Linn, Mike Kairis, Vince Murray, Clive Bunn, Keen Rafferty, Bob Rosenberger, Khash Khazaie, Alec Morley, Simon Cox, Robert Stellwagen, John Menninger, and Suresh Rattan.

ROBIN HOLLIDAY

CONTENTS BY AUTHOR

GENES, PROTEINS, AND CELLULAR AGING

INTRODUCTION

The origin of the scientific study of aging, or gerontology, can be attributed to the German zoologist, August Weismann. He first made clear the distinction between the germ cells, which are transmitted from generation to generation and are therefore potentially immortal, and the somatic cells, which in most higher organisms have finite longevity. Weismann's important contributions to the field have been recognized in a review by Kirkwood and Cremer (1982).

The many studies of aging in the hundred or so years since Weismann's publications can be assigned to fairly well-defined areas. First, a number of experimental systems have been established, including rodents, fish, insects, nematodes, cultured human and chick cells, as well as simpler organisms such as fungi and protozoa. In addition, biomedical studies of human beings of different age provide, in many cases, a valid means of advancing our understanding of aging.

Second, a variety of age-related changes have been documented using these different biological systems. The range of experimental studies is extremely wide and includes the methods of biochemistry, genetics, physiology, immunology, endocrinology, neurobiology, oncology, histology, and even psychology. Age-related changes have been documented in cells and organelles, in organ and tissue structure and function, in hormones and receptors, in the immune response, in membranes, in intracellular and extracellular proteins, and in chromatin or chromosomes, and so on. The origins of most of these changes remain obscure, however.

Third, there has been much theoretical work on aging, and it has been said that more than two hundred theories of aging have been suggested at one time or another. Although this may well be true, it is doubtful that many of them get to grips with the real biological problem: what determines the intrinsic time-dependent changes that occur during the lifetime of an organism that will finally result in its senescence, physical degeneration, and death?

Most of the contemporary theories that attempt to explain the primary causes of aging fall into two major groups: program theories and error theories. All gerontologists must accept that the maximum lifespan of a species is genetically determined, but many have assumed that this must also mean there is a *genetic program.* In this context, program usually implies that aging is the terminal stage of the normal developmental program that leads to adulthood; that is, it is part and parcel of the whole complex process of development, which ultimately depends on the temporal control of gene activity. Alternatively, a program theory of aging may be more specific, perhaps invoking a clock or time-measuring device that will trigger cell and tissue degeneration in much the same way that cell death occurs during certain developmental stages. The assumption that the genetic determination of lifespan means that aging is programmed is, however, quite false. Organisms are made up of cells, the properties of which depend on the macromolecules they contain. The structure of these is determined by the DNA. If cells survive in one species for a long period and in another for a shorter period, then this difference ultimately resides in their DNA. The genetic determination of aging is compatible with error theories that state that macromolecules accumulate defects with time and that the rate of accumulation is species specific. In fact, a central problem in gerontology is that the same kinds of biochemical changes in the same kinds of molecules occur at very different rates in short- and long-lived species. To give just two examples, 3-year-old rats contain as many cross-links in the collagen of their connective tissue as 90-year-old humans, and both aging organisms may have similar amounts of the "age pigment," lipofuscin, in their brain neurons.

Error theories, as we know them today, are derived from several sources. Following the discovery of genes, chromosomes, and mutations, it was widely realized that the accumulation of random mutations with time would have an increasingly deleterious effect on adult tissues. Several versions of this somatic mutation theory have been proposed, but when they have been put on a sound quantitative basis, it has been difficult to reconcile them with the basic biological facts on aging (Part IV). After the elucidation of the major features of protein synthesis, it was realized, initially by Medvedev, that errors in the synthesis of the RNA message and the translation of the message may also be important components of the aging process, and that there may well be many interactions between defects in DNA and proteins (see Paper 1). Later, it was pointed out by Orgel (Paper 2A) that the accuracy of the synthesis of the protein depends not only on the sequence of DNA, but also on the specificity of enzymes and proteins that are themselves necessary for protein synthesis. He concluded that errors may be perpetuated by a feedback mechanism, which could lead ultimately to an irreversible "error catastrophe" in protein synthesis. The distinction between somatic mutation and protein error theories is very clear, but unfortunately they have often been confused in the literature.

Orgel's theory has frequently been misunderstood because errors at the level of proteins and DNA have been confused. It has often been incorrectly assumed that the error catastrophe is initiated by gene mutations (e.g., Burnet, 1974; Pantelouris, 1978; Gupta, 1980; Bremermann, 1982; see Holliday and Kirkwood, 1983). However, with the discovery that the accuracy of DNA synthesis depends not only on Watson-Crick base pairing but also on the DNA polymerase and associated proteins (Kornberg, 1980), it is now widely realized that errors in proteins will probably also result in DNA replication errors. Thus, what is now referred to as the *general error theory* suggests that error propagation in proteins will lead to deleterious somatic mutations as well as gradual deterioration of organelles, membranes, and other cellular structures and that all these changes may contribute to senescence and subsequent death.

Until recently, the molecular basis of epigenetic mechanisms, which must be involved in the development and maintenance of complex organisms, has remained quite obscure. However, an important advance in this area stems from the realization that modifications to DNA structure, particularly the methylation of cytosine, can occur, and these may provide the basis for the specific control of gene activity. Evidence has accumulated that the absence of 5-methyl cytosine at particular positions in DNA is associated with transcription and its presence, with lack of transcription (for reviews, see Doerfler, 1981, 1983; Riggs and Jones, 1983). If specific control of gene activity depends on the pattern of DNA methylation, then alterations or defects in maintaining that pattern may have serious phenotypic consequences. Thus, as well as errors in DNA sequence and errors in RNA and proteins, one must now take into account "epigenetic errors" in the control of gene activity (see Part IV).

It is important to realize that there is a close connection between the possible accumulation of defects in macromolecules and the various mechanisms for maintaining the integrity of normal structure and function in the adult. Maintenance can keep in check the continual occurrence of errors and defects, but it can break down if they become too frequent. Some theories of aging do not seem to get to grips with the problem of prolonged maintenance followed by its breakdown. For example, the free radical theory does not explain why the many defenses against damage from reactive oxygen radicals are successful for most of adult life, but then become inadequate during senescence.

With regard to the accuracy of biological processes and the maintenance of normal adult functions, a distinction can be made between cellular DNA and proteins that is somewhat analogous to Weismann's distinction between germ line and soma. It is essential that the integrity of DNA sequences is maintained, and this is achieved by two means. First, replication is highly accurate since the frequency of spontaneous mutations measured in a variety of organisms indicates an error level of 10^{-9} to 10^{-11} per nucleotide per cell division (Drake, 1970). Second, all organisms contain an astonishing variety of repair mechanisms that can remove either spontaneous or induced lesions in DNA. Far less information is available about the fidelity of RNA and protein synthesis (for reasons that will be explained in Part III). Several estimates indicate, however, that errors are approximately 10^{-4} per nucleotide for mRNA (messenger RNA) synthesis and translational errors are 10^{-3} to 10^{-4} per amino acid (Kirkwood, Holliday, and Rosenberger, 1984). Thus, there is at least a ten-thousand-fold difference between the accuracy of DNA replication and polypeptide assembly. Since there are usually many molecules of any one protein, an error that inactivated one of these would have no serious physiological effect. Conversely, accuracy must be high enough to ensure that the larger polypeptide chains are usually free of errors.

Another crucial factor in determining accuracy is the need for each cell to "inherit" a functional set of that group of proteins necessary for DNA to transmit its information to new DNA molecules and to new protein molecules. For cells capable of indefinite replication, the integrity of a translation apparatus must be preserved by the avoidance of escalating protein-error propagation. It is likely that the editing and proofreading steps in protein synthesis are necessary to preserve the stability of translation (Fersht, 1981; Kirkwood, Holliday, and Rosenberger, 1984). Indeed, the complex structure of the ribosome (which has not been elucidated) is probably essential not only for polypeptide assembly but also for accuracy in decoding the message. With regard to the error theories of aging, a central issue is the possibility that accuracy is relaxed in somatic cells of higher organisms to a point where the translation apparatus loses stability or genetic and epigenetic defects accumulate. This raises the very important problem of the evolution of aging. The argument has been made that the continuity of the germ line is maintained by the investment of resources in proofreading or other processes that preserve the integrity of macromolecules, whereas it is not advantageous for the organism to maintain the soma indefinitely (Kirkwood, 1977; Kirkwood and Holliday, 1979). Most organisms die in a natural environment from disease, predation, or starvation; therefore an organism will benefit from diverting valuable metabolic resources into efficient reproduction rather than the avoidance of aging (Medawar, 1952). Thus, error theories are able to explain, in principle, the evolution of aging, whereas it has always been hard to explain why a specific program for aging should evolve.

At first sight, it appears that the selection of papers is heavily biased toward those published by the Genetics Division of the Mill Hill laboratories, but in reality this is simply the result of bringing together in this volume papers that are specifically concerned with testing the possibility that errors in proteins and DNA accumulate during aging. This work began using *Drosophila* (Harrison and Holliday, 1967) and fungi (Holliday, 1969; Paper 4) and then progressed to cultured human fibroblasts or other cells such as lymphocytes. In addition, it was important to examine in detail the conditions under which error propagation might occur using *Escherichia coli,* which can be studied with the widest range of experimental techniques. In all these investigations there is a continual interaction between experiment and theory. Several other laboratories have tried to test error theories as well, and publications from these have, of course, been included. For the most part, however, these studies were "one off" projects, not necessarily in the mainstream of their experimental program. At present, the case for the validity of the general error rests in large part on the work published from Mill Hill, and it is sometimes maintained elsewhere that its predictions have already been "disproved." The main purpose of this volume is to allow the informed reader to judge for him- or herself after examining the evidence for and against, much of which is presented in these papers.

Part I includes theoretical papers on error propagation in protein synthesis and experimental papers in which the accuracy of protein synthesis is manipulated in *Neurospora crassa* and *E. coli.* It is much easier to study cells in culture than whole animals, and it is not surprising that many investigators have taken advantage of the fact that human fibroblasts have a finite lifespan in culture. Part II begins with Hayflick's classical paper on the in vitro aging of human fibroblasts and continues with both theoretical and experimental studies of the cell biology of the "Hayflick system." Part III includes attempts that have been made to detect and characterize altered protein molecules that appear in aging fibroblasts, and it is made clear in the commentary that adequate methods have not been developed to measure the accuracy of protein synthesis. It also includes two papers on the important topic of protein turnover in aging cells. These are relevant to the likelihood that cells protect themselves by scavenging abnormal protein molecules that might otherwise have deleterious effects. Finally, the possible relationship of aging and the accumulation of genetic defects is documented in Part IV. These publications are also concerned with human fibroblasts, but also included is the first documentation of somatic mutation frequency during aging in any mammalian species, using human T lymphocytes as experimental material.

It is evident that this selection of reprinted papers raises more questions than it answers. In some cases tests of the error theory appear to have given negative results, yet it is far from clear whether the procedures used have been sensitive enough. Therefore, more sophisticated techniques are required for further studies on cultured mammalian cells. Investigations on the molecular basis of aging of organisms are even more difficult, especially as the changes that produce the phenotype of an aged individual may be far subtler than those that lead to the dying of whole populations of cells. The quest for better methods to test error and other theories of aging remains a challenge. The new techniques of genetic manipulation will undoubtedly open many possibilities. It is hoped that many different laboratories will contribute to these future investigations.

REFERENCES

Burnet, F. M., 1974, *Intrinsic Mutagenesis: A Genetic Approach to Aging,* Wiley, New York.

Bremmerman, H. J., 1982, Reliability of Proliferation Controls. The Hayflick Limit and Its Breakdown in Cancer, *J. Theor. Biol.* **97:**541-662.

Doefler, W., 1981, DNA Methylation-a Regulatory Signal in Eukaryotic Gene Expression, *J. Gen. Virol.* **57:**1-20.

Doefler, W., 1983, DNA Methylation and Gene Activity, *Annu. Rev. Biochem.* **52:**93-124.

Drake, J., 1970, *The Molecular Basis of Mutation,* Holden Day, San Francisco.

Fersht, A. R., 1981, Enzyme Editing and the Genetic Code, *Proc. R. Soc.* **212B:**351-379.

Gupta, R. S., 1980, Senescence of Cultured Human Diploid Fibroblasts-are Mutations Responsible? *J. Cell. Physiol.* **103:**209-216.

Harrison, B. J., and R. Holliday, 1967, Senescence and the Fidelity of Protein Synthesis in *Drosophila, Nature* **213:**990-992.

Holliday, R., 1969, Errors in Protein Synthesis and Clonal Senescence in Fungi, Nature **221:**1224-1228.

Holliday, R., and T. B. L. Kirkwood, 1983, Theories of Cell Aging: A Case of Mistaken Identity, *J. Theor. Biol.* **103:**329-330.

Kirkwood, T. B. L., 1977, Evolution of Ageing, *Nature* **270:**301-304.

Kirkwood, T. B. L., and T. Cremer, 1982, Cytogerontology since 1881: A Reappraisal of August Weismann and a Review of Modern Progress, *Hum. Genet.* **60:**101-121.

Kirkwood, T. B. L., and R. Holliday, 1979, The Evolution of Ageing and Longevity, *Proc. R. Soc.* **205B:**531–546.

Kirkwood, T. B. L., R. Holliday, and R. F. Rosenberger, 1984, Stability of the Cellular Translation Process, *Int. Rev. Cytol.* **92:**93–132.

Kornberg, A., 1980, *DNA Replication,* Freeman, San Francisco.

Medawar, P. B., 1952, *An Unsolved Problem in Biology,* Lewis, London. (Reprinted in *The Uniqueness of the Individual,* 1957, Methuen, London.)

Pantelouris, E. M., 1978, Involution of the thymus and ageing, in The Nude Mouse in Experimental and Clinical Research, J. Fogh and B. C. Giovanella, eds., Academic Press, New York, pp. 51–73.

Riggs, A. D., and P. A. Jones, 1983, 5 Methyl Cytosine, Gene Regulation and Cancer, *Adv. Cancer Res.* **40:**1–30.

Part I

THE CONCEPT OF ERROR ACCUMULATION

Editor's Comments
on Papers 1 Through 7

The discovery of the major features of protein synthesis was soon followed by proposals that defects in the assembly of polypeptide chains might be an important component of the aging process. Medvedev (1961, 1962) was the first to suggest explicitly that the accumulation of protein errors with time could eventually lead to degenerative changes in cells. His view was broad in that he considered all possible types of errors in macromolecules, not only miscoding at the level of transcription and translation but also posttranslational changes in protein, the synthesis of abnormal purines or pyrimidines, and so on. He was also well aware of the many possible interactions between defects in DNA and those in proteins. These ideas were originally proposed in several publications, and we reprint here in Paper 1 an updated survey, which also includes a brief history of error theories of aging.

Orgel was not aware of Medvedev's work when he published his much more specific error-feedback theory. In Paper 2A he points out that the machinery for protein synthesis is potentially unstable in the following sense. The complex series of chemical reactions that are necessary for decoding the information in DNA cannot be completely specific, so it is therefore inevitable that errors in polypeptide assembly will occur. A subset of cellular proteins is itself required to carry out protein synthesis, and an erroneous molecule in this subset may lose specificity without loss of activity. It may therefore be responsible for further errors in protein synthesis. Orgel states that this result would lead to a paradox: starting with an error-free system, errors would be expected to accumulate exponentially, and this would lead eventually to a lethal error catastrophe in protein synthesis. Yet in innumerable biological systems it is clear that error catastrophes do not inevitably occur; otherwise, organisms could not have evolved in the first place. Orgel speculates that there may be special protective or selective mechanisms that prevent error catastrophes from occurring, for example, in the germ line. In the somatic cells of higher organisms, however, the accumulation of protein errors may be "one source of progressive deterioration of cells."

Seven years later Orgel published an important correction to his original theory (Paper 2B). In this he makes it clear that it is by no means inevitable that errors will increase exponentially. He introduces a parameter, α, which can be referred to as the feedback factor. Starting with an error-free protein-synthetic apparatus, the residual, or intrinsic, error frequency (R) will generate by feedback additional errors in the next generations of proteins. If $\alpha > 1$, then errors from feedback are greater than R, and in this case errors will increase exponentially to an error catastrophe. However, if $\alpha < 1$, then errors will necessarily stabilize to a steady state. This new formulation of the protein-error theory makes it more powerful, because it can resolve the paradox mentioned in the earlier paper, if α varies in different biological situations. The theory proposes that $\alpha < 1$ in cells that show no sign of aging, whereas $\alpha > 1$ in cells that do age. Orgel was well aware that no estimates of α in any biological system were available at that time, however, and even today, only two estimates have been made in *Escherichia coli,* the organism that has been studied in more detail than any other (Paper 7; Gallant and Palmer 1979, and see pages 14–16).

A significant advance in modeling error propagation was made by Hoffman (1974). He considered the difficult problem of the evolution of an accurate protein synthetic apparatus. Since primitive organisms must have been intrinsically inaccurate, how did they manage to improve fidelity to achieve the accuracy of protein synthesis we see today? From his model he concludes that even an intrinsically inaccurate translation machinery can avoid progressive error feedback and therefore remain stable. If this is the case, then it is very unlikely that error feedback could be significant in present-day organisms, and he therefore rejects the possibility that aging could be associated with the development of an error catastrophe in protein synthesis.

Hoffman's paper does not make for easy reading, and it is not readily evident that he makes a crucial assumption that is of doubtful biological validity. He assumes that a molecule that loses fidelity in information transfer also loses most of its activity. Kirkwood and Holliday (1975) demonstrated that the assumption is unjustified by citing several examples of mutants that reduce an enzyme's specificity without substantial loss of its activity. They also modified Hoffman's model by introducing a parameter (R) that takes account of the proportion of activity retained by an enzyme required for protein synthesis (an "adaptor" in Hoffman's terminology) that has lost specificity. They show that when reasonable estimates are made about the value of R, it is perfectly feasible for progressive error feedback to occur, and they conclude that Orgel's protein error theory is based on justifiable assumptions.

Models of error propagation are comprehensively reviewed in Paper 3. The major conclusion is that Orgel's argument is essentially correct but that a more flexible formulation is provided by the Hoffman-Kirkwood-Holliday (HKH) model. It can account for the existence of either stable or unstable translation in cells, and it predicts that a normally stable translation apparatus has a threshold error level above which stability cannot be regained. It would therefore be possible for a cell or cell lineage to exist for long periods, but with a given risk that critical errors (for example, the synthesis of an inaccurate RNA polymerase molecule or errors induced by an environmental perturbation) will initiate error propagation. The chance of crossing onto an irreversible path to cell death is determined by the distance between the stable and the threshold error levels. A more recent review, which also includes an assessment of evidence for and against protein-error accumulation in biological systems, has recently been published (Kirkwood, Holliday, and Rosenberger, 1984).

The most direct experimental tests of the theory of error accumulation have been carried out with microorganisms. Bacteria and many algae, fungi, and protozoa can propagate themselves indefinitely by asexual means, but some species of fungi or protozoa have asexual cells with limited growth potential. It was established by the classical work of Sonneborn (1954) that clones of *Paramecium aurelia* eventually cease cell division and degenerate. This organism survives in a natural environment by periodic sexual reproduction or by self-fertilization (autogamy); both processes appear to rejuvenate the cells. Another classic investigation of clonal aging was carried out by Marcou (1961) using the fungus *Podospora anserina*. Individual haploid spores give rise to populations of multinucleate cells, or hypae, which grow at a constant rate for a given period, but these invariably become senescent and further growth ceases. These studies with *Podospora* were of particular importance, since they demonstrated that senescence is dominant or invasive in heterokaryons between normal and aged cells. As in the case of *Paramecium,* the organism survives by sexual reproduction, which bypasses the aging process. The behavior of these and some other species of fungi or protozoa, which also exhibit clonal aging, is therefore comparable to that seen in higher organisms, where the soma is mortal and the germ line is immortal. In this situation, two important questions arise. First, is the finite growth of vegetative clones of protozoa or fungi due to the accumulation of errors in macromolecules? Second, will genetic and environmental factors that increase error levels change a potentially immortal population of microbial cells into one that has a limited growth potential?

The ascomycete fungus, *Neurospora crassa,* which is related to *Podospora,* is one of those species that grow indefinitely. There are mutants of *N. crassa* that impose a limit to its growth, however. One of these is called *nd* (natural death), originally discovered by Sheng (1951); in effect, it produces a phenotype like that of *Podospora.* In the first test of hypothesis that microbial cultures may die through protein error accumulation (Holliday, 1969), use was made of the *nd* mutation and the phenomenon of phenotypic suppression, which had been discovered by Champe and Benzer (1962) using bacteriophage T4. Subsequently, detailed studies by Gorini and associates demonstrated that certain mutants of *E. coli* grow in the presence of error-promoting drugs such as streptomycin (Gorini, 1970). This growth is due to the correction of a genetic defect by errors in ribosomal translation, and it does not, of course, lead to a heritable change in phenotype. In *N. crassa,* some adenine auxotrophs are phenotypically suppressed by the error-inducing base analogue 5-fluorouracil (5-FU) (Barnet and Brockman, 1962). When *nd* was combined with one of these, it

13

partially suppressed the requirement for adenine, suggesting that *nd* was indeed an error prone strain (Holliday, 1969). The same study showed that the suppression of an adenine auxotroph by 5-FU led to an accelerating growth rate, followed by cell death. This would be expected if error feedback occurs, followed by a lethal error catastrophe.

More specific evidence came from studies with a second *N. crassa* mutant, *leu-5* (Paper 4). This had previously been shown to produce abnormal protein molecules, probably because a leucyl tRNA synthetase had lost specificity in charging tRNA (Printz and Gross, 1967). The *leu-5* phenotype is more extreme at 35° than at 25°, and cultures grow for only about 3 days at the higher temperature. The possibility that this limit to growth was due to an error catastrophe was tested in experiments in Paper 4, in which an antiserum was used to measure the proportion of inactive glutamic dehydrogenase (GDH) molecules. On shifting cultures from 25° to 35°, a proportion of inactive molecules appeared within a few hours, and this stayed constant during a 48-hour period of growth. Subsequently, the proportion of inactive molecules increased rapidly and growth ceased after about 75 hours. The kinetics of the appearance of an active GDH strongly suggested that an error catastrophe was responsible for the death of the culture. It was also found that *leu-5* grown at 35° was a mutator strain, which indicated that the replication of DNA was also inaccurate. (The relationship between protein errors and genetic defects will be discussed in Parts III and IV.) It was also shown that the *nd* mutant accumulated inactive GDH and that both *leu-5* and *nd* contained a significant proportion of heat-labile enzyme, which increased as the cultures ran out of growth potential.

The last three papers in Part I are devoted to experiments with wild-type *E. coli* grown in the presence of streptomycin, which binds to ribosomes and introduces ambiguity in protein synthesis. In Paper 5, Branscomb and Galas grew *E. coli* under these conditions and monitored changes in the heat stability of β-galactosidase. Langridge (1968) had previously demonstrated that about 50% of missense mutants of β-galactosidase produced active but heat-labile enzyme. Paper 5 demonstrates that there is a progressive increase in heat-labile β-galactosidase in the presence of low levels of streptomycin, which indicates that errors in proteins are accumulating. Further, it was reported that growth eventually slowed down and ceased after as many as 17 population doublings, although these results were not documented. The results of Branscomb and Galas were confirmed by Rosenberger in Paper 6. He optimized the conditions that would convert a normal population of *E. coli* to one with limited growth and was then able to demonstrate that growth ceased after 12 population doublings (more than a thousandfold increase in cell mass) and that cell viability dropped to 0.1% or less. He also studied β-galactosidase during this period, using Gorini's method for measuring errors in translation. After an initial lag, the readthrough of a nonsense codon in a structural gene increased exponentially over 10 or more generations. Since it is known that the binding of streptomycin to ribosomes is a rapid process, the results in Papers 5 and 6 strongly indicate that, under the conditions used, the antibiotic is shifting the cells from a steady-state level of errors to an unstable one in which errors progressively accumulate.

At first sight these results are in conflict with those reported in Paper 7. The misincorporation of ^{35}S-methionine into purified flagellin provides a measure of mistranslation or mistranscription, since this amino acid is not present in the normal wild-type protein. Using this method, Edelman and Gallant (Paper 7) reported that a fiftyfold increase in protein errors induced by streptomycin did not result in error catastrophe; instead, a steady-state level of errors was reached after four generations, and the population continued to grow. In other experiments, Gallant and Palmer (1979) used phenotypic suppression to monitor errors introduced by growth in streptomycin. Again, the errors increased to a steady state, approximately tenfold the spontaneous error level, and then stabilized without any significant amount of cell death. Gallant and Palmer (1979, p. 37) conclude that "the empirical data we have discussed make it unlikely that error propagation has anything whatever to do with the phenomenon of aging." This is a surprising statement, since the theory of error propagation predicts that errors will stabilize under some conditions and escalate to a lethal error catastrophe under others. It seems unlikely, therefore, that there is real contradiction between results reported in different laboratories. As previously pointed out by Rosenberger, Foskett, and Holliday (1980), the results obtained may depend on the experimental conditions used.

In principle, experiments of the type documented in Papers 4 to 7 should make it possible to measure the extent of error feedback. In Orgel's basic formulation, error propagation will continue if $\alpha > 1$, but for any population of cells that grows indefinitely, α must be < 1. Edelman and Gallant (Paper 7) carried out an experiment in which streptomycin was removed and the rate of decrease of errors in flagellin was measured. This rate of decline made it possible to measure α, and their best estimate was 0.8. In the experiments on phenotypic suppression (Gallant and Palmer, 1979), the time it took to reach a steady-state level of errors suggested that α was about 0.5. These values are not far removed from 1, and it seems reasonable that a somewhat greater perturbation of the accuracy of protein synthesis by streptomycin could easily result in α exceeding 1, which would then lead to an error catastrophe. In summary, the results with *E. coli* and *N. crassa* quite strongly suggest that error catastrophes can occur in microbial cells.

The experiments in Papers 4 to 7 are all, in a sense, dependent on artificial experimental systems where error levels are manipulated either by the use of genetic mutants or by error-promoting environmental treatments. The question was raised earlier concerning the mechanism of aging of normal microbial populations such as wild-type strains of *Podospora* or *Paramecium*. Although extensive physiological and genetic studies have thoroughly documented the main biological features of their aging, biochemical investigations have been limited to changes in mitochondria. New techniques of molecular biology have been applied to *Podospora,* and it has been unequivocally demonstrated that major changes in the mitochondrial genome occur during senescence. Substantial parts of the mitochondrial DNA are lost and a minor fraction of the sequence is amplified, or integrated into the nucleus (see, for example, Cummings, Belcour, and Grandchamp, 1979; Stahl et al., 1980; Kuck, Stahl, and Esser, 1981; Wright, Horrum, and Cummings, 1982; Wright and Cummings, 1983). Since the organism is an obligate aerobe, loss of respiration could account for senescence, and the rapid proliferation of defective mitochondrial genomes could also explain the invasive cytoplasmic spread of senescence through the whole population of cells or in heterokaryons between normal and aged cells. The experiments, however, throw little light on the origin of the defective genomes or why selection of cells with functional mitochondria is not able to sustain indefinite growth. It therefore remains possible that a breakdown of information transfer, perhaps between the nucleus and the mitochondria, is indeed responsible for this form of natural aging.

REFERENCES

Barnet, W. E., and H. E. Brockman, 1962, Induced Phenotypic Reversion by 8-Azaguanine and 5-Fluorouracil, *Biochem. Biophys. Res. Comm.* **7:**199–203.

Champe, S. P., and S. Benzer, 1962, Reversal of Mutant Phenotypes by 5-Fluorouracil: An Approach to Nucleotide Sequences in Messenger RNA, *Proc. Natl. Acad. Sci. (U.S.A.)* **48:**532–546.

Cummings, D. J., L. Belcour, and C. Grandchamp, 1979, Mitochondrial DNA from *Podospora anserina* II. Properties of Mutant DNA Multimeric Circular DNA from Senescent Cultures, *Mol. Gen. Genet.* **171:**239–249.

Gallant, J., and L. Palmer, 1979, Error Propagation in Viable Cells, *Mech. Ageing Dev.* **10:**27–38.

Gorini, L., 1970, Informational Suppression, *Annu. Rev. Genet.* **4:**107–134.

Hoffman, G. W., 1974, On the Origin of the Genetic Code and the Stability of the Translation Process, *J. Mol. Biol.* **86:**349–362.

Holliday, R., 1969, Errors in Protein Synthesis and Clonal Senescence in Fungi, *Nature* **221:**1224–1228.

Kirkwood, T. B. L., and R. Holliday, 1975, The Stability of the Translation Process, *J. Mol. Biol.* **97:**257–265.

Kirkwood, T. B. L., R. Holliday, and R. F. Rosenberger, 1984, Stability of the Cellular Translation Process, *Int. Rev. Cytol.* **92:**93–132.

Kuck, U., U. Stahl, and K. Esser, 1981, Plasmid-like DNA is Part of Mitochondrial DNA in *Podospora anserina, Curr. Genet.* **3:**151–156.

Langridge, J., 1968, Thermal Responses of Mutant Enzymes and Temperature Limits to Growth, *Mol. Gen. Genet.* **103:**116–126.

Marcou, D., 1961, Notion de longévité et nature cytoplasmique due déterminant de la sénescence chez quelques champignons, *Annu. Sci. Natl. Bot.* **12:**653–764.

Medvedev, Zh. A., 1961, Ageing of the Organism at the Molecular Level (in Russian), *Usp. Sovrem. Biol.* **51:**299–316.

Medvedev, Zh. A., 1962, Ageing at the Molecular Level and Some Speculations Concerning Maintaining the Function of Systems for Replication of Specific Macromolecules, in *Biological Aspects of Ageing,* N. Shock, ed., Columbia University Press, New York, pp. 255–266.

Printz, D. B., and S. R. Gross, 1967, An Apparent Relationship between Mistranslation and an Altered Leucyl tRNA Synthetase in a Conditional Lethal Mutant of *Neurospora crassa, Genetics* **55:**451–467.

Rosenberger, R. F., G. Foskett, and R. Holliday, 1980, Error Propagation in *Escherichia coli* and its Relation to Cellular Ageing, *Mech. Ageing Devel.* **13:**27–252.

Sheng, T. C., 1951, A Gene That Causes Natural Death in *Neurospora crassa, Genetics* **36:**199–212.

Sonneborn, T. M., 1954, The Relation of Autogamy to Senescence and Rejuvenescence in *Paramecium aurelia, J. Protozool.* **1:**38–53.

Stahl, U., U. Kück, P. Tudzynski, and K. Esser, 1980, Characterization and Cloning of Plasmid Like DNA of the Ascomycete *Podospora anserina, Mol. Gen. Genet.* **178:**639–646.

Wright, R. M., M. A. Horrum, and D. J. Cummings, 1982, Are Mitochondrial Structural Genes Selectively Amplified during Senescence in *Podospora anserina? Cell* **29:**505–515.

Wright, R. M., and D. J. Cummings, 1983, Integration of Mitochondrial Gene Sequences within the Nuclear Genome during Senescence in a Fungus, *Nature* **301:**86–88.

1

Reprinted from Mech. Age. Devel. **14**:1–14 (1980)

THE ROLE OF INFIDELITY OF TRANSFER OF INFORMATION FOR THE ACCUMULATION OF AGE CHANGES IN DIFFERENTIATED CELLS*

ZHORES A. MEDVEDEV

Division of Genetics, National Institute for Medical Research, Mill Hill, London NW7 1AA (Great Britain)

(Received January 4, 1980)

SUMMARY

While the error theory of ageing has attracted significant interest in recent times and was widely debated, in most cases the experimental tests and results (in favour or against it) were relevant only for consideration of Orgel's "error catastrophe" theory, which relates the possible mechanism of cellular ageing with self-propagation of errors in translational systems. However, Orgel's idea was one of several different "error theories", some translational and some more general. The more comprehensive theory of errors in the synthesis of macromolecules as the cause of ageing developed earlier than the "error catastrophe" explanation, and considered the infidelity of synthesis of DNA, RNAs and proteins as closely interconnected. It also considered the level of errors in protein synthesis as a balance between error-inducing and error-repairing factors and did not expect the "error catastrophe" as inevitable. The usual level of random errors of synthesis could be rather low or slowly increasing during cellular ageing, but mostly the irreversible changes of cellular structures at the functional level, which appear under the influence of errors of protein synthesis, transcription and reproduction of DNA and some other deteriorative factors, are more evident accumulators of age changes. The attempt to update this theory made in this review shows that the age-relevant changes of proteins are also much more complex than just the errors of those active in translation. Ageing at the protein level depends on the types of protein, their "half-life", structural role, origin of error, post-translational changes, *etc*. Some proteins are more "change-prone" than others and no simple test to prove or to invalidate the error theory is feasible.

---

*Paper presented at the Sixth European Symposium on Basic Research in Gerontology, Munich, F.R.G., September 4–7, 1979.

INTRODUCTION

It is well-known that the reproduction and transfer of information in cells does not take place with absolute fidelity. There are numerous studies which analyse the types and rates of errors, breaks and other alterations of DNA, RNAs and their more complex structures in functional chromatin of cell nuclei. There are also many works which demonstrate errors in protein synthesis or their post-translational changes which are reflected in alterations of the specific activity of many enzymes or in a decline of the functional quality of other intracellular and extracellular proteins. There are many theoretical works which try to relate different cases of infidelity of informational transfer with the basic causes of ageing. The "error catastrophe" theory [1], general error theory of ageing [2, 3], somatic mutation theory of ageing [4–7], are theories of ageing which consider different changes at the RNA level as the most important factor in the age-associated deterioration of cellular functions [8–10]. These theories and some others consider in fact parts of essentially the same general picture.

Errors, although inevitable, are induced or introduced into the processes of synthesis by many factors and agents which may act at different levels of the life process. The theories of ageing which pay the attention mainly to damaging factors rather than to the actual changes in the molecular population of cells – the free radical theory of ageing [11–13], the radiation theory of ageing, cross-link theories [14–16], the deamidation theory of ageing [17, 18] and similar approaches for explaining age changes – do not really contradict, but complement the error and mutation theories. The same is certainly true for some recent theories of ageing which consider the efficiency or failures of molecular repair systems to be responsible for time-related deterioration of reproduction and functional use of genetic information [19].

The infidelity of informational transfer might occur not only at the level of macromolecule synthesis, but at the level of transcription of genetic information in chromatin. There have been several attempts to explain age changes by the alteration of interaction of chromatin DNA with histones and non-histone nuclear proteins and by repression of some functional activities of cells at the transcriptional stage [20–22]. There are some other age changes that can reduce the fidelity and quality of transfer of biological information; they include changes of membranes, ribosomes, mitochondria, lysosomes and they were also suggested as possible causes of ageing. The number of experimental works which made such theories possible is so high that it is difficult to mention even the main ones in this short review. All these age changes are, as a rule, interrelated, but this does not mean that it is not possible to single out proteins as the most vital indicators of age changes of informational transfer. DNA or RNAs can accumulate many deleterious changes, which can stay latent and do not interfere with normal cellular functions. Such "invisible" changes in DNA or RNA are located within the parts of the genome that are repressed in differentiated cells anyway.

The protein population of any cell is directly connected with the cellular functions, and changes in the proteins should have an immediate effect; however, such changes are often random and therefore difficult to detect by direct biochemical methods. DNA can be compared with the complex computer memory which contains very many different programmes. For individual differentiated cells only one programme from many is really relevant. Its final products, the mixture of different proteins, accumulates the consequences of errors or repair processes at very many stages of transcription and translation and the alterations which can happen to the particular programme itself. These changes in proteins are mostly relevant for somatic tissue functions (for cells which do not divide or are used for reproduction).

Intracellular metabolism, cellular or tissue interactions and most other visible functions of life directly depend on the qualities of proteins. If the protein molecule of a certain enzyme is altered and the enzyme rendered inactive it will be difficult to determine immediately where the mistake was made, whether it was post-synthetic damage, failure of the ribosomes, or tRNA synthetase, or messenger RNA, or DNA. Only special research can answer the question. The translational errors are temporary; the errors related to tRNA or mRNA could be repeated many times; the mistakes in the DNA programme are mostly irreversible. In all these cases, however, a study of the proteins should be carried out to discover what was wrong where. This is why the relationship between the fidelity of informational transfer and molecular aspects of ageing must be started from a consideration of errors in protein biosynthesis. Some years ago this task could only be a theoretical exercise. Now we can make a review of some error theories of ageing and a condensed review of experimental results in this field as well.

1. BRIEF HISTORY OF ERROR THEORIES OF AGEING

Protein biosynthesis is not perfect from the very beginning of an individual life. There is always some "load" of mutations inherited from previous generations and, although most of them are recessive, they are reflected in, at least, partial infidelity of synthesis of a certain number of proteins. Somatic mutations which start to hit cells randomly from early embryonic stages of development permanently contribute to this "load". However, these genetic background factors introduce errors into protein synthesis in a rather general way – not all cells in all tissues and organs make errors connected with mutations. Differentiated cells realize only one genetic programme from many, and different types of cells have different patterns of functionally active genes. It is not impossible that some of such programmes are "mutation-free", when inherited mutations are under consideration. Somatic mutations are random, they hit different individual genes in the individual cells, and not all cells, at least early in life, may be affected. It is feasible to suggest that some cells in different tissues could have an *absolutely perfect* genetic programme and all errors that appear are errors of translation. Could such errors be irreversible, cumulative or self-propagative? Or maybe, they are just transitory, and appear and disappear during protein turnover? Are some errors more stable than others, or are some proteins more vulnerable than others? The hypothetical answers to these and some similar questions represent the so-called "error theories of ageing".

Error theories of ageing started to appear before the discovery of the template mechanism of protein biosynthesis. In 1951–1952 the double helix structure of DNA was not yet known and neither messenger RNAs, nor transfer RNAs had been identified. The idea of the genetic code was not yet clearly formulated. The dominating explanation of protein synthesis and turnover considered the synthesis and degradation of proteins as interreversible processes. Because protein degradation was found to be a stepwise process with polypeptides and peptides as intermediate products, protein biosynthesis was also seen as an assembly through intermediate peptide and polypeptide stages with the possibility of re-utilizing peptides and polypeptides for synthesis before their complete degradation into the amino acid pool. It had already been found that peptide bonds between different amino acids are not equal, the formation of some bonds releasing more energy than others [23]. Such peptide bonds also need more energy for their degradation. Pasynsky [24] suggested in 1951 that peptide bonds which need more energy for their degradation should have lower turnover rates. In this case the inevitable erroneous substitution of one amino acid by another during turnover can produce the accumulation of peptide bonds with a lower level of energy – a change of a thermodynamic nature. This could lead to lower and lower rates of turnover and to a reduction in the functional abilities of proteins, which could mean their ageing [24].

This idea was a starting point of my own attempt to explain the molecular mechanism of ageing [25]. Pasynsky, as a biophysicist, tried to suggest the energy (entropy) approach to the possible result of erroneous amino acid substitution. I, as a biochemist, suggested that the errors in protein biosynthesis should lead to the increased stability of proteins against degradation by proteolytic enzymes. The intracellular proteases break peptide bonds selectively. The peptide bonds between some specific amino acid residues are more vulnerable to the action of certain enzymes. The stable bonds could have, therefore, a lower rate of turnover. In the case of peptides as intermediate products of both degradation and synthesis (concept of turnover in 1945–1954), the errors of restoration of amino acid sequence specificity should lead to the accumulation of more stable bonds because they participate in the cycles of degradation and resynthesis less frequently and have, therefore, fewer chances of erroneous substitution. At this time some experimental works indicated that the total proteins of old tissues were more stable for proteolytic digestion and there were attempts to show the same for individual proteins as well (serum proteins, myosin, collagen). Later it was shown that these changes were related more with the structural changes of proteins (cross-links in collagen), but in 1952 the nature of the increased stability of some proteins was not yet clear.

Both schemes of "inactivation" of proteins were based on the assumption that the specific proteins increase in time the number of sequence alterations up to the point when this can be found by physical or chemical methods. There were several attempts made in 1940–1955 to find the age-related changes in amino acid composition of proteins in different tissues (see reviews [26, 27]), but most of them did not show definite changes, when individual specific proteins had been analysed.

In 1960, when the genetic mechanism of the control of protein synthesis was already well studied and the transfer of genetic information through the system

21

                    tRNAs

DNA → mRNAs → proteins was found to be the permanent functional cellular process
                    rRNAs

(with the genetic code still under theoretical consideration), I made an attempt to develop a more general error theory of ageing, which did not isolate errors of protein bio-synthesis from errors of transcription and changes and errors at the level of DNA [2, 3]. The possibility of the accumulation of errors at different stages of transfer of genetic information and the rate of this accumulation was suggested to be dependent on the com-parative activity of error-inducing factors (free radicals, analogues, cross-links, sequence substitutions, *etc.*) and repair and turnover processes which could reduce the final out-put of functionally relevant changes. It was also postulated that, because the errors in most cases are random and could be unique at the translational stage for individual molecules of proteins, it should be extremely difficult to determine the accumulation of errors by conventional biochemical procedures. I also suggested that even in the case of a permanent level of errors (regulated by "error-eliminating processes") the *accumulation* of deleterious consequences of random errors can occur at the higher levels of cellular systems, where some protein-related changes could be irreversible. The finding of age changes of different types (age pigments, changes of membranes, chromatin, *etc.*) was not considered as something completely independent or an alternative to the error theory of ageing. They could be directly or indirectly induced by random translational errors and the temporary appearance of faulty protein molecules (the background informational "noise"). The somatic mutations were part of the general picture and the genetic message in DNA was considered to deteriorate with time. This deterioration and the accumulation of age changes at the higher levels of the cellular system could reduce the fidelity of translation as well, but I did not postulate that this interrelation might lead to a perma-nent acceleration of the ageing rate.

The possibility of a dramatic acceleration of error frequency had been suggested in 1963 by Orgel [1] in his well-known theory of "error catastrophe". It was a more simple "translational" approach and the theory offered a more straightforward explanation of imminent and accelerated ageing and death. It also promised easier approaches for experi-mental tests. (Most studies to test the error theory of ageing, which we consider later, had been designed to test the "error catastrophe" theory.) Orgel's theory divided indivi-dual proteins of living systems into two groups, not equally important for accumulation of errors. The first group of structural proteins (and enzymes) was responsible for inter-mediate metabolism and different intracellular and extracellular formations, the second with direct processing of genetic information (determination of the sequences of nucleic acids and polypeptides). The consequences of errors for these two groups of proteins were expected to be rather different. If, for example, a small proportion of enzyme mole-cules responsible for some metabolic function (glucose or other metabolism) were in error, the theory expected a slight reduction of the *average* specificity of the reactions. This effect would not be cumulative; once the faulty mRNA or protein had been degraded, all memory of the error would soon be lost. However, errors that result in a reduced specificity of an information-handling enzyme load, according to Orgel's theory, to an increasing error frequency. Such processes should be cumulative and, in the absence of an imposed selection for "accurate" protein-synthesizing units, must lead ultimately to an error catastrophe. Error catastrophe means that "the error frequency must reach a value at which one of the processes necessary for the existence of viable cell becomes critically inefficient...". The cumulative effect is not related to the preservation and accumulation of errors, but to the possibility that altered proteins (like ribosomal, tRNA synthetases, polymerases) "may contribute to the synthesis of further altered proteins and the mistakes would then tend to become self-propagating. The amount of altered, malfunctional protein would thus increase in cells with every division until a state would finally be reached when the cell could no longer function normally and degenerative changes would be evident."

This expectation (that the level of accumulation of altered proteins can reach rather a high proportion) induced many attempts to find such altered proteins in ageing cells or to reduce the life span by temporary misincorporation of amino acid analogues. In the latter case the prediction of self-propagation of errors was tested. These experimental approaches had been suggested as possible in Orgel's paper. However, he made clear that his idea of the possibility of an "error catastrophe" is not an explanation of the "mechanism of ageing". He realized that ageing is a very complex process and may depend on different factors (including selective scavenging of incorrect proteins by hydrolytic enzymes). Orgel wanted to point out one possible source of progressive deterioration of cells and cell lines and to show that this kind of theoretical speculation "... can suggest experiments which should show where, if anywhere, it contributes to the ageing process in higher organisms." We can see that Orgel originally did not offer his idea as a special "error catastrophe theory of ageing"; the idea was only later transformed into the "error catastrophe theory of ageing" when preliminary experiments along the lines suggested by Orgel seemed to support the theoretical expectations. Later experiments, however, were contradictory and showed that real ageing is a much more complex and multi-channel process.

2. AGEING AND THE FIDELITY OF PROTEIN SYNTHESIS. EXPERIMENTAL APPROACHES

The problem of the fidelity of protein biosynthesis is a part of a more general pattern of research into molecular mechanisms of translation. A significant proportion of experimental works on the fidelity of protein synthesis, misincorporation of amino acid analogues or selective degradation of altered proteins does not have direct links with gerontology and many experiments have been carried out with bacterial systems. The reproduction of sequence specificity of proteins was probably the event when the origin of life and early evolution of protein biosynthesis started, and the selection process worked later towards the development of a more and more error-proof system of translation. It is impossible, however, to reach one hundred per cent perfection in any system of informational transfer and the estimation of the error rates (which might be species-specific) has many different implications. The importance of the fidelity of translation for cellular ageing is only one of them.

23

The first experimental attempt to find the actual level of errors of protein biosynthesis in specialized cells was made by Loftfield [28] without any purpose of testing a theory of ageing. Before his research, carried out in 1963, the search for altered proteins was a task of "molecular pathology", when about a hundred genetically determined abnormal hemoglobins and some other proteins had already been discovered. The defects in these proteins were, as a rule, related to amino acid sequence substitutions. Loftfield's experiments were designed to discover if similar substitutions can occur as spontaneous, random, non-genetic errors "owing to the finite ability of macromolecular surfaces to distinguish between closely related molecules". The experimental system to test the frequency of errors consisted of the *in vitro* synthesis of oviduct chicken ovalbumin and the possibility of misincorporation (or substitution) of three chemically similar amino acids – leucine, isoleucine and valine – which were labeled with ^{14}C. The use of highly radioactive amino acids is much more sensitive than any chemical method. Loftfield's scheme was simple and was repeated later by others. He tried to demonstrate incorporation of a labeled amino acid into peptide sequences where this amino acid should not normally be present. Because ovalbumin has all the amino acids represented in its polypeptide chains, Loftfield tried to isolate individual peptides after partial degradation of the protein and to find, for example, labeled valine in a peptide that would normally contain isoleucine but not valine. Results with different peptides were not identical, but the preliminary estimation (on the basis of better purified peptides) put the frequency of mistranslation at the level of one error per 3000 correct amino acid incorporations (in the case, for example, of valine–isoleucine substitutions). The absolute purification of ovalbumin or its peptides was rather difficult and this limitation could influence the final figures. The ratio 1/3000 was, therefore, indicated as a maximum probable error level, rather than the actual biologically determined error frequency. About ten years later Loftfield and Vanderjagt tried to improve the sensitivity of the method by using the same approach to study the synthesis of hemoglobin in rabbit reticulocytes [29]. Hemoglobin with its more distinctive molecule could be more highly purified than ovalbumin.

The use of amino acids with higher specific activities and better procedures for purification of peptides has lead to four almost identical estimates of error frequency. These results made authors believe that they were observing a true error frequency rather than a maximum probable value. Labeled valine was again used as a substitute for leucine or isoleucine, but authors did not exclude the possibility that erroneously incorporated valine was substituted for amino acids other than isoleucine. The error frequency (3 mistakes per 10 000 incorporations) was close to the figure obtained for ovalbumin, and was considered to be an indication of a very high precision of the peptide assembly system. It was not, of course, clear whether this figure reflected the error rate of aminoacyl-tRNA synthetases or ribosomal decoding of mRNA.

A similar method to measure the frequency of misincorporation may be used when working with specific proteins which lack certain amino acids. The incorporation of amino acids (in labeled form) that are normally absent can be an indication of misincorporation (theoretically random, but possibly preferential depending on the chemical resemblance of amino acids especially at the level of activation), or similarities of codons (for example, more probable substitutions between codons AGU(Asp) and AGA(Glu), than for AGU and CCC(Pro) in the case of tRNA—mRNA interactions). Several studies of this kind have been carried out in recent years. The subjects of study (specific protein and absent amino acids) varied considerably. Flagellin — the protometric subunit of *Escherichia coli* which does not contain cysteine [30] — showed about 2—6 erroneous incorporations of [^{35}S]-cysteine per 10 000 molecules. H1 histone from mouse tissues [31] and from cultured human fibroblasts [32] which does not contain either cysteine or methionine, showed maximum error frequency (misincorporation of [^{35}S]-methionine) at the level of 10^{-5} for mouse histone and 10^{-4} for fibroblasts (per amino acid residue). Old fibroblast cultures [32] and the old animals in our work [33] showed higher levels of [^{35}S]-methionine misincorporation into H1 histone, but the increase (by a factor of about 2) can not yet be definitely considered as related to mistranslation and it was not of the order which may be expected as "error catastrophe".

Popp *et al.* [34] have made an attempt to measure the possible error frequency in the synthesis of human hemoglobin (by finding isoleucine which normally is not present in this human protein). The average error frequency was about 3×10^{-5}, but the individual variations were significant and a small age-associated increase could be postulated only as a tendency, not as a clearly visible change. (The authors made their measurements for persons living on the Marshal islands and there were no really old persons among those whose blood samples had been taken for analysis.) The same method was also tested by those authors to study the error frequency in young and old marmosets [35]. The error rate of hemoglobin synthesis in reticulocytes from the young animals was about 10^{-5} and reached 4×10^{-5} in reticulocytes from old animals.

In all these attempts to estimate the fidelity of protein synthesis it was practically impossible to determine when the errors were a result of real mistranslation, and when they might reflect the presence of mutated genes in a few cells. It is, however, a problem of theoretical interest; from a functional point of view the presence of faulty or altered protein molecules is what really matters, not so much the cause of the alteration. Nobody was yet able to design the method which could serve both purposes: to estimate the frequency of errors of mistranslation and to show that the estimated level of errors is directly responsible for an age-associated decline in functional abilities. The problem of relations lie between translational errors and ageing and was tested by indirect methods only — by the attempts to reduce the life span by amino acid analogues which induce the experimental formation of altered or deficient proteins. The first experiment of this kind had been carried out by Harrison and Holliday [36], who tried to reduce the life span of *Drosophila melanogaster* by feeding them during the third instar larval stage with a mixture of five amino acid analogues. The results were positive; in some experiments the life span of adult flies was reduced from 55 to 42—43 days. However, the authors pointed out that this effect can also be explained by interference with protein synthesis immediately before metamorphosis. The deterioration of the fidelity of protein biosynthesis may have morphological consequences, and the shortened life may result from morphological abnormalities. This "non-error" explanation of shorter life span was later supported by Dingley and Maynard Smith [37] who were not able to show any effect of amino acid analogues on the life span of flies when analogues were added in sub-lethal doses to the food of young adult flies (the incorporation of analogues into proteins was shown in experiments with labeled analogues). The results were confirmed in 1976 by Bozcuk [38]. The young adult males of *Drosophila melanogaster* did not show any accelerated ageing when their food contained three amino acid analogues in just sub-lethal doses. At the same time the abnormal proteins with analogues showed a faster turnover than their respective normal proteins — an indication that there is some system of preferential degradation of abnormal proteins in differentiated cells. Earlier the existence of preferential degradation of abnormal proteins had been shown for bacterial systems as well.

In subsequent experiments Holliday and his colleagues were able to show that amino acid analogues can accelerate the clonal senescence in fungi (*Podospora*) and in cell cultures of human fibroblasts [39, 40]. Ryan *et al.* [41], however, were not able to confirm the age-accelerating effects of amino acid analogues for human fibroblasts. Recent experiments on the effect of amino acids analogues on the life span of mice [42] are interesting but are open to different interpretations.

The new, but also indirect, approach to test the possible relationship between ageing and the fidelity of protein synthesis was based on attempts to show that the systems of protein biosynthesis in old cells reduce their ability to discriminate between normal amino acids and their analogues [43].

After several pioneering studies [44-47] which showed that there is a certain reduction associated with cellular ageing in the *specific activity* of some enzymes and the parallel accumulation of altered or inactive enzyme molecules, the investigation of the quality of enzymes during ageing became the purpose of many studies. In this short review we are not able to discuss the contradictory results with different systems and different enzymes obtained from this approach. However, it is now clear that the alteration to an enzyme (or isoenzyme group), when it is shown, does not as a rule have the character of accumulated random errors [48–51]. This is a more complex and probably secondary age-associated change which cannot be considered without analysis of the metabolic or functional activity related to one or other enzymatic reaction.

3. THE COMPLEX ORIGIN OF AGE CHANGES OF PROTEINS IN DIFFERENTIATED CELLS

Age-associated changes of proteins have very many different connections with many independent and interdependent causes of potential ageing. I would like especially to underline the term *potential ageing,* because in real systems not all deteriorative changes are irreversible and not all local ageing processes contribute to the general ageing of the system, or the ageing of different types of specialized cells. Red blood cells and lymphocytes, for example, age and die permanently in the blood circulation of vertebrate animals, but these local cases of ageing are not connected with the ageing of the whole organism. Animals with shorter life spans could have very long-lived erythrocytes (frogs and other amphibians), while animals with long life spans could have red blood cells with very short individual lives in circulation (most birds). The discrepancy between life span possibilities of red blood cells and the ageing rate of the organism as a whole exists among mammalian species as well, which have the simplest anuclear forms of erythrocytes (see review [52]). There are many studies of changes of proteins in erythrocytes which age while in the blood circulation (see review [53]). The pattern of molecular ageing is familiar — reduction in the specific activity of many enzymes, modifications of protein molecules, inactivation of nuclei (amphibian and bird species) — but all these changes are not seriously relevant for the ageing rates of other tissues. Non-dividing *irreplaceable* cells (nerves, muscle, *etc.*) are much more important for the general ageing, and species in which all types of cells are non-dividing and irreplaceable (insects, nematodes, *etc.*) live, as a rule, for a few weeks or months only.

These examples (many more could be given) indicate that even in the case when one discusses the molecular level of ageing, the tissue specificity of ageing and different relevance of tissues for general ageing should be taken into consideration. However, similar tissues (or similar types of specialization) in species which are closely linked in evolution can have rather different rates of ageing. Among different species of rodents the normal maximum biological life span can vary from 1 to 20 years and the pattern of age changes, for example, in almost identical non-dividing nervous or muscle cells is similar in all aspects but speed. The maximum life span within primate species varies between 10 and 100 years, but the ageing pattern is similar, including the ageing of collagen. How can it be that the accumulation of cross-links in collagen or elastin proceeds with such different speeds? The different rates of ageing in the differentiated cells of the same type (or the same design) is the crucial problem which faces any attempt to explain the mechanism of ageing.

The theoretical approaches which were briefly reviewed in the first part of this paper tried to explain the molecular or cellular *picture* of age alterations, not the speed or evolution of ageing and individual forms of age changes. There were some molecular theories which have paid most attention to an explanation of the rate of age changes in similar cells of different species; the suggestion of the importance of the redundancy of vital genes in cellular DNA [54–56] and the suggestion of the existence of special genetic programmes of ageing for all tissues — the latter idea is especially popular among plant gerontologists [57, 58] — are two examples of this approach. Both control systems (repetition of vital genes and the genetic "clocks" programmes) may act, however, through the existing channels of transfer of information and their realization does not eliminate the importance of studying all forms of age changes in proteins, RNA and DNA.

In any possible system of transfer of information (either programmed or not) the proteins act as intermediates between the stored information and the actual functional structure of any cell. The normal activity of a differentiated cell is related to several thousands of different types of proteins. About 2000 individual proteins can be identified now simply by two-dimensional electrophoresis of cytoplasm or chromatin material [59]. A protein biochemist immediately realizes that all the protein alterations that have been suggested as age-important (errors, cross-links, post-synthesis modifications) cannot be

TABLE I

DIFFERENT TYPES OF AGE CHANGES IN PROTEINS

Errors of synthesis (misincorporation)
- a. Errors in aminoacyl-tRNA synthetases
  - 1. Translational
  - 2. Transcriptional
  - 3. Mutational (inherited or somatic)
- b. Miscoding (ribosomal or mRNA)
  - 1. Translational
  - 2. Transcriptional
  - 3. Mutational (inherited or somatic)

Errors of synthesis (formation of abortive or altered molecules)
- a. Errors of termination of synthesis
- b. Errors of initiation of synthesis
- c. Incorporation of amino acid analogues
  - 1. External
  - 2. Internal (errors of amino acid metabolism)
- d. Incorporation of *d*-amino acids
- e. Incorporation of peptides

Changes in enzymes
- a. Inactivation
- b. Decrease of specific activity
- c. Changes in isoenzyme pattern
- d. Increased thermolability
- e. Decrease in the fidelity of reactions
- f. Changes in induction (adaptation) sensitivity
- g. Others: random errors, selective increase in activity, *etc.*

Changes in turnover rates and life spans of molecules

Post-synthesis changes and alterations
- a. Cross-links
- b. Acetylation
- c. Phosphorylation
- d. Methylation
- e. Mineralization
- f. Protein–metal complexes
- g. Deamidation
- h. Changes of conformation (secondary or tertiary structure)
- i. Others (dimerization, aggregation, amyloidization, complexes with lipids, oxidation, denaturation, *etc.*)

Changes in number, pattern, individual concentrations, localization, interactions
  - 1. Random
  - 2. Transcriptional
  - 3. Mutational

equally applied to different types of proteins. The changes which are usual for collagen or elastin do not happen in most enzymes. Post-synthetic modifications (acetylation, phosphorylation, *etc.*) are more typical for long-lived histones than for short-lived (high turnover rate) albumins of cytosol. At the same time, one may realize that, even with all protein molecules in a perfectly unchanged condition, the change in the pattern of enzymes only — increase of these with proteolytic activity, or the increase of some which intensify or delay special functions, or the appearance of some proteins which are not "normal" for one cellular type, but quite normal for another — may lead to the malfunction, deterioration and death of cells. The attempts to find some age-related changes in the number and pattern of individual proteins in some cellular or nuclear fractions [60, 61] have the same relevance for the molecular picture of ageing as the experiments designed to find the errors in proteins or DNA.

If some of the age changes in proteins which have already been experimentally studied as age-related were simply listed here (the review of the experimental works to show that this is really the case should include more than 200 publications), the complexity of ageing at the level of proteins will be quite evident. At the same time each individual form of the possible age changes in proteins listed in Table I may have a rather complex origin and multiple connections with the informational "memory" of the cell (DNA) and its transcription. At these levels of reproduction and transfer of information the pictures of age-associated alterations are also extremely complex. Within the limits of a single paper it is impossible to discuss the protein and the RNA and DNA levels of informational transfer. One should remember, however, that these levels do exist and are important. It is not easy yet to decide whether DNA—RNA levels have a primary or a secondary role in the accumulation of age changes. Transfer of information during normal cellular function is so interrelated that at any level one can find some primary and some secondary alterations. In the same way as one can not yet clearly distinguish between errors of protein synthesis of translational, transcriptional or mutational origin, it is not yet easy to give a universal explanation of what causes of informational changes in the specific genetic message play a leading role. The same type of nucleotide sequence alteration may occur as a result of misincorporation, action of free radicals, modification, faulty repair (many repair enzymes), mistakes of DNA polymerases, incorporation of analogues, breaks, dimers, and others. This complexity of all interrelations in the transfer of genetic information gives the selection (and the evolution) the possibility to manipulate the actual durability of the integrity of the differentiated cell. Evolution may play different molecular "instruments" to change the life span, acting through gene amplification, efficiency of repair systems, level of damaging factors (free radicals, the fidelity of metabolism, metabolic origin of analogues, mutability, mutator genes, turnover, cellular replenishment, mitochondria, ribosomes, enzymes, and many others). The simplistic explanation of ageing by one or other dominant factor of a molecular nature is hardly possible. The molecular biology of ageing is still very much a descriptional, not an explanational, branch of gerontology.

REFERENCES

1 L. E. Orgel, The maintenance of the accuracy of protein synthesis and its relevance to ageing. *Proc. Natl. Acad. Sci. U.S.A., 49* (1963) 517–521.
2 Zh. A. Medvedev, (Ageing of organism at the molecular level). *Usp. Sovrem. Biol., 51* (1961) 299–316.

3 Zh. A. Medvedev, Ageing at the molecular level and some speculations concerning maintaining the functioning of systems for replication of specific macromolecules. In N. Shock (ed.), *Biological Aspects of Aging*, [*Proc. 5th Int. Congr. Gerontol.*, Vol. 3], Columbia University Press, New York, 1962, pp. 255–266.

4 L. Szilard, On the nature of aging process. *Proc. Natl. Acad. Sci. U.S.A., 45* (1959) 30–45.

5 G. Failla, The aging process and somatic mutations. In B. L. Strehler (ed.), *The Biology of Aging*, American Institute of Biological Sciences, Washington, 1960, pp. 170–175.

6 H. J. Curtis, Biological mechanism underlying the ageing process. *Science, 141* (1963) 686–694.

7 H. J. Curtis, *Biological Mechanisms of Aging*, C. C. Thomas, Springfield, 1966.

8 Zh. A. Medvedev, The nucleic acids in development and aging. *Adv. Gerontol. Res., 1* (1964) 181–206.

9 V. J. Wulff, Age-associated changes in the metabolism of ribonucleic acids. In N. W. Shock (ed.), *Perspectives in Experimental Gerontology*, C. C. Thomas, Springfield, 1966, pp. 69–82.

10 B. L. Strehler, G. Hirsch, D. Gusseck, R. Johnson and M. Bick, Codon restriction theory of ageing and development. *J. Theor. Biol., 33* (1971) 429–474.

11 D. Harman, Aging: Theory based on free radicals and radiation chemistry. *J. Gerontol., 11* (1956) 298–300.

12 D. Harman, Free radical theory of aging: effect of free radical inhibitors on the lifespan of LAF_1 mice. *J. Gerontol., 23* (1968) 476–482.

13 D. Harman, Free radical theory of aging: effect of amount and degree of unsaturation of dietary fat on mortality rate. *J. Gerontol., 26* (1971) 451–456.

14 F. M. Sinex, Cross-linkage and aging. *Adv. Gerontol. Res., 1* (1964) 165–180.

15 J. Bjorksten, The crosslinkage theory of aging. *J. Am. Geriatr. Soc., 16* (1968) 408–427.

16 F. Verzar, Aging of connective tissue. *G. Gerontol., 12* (1964) 915–921.

17 A. B. Robinson, J. W. Scotchler and J. H. McKerrow, Rates of nonenzymatic deamidation of glutaminyl and asparagynil residues in pentapeptides. *J. Am. Chem. Soc., 95* (1973) 8156–8159.

18 A. B. Robinson, R. Willoughby and L. R. Robinson, Age dependent amines, amides, and amino acid residues in *Drosophila melanogaster*. *Exp. Gerontol., 11* (1976) 113–120.

19 R. W. Hart and R. B. Setlow, Correlation between deoxyribonucleic acid excision–repair and lifespan in a number of mammalian species. *Proc. Natl. Acad. Sci. U.S.A., 71* (1974) 2169–2175.

20 H. P. von Hahn, Structural and functional changes of nucleoproteins during the ageing of the cell. *Gerontologia, 16* (1970) 116.

21 H. P. von Hahn, Failures of regulation mechanisms as causes of cellular ageing. *Adv. Gerontol. Res., 3* (1971) 1–38.

22 A. I. Klimenko, (Nucleic acids and histones of cell nuclei in ontogenesis). In V. I. Machinko (ed.), *Molecular and Functional Basis of Ontogenesis*, Medicina, Moscow, 1970, pp. 89–109.

23 F. Lipmann, Mechanism of peptide bond formation. *Fed. Proc., 8* (1949) 597–602.

24 A. G. Pasynsky, (The role of unequal peptide bonds in proteins). *Dokl. Akad. Nauk SSSR, 77* (1951) 863–866.

25 Zh. A. Medvedev, (Problem of turnover and ageing of intracellular proteins). *Usp. Sovrem. Biol., 33* (1952) 202–217.

26 A. V. Nagorny, V. N. Nikitin and I. N. Bulankin, (*Problem of Ageing and Longevity*), Medgiz, Moscow, 1963.

27 Zh. A. Medvedev, *Protein Biosynthesis and Problems of Heredity, Development and Ageing*, Plenum Press, New York, 1966.

28 R. B. Loftfield, The frequency of errors in protein biosynthesis. *Biochem. J., 89* (1963) 82–92.

29 R. B. Loftfield and D. Vanderjagt, The frequency of errors in protein biosynthesis. *Biochem. J., 128* (1972) 1353–1356.

30 P. Edelman and J. Gallant, Mistranslation in *E. coli*. *Cell, 10* (1977) 131–137.

31 Zh. A. Medvedev and M. N. Medvedeva, Use of H1 histone to test the fidelity of protein biosynthesis in mouse tissues. *Biochem. Soc. Trans., 6* (1978) 610–612.

32 J. H. Buchanan and A. Stevens, Fidelity of histone synthesis in cultured human fibroblasts. *Mech. Ageing Dev., 7* (1978) 321–334.

33 Zh. A. Medvedev and M. N. Medvedeva, Effect of age on the fidelity of H1 histone biosynthesis in mouse tissues. *Gerontologist, 18* (II) (1978) 100.

30

34 R. A. Popp, E. G. Bailiff, G. P. Hirsch and R. A. Conrad, Errors in human hemoglobin, as a function of age. *Interdiscip. Top. Gerontol., 9* (1976) 209–218.

35 G. P. Hirsch and R. A. Popp, Somatic base substitutions and errors in hemoglobin from aging mammals. *Gerontologist, 13* (1973) 47.

36 B. J. Harrison and R. Holliday, Senescence and the fidelity of protein synthesis in *Drosophila. Nature, 213* (1967) 990–992.

37 F. Dingley and J. Maynard Smith, Absence of a life-shortening effect of amino-acid analogues on adult *Drosophila. Exp. Gerontol., 4* (1969) 145–149.

38 A. N. Bozcuk, Testing the protein error hypothesis of ageing in *Drosophila. Exp. Gerontol., 11* (1976) 103–112.

39 R. Holliday, Errors in protein synthesis and clonal senescence in fungi. *Nature, 221* (1969) 1224–1228.

40 C. M. Lewis and G. M. Tarrant, Error theory and ageing in human fibroblasts. *Nature, 239* (1972) 316–318.

41 J. M. Ryan, G. Duda and V. J. Cristofalo, Error accumulation and aging in human diploid cells. *J. Gerontol., 29* (1974) 616–621.

42 R. Holliday and A. Stevens, The effect of an amino acid analogue, *p*-fluorophenylalanine, on longevity of mice. *Gerontology, 24* (1978) 417–425.

43 J. P. Ogorodnik, J. H. Wulf and R. G. Cutler, Altered protein hypothesis of mammalian ageing processes. Discrimination ratio of methionine *vs.* ethionine in the synthesis of ribosomal protein and RNA of C57BL/6J mouse liver. *Exp. Gerontol., 10* (1975) 119–136.

44 H. Gershon and D. Gershon, Detection of inactive enzyme molecules in ageing organisms. *Nature, 227* (1970) 1214–1217.

45 H. Gershon and D. Gershon, Inactive enzyme molecules in ageing mice: Liver aldolase. *Proc. Natl. Acad. Sci. U.S.A., 70* (1973) 909–913.

46 P. Zeelon, H. Gershon and D. Gershon, Inactive enzyme molecules in ageing organisms. Nematode fructose 1,6-diphosphate aldolase. *Biochemistry, 12* (1973) 1743–1750.

47 R. Holliday and G. M. Tarrant, Altered enzymes in ageing human fibroblasts. *Nature. 238* (1972) 26–30.

48 P. D. Wilson, Enzyme changes in aging animals. *Gerontologia, 19* (1973) 79–125.

49 M. Rothstein, Aging and the alteration of enzymes: A review. *Mech. Ageing Dev., 4* (1975) 325–338.

50 D. Gershon and H. Gershon, An evaluation of the "error catastrophe" theory of aging in the light of recent experimental results. *Gerontology, 22* (1976) 212–219.

51 R. J. S. Reis, Enzyme fidelity and metazoan ageing. *Interdiscip. Top. Gerontol., 10* (1976) 11–23.

52 N. I. Berlin, Life span of the red blood cell. In Ch. Bishop and D. M. Surgenor (eds.), *The Red Blood Cell,* Academic Press, New York, 1964, pp. 423–450.

53 Zh. A. Medvedev, (Biochemical mechanisms of ageing of nuclear and anuclear erythrocytes). *Cytologia, 15* (1973) 963–975.

54 Zh. A. Medvedev, Possible role of repeated nucleotide sequences in DNA in the evolution of life span of differentiated cells. *Nature, 237* (1972) 453–454.

55 Zh. A. Medvedev, Repetition of molecular–genetic information as a possible factor in evolutionary changes of life span. *Exp. Gerontol., 7* (1972) 227–238.

56 R. G. Cutler, Redundancy of critical genes in mammalian species of different maximum life spans. *Proc. 25th Ann. Meet. Am. Gerontol. Soc.,* Puerto Rico, 1972, p. 40 (abstract).

57 H. W. Woolhouse, The nature of senescence in plants. In H. W. Woolhouse (ed.), *Aspects of the Biology of Ageing,* [*Symp. Soc. Exp. Biol.,* No. XXI], Cambridge University Press, 1967, pp. 179–214.

58 P. F. Wareing and A. K. Seth, Ageing and senescence in the whole plant. In H. W. Woolhouse (ed.), *Aspects of the Biology of Ageing,* [*Symp. Soc. Exp. Biol.,* No. XXI], Cambridge University Press, 1967, pp. 543–558.

59 J. L. Peterson and E. H. McConkey, Non-histone chromatin proteins from HeLa cells. *J. Biol. Chem., 251* (1976) 548–554.

60 D. L. Wilson, M. E. Hall and G. C. Stone, Test of some ageing hypotheses using two-dimensional protein mapping. *Gerontology, 24* (1978) 426–433.

61 Zh. A. Medvedev, M. N. Medvedeva and L. Robson, Age-related changes of the pattern of non-histone proteins from rat and mouse liver chromatin. *Gerontology, 25* (1979) 219–227.

2A

Reprinted from *Proc. Natl. Acad. Sci. (U.S.A.)* **49:**517-521 (1963)

THE MAINTENANCE OF THE ACCURACY OF PROTEIN SYNTHESIS AND ITS RELEVANCE TO AGEING

By L. E. Orgel

UNIVERSITY CHEMICAL LABORATORY, CAMBRIDGE, ENGLAND

Communicated by Lord Todd, F. R. S., February 15, 1963

The ways in which the accumulation of mutations might contribute to the process of ageing in higher organisms or in individual clones of cells has been discussed at length.[1]   No corresponding treatment of the consequences of transcription errors in the translation of the DNA message into RNA and protein sequences seems to be available.   Here I show that a consideration of the rate of accumulation of such errors leads to a paradox, the resolution of which may be relevant to the problem of ageing, and that there are a number of simple experiments which should decide whether this is the case.

The basic idea is a simple one, namely, that the ability of a cell to produce its complement of functional proteins depends not only on the correct genetic specification of the various polypeptide sequences, but also on the competence of the protein-synthetic apparatus.   A cell inherits, in addition to its genetic DNA, the enzymes necessary for the transcription of that material into polypeptide sequences; the inheritance of inadequate protein-synthesizing enzymes can be as disastrous as the inheritance of a mutated gene.   Similarly, a cell may deteriorate through a progressive decrease in the adequacy of its transcrption mechanism, just as it may through the accumulation of somatic mutations.

As a basis for further discussion we shall suppose:[2] (*a*) that the sequence of amino acids in a protein is determined by the sequence of nucleotides in a corresponding region of the DNA; (*b*) that the immediate functions of DNA include the direction of the synthesis of transfer, microsomal, and messenger RNA but not of proteins; and (*c*) that the information for peptide sequence determination is carried by messenger RNA, but that other forms of RNA may affect the accuracy of protein synthesis.

There is little direct evidence concerning the accuracy of polypeptide-sequence determination.   The accuracy of synthesis of messenger RNA is unlikely to exceed that of DNA replication, for which an error frequency of 10^{-8} per base has been estimated.[1]   This would contribute an error frequency of about 3×10^{-8}

to protein synthesis if a degenerate three-letter code is assumed. However, the process of sequence determination also involves the specific reaction of the amino acids with their activating enzymes; this is likely to be a more important source of error. It must be difficult to distinguish a pair of amino acids as similar as valine and isoleucine even with an error level of 10^{-4}.

Our lack of knowledge of the error-frequency in protein synthesis makes it impossible to say, *a priori*, whether or not the accumulation of errors of protein sequence is relevant to the processes of ageing in higher organisms; this point must be decided experimentally. The nature of the mechanisms by means of which the accuracy of protein synthesis is maintained remains of interest, however low the error frequency. The first part of this paper will be concerned with these mechanisms.

We consider the following problem: suppose we could obtain a cell in which initially every polypeptide chain was in exact correspondence with its DNA sequence, the correspondence being assumed unambiguous. What would be the consequences of errors in protein transcription? We shall concentrate on what must be one of the main types of transcription error, namely, the replacement of a given amino acid by an incorrect amino acid. Let us roughly (and necessarily somewhat arbitrarily) divide the proteins of the cell into two sets, the first concerned with physical structure and intermediary metabolism, and the other concerned with the processing of genetic information, that is, the determination of the sequences of nucleic acids and polypeptides. The consequences of transcription errors for the two classes are very different.

If a small proportion of the protein molecules responsible for some metabolic function, say a part of glucose metabolism, were in error, then we might expect a somewhat lower *average* turnover number for the enzyme, a slightly reduced *average* specificity of the reaction, or perhaps a slight loosening of the *average* "control" of the reaction by feed-back inhibition, etc. These effects would not be cumulative; once the faulty messenger RNA or protein had been degraded, all memory of the error would soon be lost.

The situation is very different for the second class of proteins. Errors leading to complete loss of function would again affect only the efficiency of the information-transfer process in terms of average number of polymers replicated or synthesized per enzyme molecule. However, errors which lead to a reduced specificity of an information-handling enzyme lead to an increasing error frequency. Such processes are clearly cumulative and arguments which we shall next give in some detail suggest that, in the absence of an imposed selection for "accurate" protein-synthesizing units, must lead ultimately to an error catastrophe; that is, the error frequency must reach a value at which one of the processes necessary for the existence of viable cell becomes critically inefficient.

Under any given conditions the initial error frequency corresponding to the replacement of a particular amino acid in a particular position in a protein must be well-defined. Naturally it depends both on the amino acid replaced, the new and incorrect amino acid substituted (or, more correctly, on their base representations in the nucleic acid), and perhaps also on the environment in the peptide. Some replacements must certainly be more frequent than others on account of specificity at the transfer-enzyme level; further variety is introduced by any nonrandom distribution of errors of RNA synthesis. Similarly, each error in the protein-synthesizing system will induce a quite specific family of errors in the material synthesized; a loss of amino acid specificity of the phenylalanine activating enzyme, for example, could only effect substitutions involving the replacement of phenylalanine.

One feature of the dependence of the frequency of induced errors on the concentration of errors already present seems fairly certain, namely, that for small enough concentrations of errors the principal term is linear; that is, if we double the number of each kind of error already present in the protein-synthesizing system, we shall double the excess of induced errors over that in the initial error-free system. We shall not lose sight of the general features of the problem if, instead of considering the development of errors class by class, we lump them together and consider the error frequency p measured as the proportion of errors per amino acid present in polypeptide.

New protein synthesized by the initially correct enzymes will have some characteristic error concentration, say p_0. In the absence of any influence of errors already present on the frequency of error production, the error-level would settle down to p_0 once an amount of protein, large compared with that initially present, had been synthesized. If a slight linear dependence of the frequency of error production on the concentration of errors already present must be taken into account, we may, as a first approximation, write

$$\frac{dp}{dt} = \alpha p \tag{1}$$

where we consider the development of the system only after a time t_0 at which the error frequency p_0 has become established. This equation is only valid if p is small.

The solution of equation (1), $p = p_0 e^{\alpha t}$, shows that the error frequency initially increases exponentially and hence we predict the error catastrophe already mentioned. A proper treatment of this problem would take account of the time delay in the expression of errors. More importantly, the methods of probability theory should be used to determine the variance of p, etc.; the use of the differential equation (1) may be a rather poor approximation here, since the number of errors per cell may be small. These elaborations do not seem justified in the absence of more detailed experimental evidence.

How could an error catastrophe be avoided? Genetic selection for these proteins of DNA sequences, which in the course of transcription give particularly low error frequencies, may be possible within limits; selection for sequences which are completely inactivated by amino acid substitution would be more powerful. Selective scavenging of all incorrect proteins by hydrolytic enzymes might also contribute. However, in the light of recent work on ambiguous mutants,[3] it seems unlikely that these methods are powerful enough. What is needed is a selection based on the accuracy of protein synthesis, that is, a selection which rejects enzymes which lead to too many errors in protein synthesis. This could be achieved within a single cell only by a partial or complete segregation of the products of one piece of "protein-synthetic apparatus." All the available evidence argues against such segregation. We must presume then that the selection works at a cellular or higher level.

In any population of cells there must be a certain variance in the accuracy of the protein-synthetic apparatus so that a sufficiently severe selective pressure could eliminate the least accurate cells; otherwise, all cell lines would ultimately deteriorate. I shall try to put this result in a paradoxical form as follows: suppose we take a single bacterial cell and culture it under "ideal" conditions. After the first division we choose one daughter cell at random, allow it to divide again, choose one daughter cell at random, and proceed in this way in the absence of any "competition" between cells.[4] Then the arguments given above show that we finally get a nondividing cell even in the absence of mutation (naturally, the mutation frequency would also increase when the error frequency rose sufficiently). If these arguments are correct, one of the important functions of selection at cellular level is the maintenance of the accuracy of protein synthesis. Such a function would be additional to that usually discussed by geneticists and might quite possibly be equally important.

These speculations raise two important questions when applied to higher organisms. Firstly, does a cumulative inaccuracy in protein synthesis, connected with mutation only in so far as it is likely to lead to an increased somatic-mutation rate, play any role at all in the clearly extremely complex phenomenon of ageing? Secondly, what are the protective and selective mechanisms which prevent the error catastrophe in higher organisms?

There seems a fairly obvious experimental approach to the first of these questions. If we wish to study the effect of errors in protein synthesis in the absence of complications due to *primary* nucleic acid changes, we must increase such errors specifically, that is, without affecting nucleic acid synthesis. This could now be done by incorporating appropriate amino acid analogues. In the case of microorganisms of fairly representative selection of amino acid, errors could be induced by adding subtoxic concentrations of, say, p fluorophenylalanine, or, better, a mixture of p fluorophenylalanine and ethionine,[5] to the growth medium. A related experiment could almost certainly be done with mice. The level of incorporation of the analogue in these experiments should be far below that which causes death directly by gross inactivation of enzymes. In principle a pulse of analogue should produce consequences which cannot be reversed (except by selection) even after elimination of the analogue. It is perhaps worth remarking that it should be possible to study the effects of point mutations and primary errors of RNA synthesis by using bromouracil and fluorouracil, respectively, instead of amino acid analogues.

In higher organisms the situation is complicated by the existence of dividing and nondividing cells. The accumulation of transcription errors is likely to be particularly serious for the latter since, in the absence of division, selection at the cellular level is impossible. In slowly dividing tissue, selection may or may not be able to maintain the accuracy of protein synthesis; this can only be determined by experiment.

Now we come to the mechanism of protection or selection in higher organisms. If any part of the ageing process has to do with the accumulation of errors of polypeptide sequence, we need to know how it is that each new organism comes to have as clear a start as its parents. Separation of the germ line early may help in some way, but we have no reason to believe that the *enzymes* initially present in the egg have been subjected to a significantly reduced possibility of error. Various solutions seem possible; for example, selection in the growing embryo may be strong enough or the process of embryogenesis may demand such a high accuracy that its successful completion guarantees the necessary accuracy of protein synthesis. One can also conceive of special mechanisms of quality control; for example, special proteins might be synthesized which are converted by a certain class of errors into lethal polypeptides. This would guarantee that the frequency of this class of errors in viable cells is kept low. At present there is no evidence available which enables one to select among these possibilities.

I wish to make it quite clear that I am not proposing here that the accumulation of protein transcription errors is "the mechanism of ageing." My object is the more modest one of pointing out one source of progressive deterioration of cells and cell lines. Since I am unable to estimate the time scale of this process, I can only suggest experiments which should show where, if anywhere, it contributes to the ageing process in higher organisms.

I am indebted to Professor H. C. Longuet-Higgins and Dr. F. H. C. Crick for valuable criticisms of my original manuscript.

[1] Szilard, L., these PROCEEDINGS, **45,** 30 (1959); Maynard Smith, J., *Proc. Roy. Soc.,* **B157,** 115 (1962).

[2] Perutz, M., *Proteins and Nucleic Acids: Structure and Function* (Amsterdam: Elsevier, 1962).

[3] Benzer, S., and S. P. Champe, these PROCEEDINGS, **47,** 1025 (1961); **48,** 532 (1962).

[4] Note that in an exponentially growing culture there is a strong selection for cells with a short division time.

[5] See Kempner, E. S., and D. B. Cowie, *Biochim. et Biophys. Acta,* **42,** 401 (1960) and references therein.

2B

Reprinted from *Proc. Natl. Acad. Sci. (U.S.A.)* **67**:1476 (1970)

The Maintenance of the Accuracy of Protein Synthesis and Its Relevance to Ageing: A Correction

Leslie E. Orgel

SALK INSTITUTE FOR BIOLOGICAL STUDIES, SAN DIEGO, CALIFORNIA 92112

Communicated by Robert W. Holley, September 21, 1970

An argument[1] purporting to show that the accuracy of protein synthesis would deteriorate in the absence of cellular selection, thus leading to an "error catastrophe," contains a hidden assumption that no longer seems justified. I supposed that the error frequency in protein synthesis could be approximated as the sum of a residual error frequency (applicable where the protein-synthetic apparatus contains no errors) and a term dependent linearly on the number of errors already present in the protein-synthetic apparatus. I deduced that the error frequency would increase exponentially.

To clarify the nature of the hidden assumption I now consider a simpler model in which successive generations of the protein-synthetic apparatus are discrete and distinguishable. Let c_n be the error frequency in the nth generation, R the residual error frequency, and α the proportionality constant between errors in the synthetic apparatus and errors in freshly synthesized protein. Then

$$c_{n+1} = R + \alpha c_n$$

and if $c_0 = 0$

$$c_n = R \, (1 + \alpha + \alpha^2 + \ldots + \alpha^{n-1}).$$

If $\alpha > 1$, c_n increases indefinitely and we get an error catastrophe; if $\alpha \gg 1$, the error frequency increases exponentially. Otherwise, a steady-state error frequency of $R/(1 - \alpha)$ will be reached.

Arguments concerning the value of α turn out to be more subtle than I appreciated and it is not clear that $\alpha > 1$ under all circumstances. Thus, while an error catastrophe can occur, and apparently does in certain *Neurospora* mutants,[2] it may not be inevitable even in the absence of cellular selection.

[1] Orgel, L. E., *Proc. Nat. Acad. Sci. USA*, **49**, 517 (1963).
[2] Lewis, C. M., and R. Holliday, *Nature*, in press.

3

Reprinted from *J. Theor. Biol.* **82**:363–382 (1980)

Error Propagation in Intracellular Information Transfer

T. B. L. KIRKWOOD

National Institute for Biological Standards and Control, Holly Hill, Hampstead, London NW3 6RB, England

(*Received 9 April* 1979, *and in revised form* 3 *August* 1979)

The translation of genetic information from polynucleotides to proteins is mediated by proteins themselves. The cyclic nature of this process admits the possibility of a feedback of errors which may become lethal to the cell. During ageing, it is known that cells in some organisms show increased levels of altered or defective protein, and it has been suggested that the propagation of macromolecular errors may play a causative role in the progressive loss of homeostasis with increasing age. Experimental studies of this hypothesis have so far been inconclusive, and it is shown that theoretical models of intracellular error propagation throw important light on the determinants of stability within the translation apparatus and can improve the design of future experiments, as well as aid in their interpretation.

   Critical features of any model are its assumptions about the amino acid sequence changes required for a component of the translation apparatus to become error-prone and about the magnitude of any resultant change in activity. Existing models, which differ in these respects, are critically compared, and one is shown to be more flexible than the rest. In common with others, this model predicts that a normally stable translation apparatus has a threshold error level above which stability cannot be regained. The risk of crossing onto an irreversible path to cell death is determined by the distance between the stable and threshold error levels, and experiments to estimate this "safety margin" are suggested. Evolutionary modification of translational stability is also discussed.

1. Introduction

The fundamental characteristic of a living cell is its ability to translate into protein, and thereby implement, the information which is encoded in the polynucleotide sequences of its genes. The viability of a cell continues only as long as the integrity of its information-handling systems is preserved since, otherwise, the cell loses the capacity to repair molecular damage, for example due to thermal noise, or to multiply.

   Orgel (1963) first drew attention to the possibility that a cell might lose viability through a progressive breakdown in the accuracy of macromolecular information transfer. He pointed out that when information from DNA is

38

translated into protein a fraction of the newly-synthesized protein is destined itself to become a part of the information processing, or translation, apparatus. Errors in the synthesis of this fraction of protein might lead to the creation of erroneous but active units of the translation apparatus which would, by participating in fresh translation, generate further errors. In this way errors of protein synthesis could be amplified by positive feedback and in time, perhaps over several cell generations, could reach a lethal level. Orgel graphically described this outcome as an "error catastrophe".

In its original formulation, Orgel's hypothesis referred specifically to errors in protein synthesis as the source of error catastrophe, but subsequently Holliday & Tarrant (1972) have remarked that a build-up of errors in the accuracy of protein synthesis may also be expected to increase the frequency of mutations due to random errors in DNA polymerases. Orgel (1973) described how the elementary protein error theory may be extended into a more general error theory in which all macromolecular information transfer systems participate.

The main information transfer pathways involved in macromolecular biosynthesis within a typical cell are well known. Information, stored as DNA, is transcribed into secondary copies (RNA), which are then translated into protein. Proteins participate in DNA replication and repair, in transcription and in translation, so there exist three principal loops around which feedback can occur. The first involves distortion of primary information through errors of mutation or misrepair that may be induced by faulty DNA polymerases, the second involves the possible corruption of this information when it is transcribed to RNA by faulty RNA polymerases, and the third involves direct feedback of errors in the synthesis of ribosomal proteins and aminoacyl-tRNA synthetases. These loops are interrelated through the protein synthesizing machinery so that initiation of error feedback about any of them could trigger the feedback of errors around the others.

Originally, Orgel suggested that errors may be expected to multiply exponentially within the protein synthesizing machinery and he conjectured that this might explain the finite lifespans of cells such as fibroblasts which exhibit clonal senescence (Hayflick & Moorhead, 1961). He supposed that, if this were true, special mechanisms must operate to protect the germ line and other non-clonally senescing cells. Later he revised this suggestion to allow the alternative possibility that the error level might converge to a stable value (Orgel, 1970). Several experiments have been performed to test the relevance of the error theory to cellular senescence. Lewis & Holliday (1970) showed that two mutants of the fungus, *Neurospora crassa*, that exhibit clonal senescence do accumulate increased levels of faulty protein as they approach death, while Holliday & Tarrant (1972) demonstrated a similar effect in human fibroblasts. Fulder & Holliday (1975) found evidence of a rapid increase in the number of mutants during the late stages of senescence of human fibroblast populations, consistent with a general accumulation of macromolecular errors, and Linn *et al.* (1976) have since shown that the fidelity of DNA polymerase is reduced in old fibroblasts relative to young ones. These results agree well with the error theory's predictions but, although suggestive, they do not prove that cellular senescence is due to a feedback of errors rather than to some other process, such as post-translational protein modification (Gershon & Gershon, 1970; Gershon & Gershon, 1973), that also generates defective or altered protein. Furthermore, Evans (1977) has claimed that there is no corresponding evidence for an accumulation of defective protein in senescing mouse fibroblasts.

Clearly it may be some time before experimental evidence is available that is adequate to determine whether or not the error theory has any relevance to present-day organisms but, in the meantime, insight into the question may be gained from studies of theoretical models of the translation apparatus. In addition, a better theoretical understanding of the factors determining stability in self-replicating macromolecular systems should make it possible to improve the design of future experiments.

2. Linear vs. Non-linear Feedback of Errors

The first formal model of error propagation within the translation apparatus was described by Orgel (1970) in a correction to his original (1963) paper. For simplicity he assumed the translation apparatus to consist of discrete and distinguishable generations. He defined the error frequency of any one generation to be the average proportion of incorrect amino acid residues incorporated by that generation in the synthesis of fresh protein, and he denoted the error frequency in generation t by c_t. He then assumed c_t to consist of two parts, one a residual error frequency, R, such as would characterise a translation apparatus containing no errors, and the other an error frequency due to feedback of errors from the preceding generation. The error frequency contributed by feedback was assumed to be directly proportional to the total error frequency of the preceding generation, the constant of proportionality, α, being a fixed parameter of the system. Thus the model is expressed by the equation

$$c_t = R + \alpha c_{t-1} \qquad (1)$$

and it may readily be shown that if $\alpha > 1$ the error frequency increases without limit, resulting in catastrophe, while if $\alpha < 1$ the error frequency asymptotically approaches a steady level, $R/(1-\alpha)$.

The model is simple to understand and generates the important conclusion that, depending on the amount of error feedback (determined by α), the cell either deteriorates to an error catastrophe or maintains a stable existence. Beyond this, however, the limited realism of its assumption of linear feedback with a constant coefficient, α, is seriously restrictive. In general, the accuracy of a component of the translation apparatus will be determined by more than one amino acid residue, so the relationship between c_t and c_{t-1} will be reasonably complex. As a very simple example, consider a hypothetical polypeptide unit of the translation apparatus whose specificity is determined by exactly two residues and suppose that the insertion of an incorrect amino acid into either of these residues results in loss of specificity but not of activity. The proportion of erroneous copies of this unit in the generation t that are produced by feedback of errors from generation $t-1$ will be proportional to $1-(1-c_{t-1})^2 = 2c_{t-1}-c_{t-1}^2$ and not to c_{t-1}, as assumed by Orgel. (Of course, for *very small* c_{t-1}, Maclaurin's well-known theorem in mathematical analysis (for example, see Scott & Tims, 1966) permits any reasonable functional relationship

$$c_t = g(c_{t-1})$$

to be written as

$$c_t \simeq g(0) + c_{t-1} \cdot g'(0)$$

where g' denotes the first derivative of g. This is similar to equation (1) if we set $R = g(0)$ and $\alpha = g'(0)$. However, without detailed knowledge of the function g there is no simple way of determining at what value of c_t the adequacy of the linear approximation starts to break down.)

It should be stressed that the difference between linear and non-linear feedback has implications beyond merely altering the rate at which errors may accumulate. Orgel's model, with its assumption of linear feedback, requires that cells belong to one of two types, those with $\alpha < 1$ which are *always* stable, and those with $\alpha > 1$ which *never* are. If $\alpha < 1$, a cell will always revert to its stable error frequency even if a temporary change in external conditions generates a doubling or even a thousandfold increase in the error level.

3. Translation by Polypeptide "Adaptors"

A significant advance in modelling error propagation was made by Hoffman (1974), who proposed a more detailed model than Orgel's while retaining the assumption of discrete and distinguishable generations of the translation apparatus. Hoffman assumed that translation of the genetic message was effected in a single stage by polypeptide "adaptors". This has the advantage that, neglecting errors of DNA mutation or misrepair, attention is restricted to a single feedback loop. Hoffman further assumed that each adaptor contained a subset of m residues whose correct insertion was vital for activity, and a further non-overlapping subset of n residues vital for specific recognition of the correct substrate. Incorrect insertion of one or more of the m residues would result in total loss of activity, while incorrect insertion of one or more of the n residues would result in a total loss of specificity. For the purposes of modelling error propagation, attention can be restricted to those adaptors which contain no errors in any of the m sites, as any containing errors in these sites do not contribute to translation and therefore cannot affect its accuracy. Note, however, that if error propagation does occur, the proportion of active adaptors may become quite small.

The average accuracy of translation by the generation t adaptors, or probability that an amino acid is correctly translated, is denoted by q_t. (Accuracy, q_t, is directly related to the error frequency, c_t, since $q_t = 1 - c_t$).

The viable adaptors in generation t consist of two types: (i) a fraction q_{t-1}^n which contain no errors in the n sites, and (ii) a fraction $1 - q_{t-1}^n$ which contain errors in one or more of the n sites and which have therefore lost their specificity. The rate at which an adaptor which retains its specificity makes correct assignments is denoted by k (assignments per adaptor per second; Hoffman used α but this clashes with Orgel's terminology), while a dimensionless specificity, S, is defined such that the average rate with which this fraction makes one of the possible mis-assignments by incorporating one of the wrong amino acids in a peptide chain is given by k/S. If the number of different amino acids in the system is λ, there are $(\lambda - 1)$ wrong amino acids that can be incorporated, and the total average rate of making errors by adaptors of this fraction is $(\lambda - 1)k/S$. For the fraction $1 - q_{t-1}^n$ which loses its specificity, Hoffman assumed the rate of making correct assignments to drop to k/S, the same as the rate for making any particular mis-assignment.

On the basis of these assumptions, the average rates for generation t adaptors of making correct or incorrect assignments may be calculated, and the average accuracy, q_t, determined as the rate of making correct assignments divided by the total rate of making assignments, both correct and incorrect. This gives:

$$q_t = \frac{(S-1)q_{t-1}^n + 1}{(S-1)q_{t-1}^n + \lambda} \qquad (2)$$

42

Hoffman noted that this equation can be represented graphically in two ways. In the first, q_t is simply plotted against time, expressed in generations of adaptors. In the second, q_t is plotted against q_{t-1} in a format which I shall refer to as a q-diagram (Fig. 1).

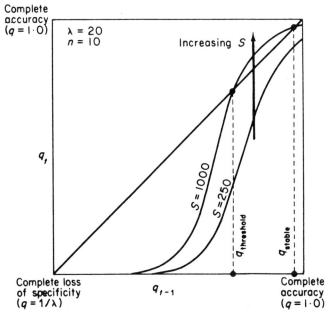

FIG. 1. The translation apparatus is assumed to consist of discrete generations of polypeptide adaptors responsible for translation of DNA into protein (Hoffman, 1974). The curves illustrate the relation [equation (2)] between the average accuracy, q_t, of generation t adaptors and the average accuracy, q_{t-1}, of the immediately preceding generation. See text for definitions of the parameters S, λ and n. This format is described as a q-diagram.

The q-diagram provides a convenient basis for discussing the stability of the translation apparatus. The diagonal line $q_t = q_{t-1}$ bisects the graph into two regions, one of increasing accuracy (upper left) and one of decreasing accuracy (lower right). On the line, accuracy remains constant. The important feature of equation (2) is that it predicts an S-shaped dependence of q_t on q_{t-1} and this allows the possibility that the curve may cross into the region of increasing accuracy and thereby define a point of stability where $q_t = q_{t-1} = q_{stable}$ and $\partial q_t / \partial q_{t-1} < 1$. Whether or not this happens depends on the specificity, S. For given λ and n, increasing S displaces the curve upwards and Hoffman showed that one can calculate a critical value of S above which translation is stable. In particular, for $\lambda = 20$ and $n = 10$ this critical value is $S = 500$ and hence he concluded that present-day translation, with measured specificity well in excess of this (Loftfield, 1963), should be sufficiently stable to rule out the possibility of error catastrophe occuring under normal conditions. (It is important to note, here and later, that most experimental data on translational specificity estimate q, not S, since it may be assumed that the experiments are performed with typical mixtures of correct and erroneous molecules. Hoffman's specificity, S, will usually be well in excess of the observed discrimination ratio).

4. A Modification of the Adaptor Model

In formulating his model, Hoffman made one crucial assumption that contributed heavily to his conclusion, but which does not seem biologically justified. For an adaptor which retains its specificity, the total activity (rate of making assignments both correct and incorrect) is $k + (\lambda - 1)k/S$, while for an adaptor which loses specificity through errors in one or more of the n sites, the total activity is assumed to be only $\lambda k/S$. This means that the activity of an erroneous adaptor is only $\lambda/(S + \lambda - 1)$ times that of an adaptor which retains its specificity. For any reasonable values of λ and S ($\lambda \leq 20$, $S \geq 1000$) this amounts to a drastic reduction in activity. Kirkwood & Holliday (1975) drew attention to this defect in Hoffman's model and cited numerous examples of enzymes with altered substrate specificity which retain most of their activity. In effect, Hoffman's model so greatly inhibits the feedback of errors that the conclusions it generates may have little meaning. A similar criticism of Hoffman's model was made independently by Ninio (1974), who noted that "The sole effect of an error in his (Hoffman's) model is to inactivate the gene product; an error does not lead to the possibility of making further errors. Thus Hoffman calculates the accumulation of error-containing, not error-producing, proteins. As a result, his model is exaggeratedly optimistic as regards the risk of error catastrophe."

Despite this limitation, Hoffman's work represented an important advance in discussing theoretical aspects of the error theory, and Kirkwood and Holliday showed how it could be modified to be more biologically realistic. As it is likely that errors will have some effect on the overall activity of an adaptor, a variable, R, was defined to be the average proportion of the total activity of an error-free adaptor which is retained by an adaptor containing errors in one or more of the n sites. Hoffman's model is a special case of this more general model corresponding to $R = \lambda(S + \lambda - 1)$.

Using Hoffman's notation, the modified equation (2) becomes:

$$q_t = \frac{q_{t-1}^n(\lambda S - (S + \lambda - 1)R) + (S + \lambda - 1)R}{q_{t-1}^n \lambda(S + \lambda - 1)(1 - R) + \lambda(S + \lambda - 1)R} \tag{3}$$

For $R = 1$, that is if all erroneous adaptors retain normal activity, equation (3) degenerates to

$$q_t = q_{t-1}^n \cdot \frac{(\lambda - 1)(S - 1)}{\lambda(S + \lambda - 1)} + \frac{1}{\lambda}. \tag{4}$$

The q-diagram of this equation (see Fig. 2) defines a curve of strictly increasing slope which lies entirely in the lower (unstable) region, so an error catastrophe is unavoidable.

For $\lambda/(S + \lambda - 1) < R < 1$, equation (3) defines an S-shaped curve which is intermediate between those given by equations (2) and (4), as illustrated in Fig. 2. For any set of values of S, λ, n, the question of stability depends on R, low R values tending to give a region of stability, so there exists a value R^*

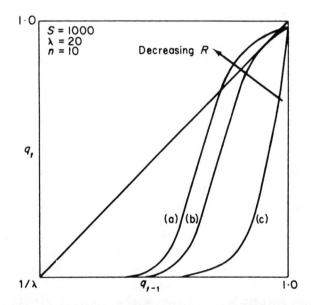

FIG. 2. q-diagrams illustrating the effect of the Kirkwood–Holliday (1975) modification of Hoffman's (1974) model. Curve (a) shows $R = \lambda/(S + \lambda - 1)$, equivalent to Hoffman's model (R is the average proportion of normal activity retained by erroneous adaptors); curve (b) shows $R = R^*$, when the curve just touches the line $q_t = q_{t-1}$; curve (c) shows $R = 1\cdot0$. See text for definitions of R^*, S, λ and n.

such that for $R > R^*$ the translation apparatus is unstable while for $R < R^*$ it is stable. Kirkwood and Holliday showed that as a good approximation,

$$R^* \simeq \frac{\lambda}{n(\lambda - 1)} \cdot \frac{S - C^{1/2}}{S + C^{1/2}} \tag{5}$$

where

$$C = (S - 1) \cdot \left(\frac{(\lambda - 1)(n - 1)}{2} - 1 \right).$$

For given λ and n, equation (5) shows that R^* increases with S to an asymptotic maximum $\lambda/(n(\lambda - 1))$. For example, if $\lambda = 20$ and $n = 10$, then however large S may be, there must be at least an 89% reduction in the activity of an erroneous adaptor if stability is to be possible. (Note that R depends not only on the intrinsic activity of the erroneous adaptors but also on their lifetime in the cell. Rapid breakdown of erroneous protein would decrease R).

The Kirkwood–Holliday modification of Hoffman's model is important for two reasons. Firstly, it avoids the assumption that erroneous adaptors necessarily lose most of their activity and, secondly, the introduction of the new variable, R, defines an additional degree of freedom in the evolution and control of the translation apparatus. For fixed λ and n, both R and S influence the shape and position of the curve in the q-diagram. Decreasing R, for example by means of a scavenging system for rapidly degrading erroneous proteins, displaces the curve upwards and to the left, while leaving the end points fixed. Increasing S causes a predominantly vertical displacement of the curve, including the right-hand end point. Changing both together results in a combination of these effects.

The model remains open to several criticisms, of which the most serious are: (i) the all-or-none loss of specificity is unrealistic and it is more likely that errors in recognition sites would progressively alter or reduce specificity; (ii) the assumption of discrete generations is certainly an over-simplification and ought, in due course, to be replaced by a continuous time model, taking account of rates of protein synthesis and turnover; (iii) by restricting attention to hypothetical polypeptide adaptors and considering only the average accuracy of these, the model fails to take account of the more complex realities of transcription and translation and cannot describe a non-uniform increase in error frequencies in different information handling enzymes.

Nevertheless, the Hoffman–Kirkwood–Holliday (HKH) model does provide a useful framework within which to discuss the stability characteristics of self-replicating macromolecular systems and, at least with regard to the elementary protein error theory, it generates a number of non-trivial results. Firstly, in common with Orgel's model, it predicts that there may exist two kinds of translation apparatus, that which is stable and that which is not. Secondly, it suggests that the stability, where it occurs, will not be absolute and that if, for some reason, the accuracy falls or is forced below a certain critical level ($q_{\text{threshold}}$), stability will not be regained and error catastrophe will occur. Thirdly, it identifies two principal variables, R and S, which determine the degree of stability.

5. Errors in Aminoacyl-tRNA Synthetases

A further model of error propagation in protein synthesis was proposed by Goel & Ycas (1975) and has been further developed in two later papers (Goel & Ycas, 1976; Goel & Islam, 1977). Like the previous models, the Goel–Ycas model describes the feedback of errors around a single information transfer loop, specifically in the synthesis of aminoacyl-tRNA synthetases. In this sense it is equivalent to Hoffman's model, and it again assumes discrete generations. The difference is that, instead of considering the *average* accuracy of whole generations of synthetases, Goel & Ycas argued that for realism it was necessary to take account of the kinetic behaviour of different synthetases. Defining S_i to be the synthetase responsible for inserting amino acid i, they assumed, like Hoffman, that a subset of amino acid residues in each synthetase needed to be correct for the synthetase to have any activity at all, and that a further subset was required to be correct for normal specificity.

Specifically, they defined x_{ij} to be the total number of residues of amino acid j in S_i necessary for normal function (i.e. normal specificity and activity) and y_{ij} to be the subset of x_{ij} needed only for specificity; in other words, the subset $x_{ij} - y_{ij}$ are those residues essential for activity. Whereas Hoffman assumed that any error in any of the y_{ij} sites (any j) would cause a total loss of specificity, Goel and Ycas demanded that *all* the y_{ij} residues must be incorrectly substituted in a *specified* way for the synthetase to have a specified new (erroneous) amino acid specificity. Like Hoffman, they did not consider changes in codon specificity. Thus they defined three types of synthetases: (i) normal synthetases, (ii) inactive synthetases, (iii) active but erroneous synthetases. All active synthetases were assumed to have the same activity. Goel & Ycas then derived a relationship between the relative frequencies of normal and erroneous synthetases in successive generations which formed the basis of their theoretical development of the model. Letting $f_i(t)$ and $f_i'(t)$ denote the fractions of synthetases S_i in the generation t that were normal and erroneous respectively, they obtained the equations:

$$f_i(t) = \prod_j \left(\frac{f_j(t-1)}{f_j(t-1)+f'_j(t-1)} \right)^{x_{ij}} \tag{6a}$$

$$f'_i(t) = \prod_j \left(\frac{f_j(t-1)}{f_j(t-1)+f'_j(t-1)} \right)^{x_{ij}-y_{ij}} \left(\frac{f'_j(t-1)}{f_j(t-1)+f'_j(t-1)} \right)^{y_{ij}} \tag{6b}$$

(N.B. Goel and Ycas used q in place of f). Dividing equation (6a) by (6b) gave the relationship

$$V_i(t) = \prod_j [V_j(t-1)]^{y_{ij}} \tag{7}$$

where $V_i(t) = f_i(t)/f'_i(t)$ is the ratio of normal to erroneous synthetases S_i. Taking the logarithm of this equation generated the matrix relationship

$$Q(t) = YQ(t-1) \tag{8}$$

where Q is a column vector with $Q_i = \log V_i$ and Y is the matrix (y_{ij}).

A striking feature of equations (7) and (8) is their total symmetry with regard to normal and erroneous synthetases; there is no change if V_i is defined instead to be the ratio of erroneous to normal synthetases, rather than the other way around. The reason for this is readily seen in the definition of an erroneous synthetase since the *requirement for a synthetase to be erroneous is as stringent as for it to be normal*; in either case the same subset of amino acid residues must be precisely specified. This requirement, when examined more closely, throws light on a highly questionable assumption in the derivation of equation (6), namely that any synthetase which contains errors in only some of the y_{ij} sites or which contains the 'wrong' errors in these sites does not feature in the model. This class of synthetases would normally far outnumber that which has been defined to be erroneous, and it seems quite unreasonable to suppose that all such synthetases would be inactive. In effect, by placing such stringent requirements on erroneous synthetases and by choosing to ignore those synthetases which are neither normal nor sufficiently wrong to be "erroneous", Goel & Ycas have constructed a model which so artificially restricts error propagation that it can have little practical use in studying the plausibility of the error theory. This point is examined more closely in the next section.

6. Comparison of the Goel–Ycas and Hoffman–Kirkwood–Holliday Models

In the second of their papers, Goel & Ycas (1976) contrast their model with the HKH model and make two main criticisms of the latter. Firstly, they question the concept of specificity being determined simply by differential rates of insertion of correct and incorrect amino acids. Secondly, since an adaptor is assumed to be erroneous if even one of the n sites determining specificity is incorrect, they point out that the HKH model allows for a very large number of erroneous adaptors. (Note that the number of adaptors here means the total number of different ones that potentially could arise, not the actual number present within the translation apparatus at any one time).

Regarding specificity, Goel & Ycas are correct in as much as the rates of insertion of correct and incorrect amino acids will depend on factors other than the molecular structure of the translation apparatus, for example on the relative concentrations of correct and incorrect amino acids in the available free pool. Nevertheless, in modelling error propagation, it is reasonable to assume that such factors remain more or less constant since they fall outside our sphere of interest. Indeed, it is hard to see how specificity in this context could be simply described in any way other than by differential insertion rates. It should also be noted that, far from implementing a more constructive alternative, Goel & Ycas, who favour instead the idea that specificity is "more like a discrete (or go–no go) quantity", assume in their model that correct synthetases never make mistakes and erroneous ones never insert the correct amino acid. This somewhat rigid approach is at variance with comtemporary understanding of enzyme-substrate interactions.

Regarding the requirement for loss of specificity, the point made by Goel and Ycas is an important one which needs careful consideration in the context of both models. Clearly it matters greatly which assumptions are made about what is needed for a change in specificity, since this directly influences the amount of error feedback that is assumed to occur. In fact, careful consideration of the parameter R (average proportion of normal activity retained by erroneous adaptors) shows that the HKH model is immune to their criticism for, since R is an *average* over the whole class of erroneous adaptors, it can take account of wide variation in activity among them. If only a small proportion of the 'erroneous' adaptors were, in fact, active, R might also be very small. However, since attention is already restricted to those sites which are responsible for specificity, rather than activity, there is no reason to suppose that this need be the case.

If we consider now the Goel–Ycas equations (6a) and (6b), we find they are consistent with a range of assumptions about the erroneous synthetases. At one extreme, they represent the case where each correct synthetase has only one 'erroneous' alternative (in other words all the y_{ij} sites must be incorrectly occupied in a unique way); at the other, the occupation of the y_{ij} sites by *any* of the incorrect amino acids produces erroneous synthetases which are completely non-specific in that they can insert *any* amino acid, *except the correct one*. However, in all but the latter extreme, the equations embody an important hidden assumption, namely that the erroneous synthetases are mutually *coherent*, in other words the new specificities of the erroneous synthetases are such that they act together to insert just those wrong amino acids that are required for a newly-synthesized synthetase itself to be erroneous. In this case, the process represented in the Goel–Ycas model is, in fact, the intracellular competition between two rival translation apparatuses, neither of which ever makes mistakes and which are mutually completely incompatible. However, the residues crucial for activity must always be inserted by the 'normal' synthetases. Whether this particular phenomenon is of scientific interest or not is debatable, but it certainly has little or no relevance to the theoretical study of whether error catastrophe is a serious threat to an established translation apparatus.

The remaining possibility within the framework of the Goel–Ycas model, namely that any set of incorrect y_{ij} residues generates complete loss of specificity, is not very different from the HKH assumption, except that it does not allow changes in specificity to result from only minor errors such as a single incorrect residue among the y_{ij} sites. This seems unnecessarily restrictive. (Note that if the error frequency is 10^{-3}, a single residue has a probability 10^{-3} of being incorrect, while a pair of residues only has a probability of 10^{-6}, and so on.) The reason for retaining this restrictive assumption is not hard to surmise, since without it the symmetry of equations (7) and (8) disappears and the mathematics of the model become enormously less tractable.

In summary, the main advantage of the Goel-Ycas model over the HKH model is that it considers individual synthetases rather than averages, and in this respect has the potential to be more realistic. However, this advantage is heavily outweighed by the highly restrictive nature of its assumption about erroneous synthetases.

7. Evolution of Stability in the Translation Apparatus

Eigen (1971) observed that the nucleation of the genetic code probably involved two sets of adaptors which made the same amino acid-codon assignments. The first resulted from random polymerisation and led to the initial translation of those nucleic acid sequences which happened, by chance, to be present. This translation led to the production of a second set of adaptors which then took over the translation process. Hoffman (1974) concurred with this view and pointed out that, in terms of his model, the initial specificity of the second set of adaptors would have needed to exceed the threshold value required for stability and that the accuracy of the preliminary translation by the random precursor set of adaptors would also have had to be greater than the corresponding accuracy threshold.

Ninio (1974) criticised Hoffman's use of his model in this context, on the grounds that it was unrealistic to consider the evolution and stability of the translation apparatus in terms of errors expressed as deviations from an archetypal norm. Instead, he argued "a correct treatment would have to give a complete description of the recognition processes in the systems and, by evaluating their discriminatory capacity, would model their evolution." He predicted that, because of its complexity, it would be some time before such an approach could generate biologically useful results, a view with which it is hard to disagree.

While agreeing with Ninio (1974) that ultimately it is unrealistic to model the evolution of the translation apparatus in terms of deviations from a norm because this precludes the possibility of the norm itself progressing, I suggest that the HKH model contributes to an understanding of how the translation apparatus may have developed following its original genesis. Once the genetic code was established (perhaps in a significantly more primitive and inexact form), the frequency with which the translational norm changed through the establishment of beneficial mutations would have been relatively small compared with the frequency of disadvantageous or "erroneous" departures from the norm. To a first approximation, therefore, the norm could be taken to be static over any short period of evolutionary time and the HKH model permits discussion of the stability of the translation apparatus at that time. In particular, it allows identification of two parameters of the translation apparatus, R (average proportion of normal activity retained by erroneous adaptors) and S (specificity), whose evolutionary modification could control its stability.

The notion that R and S, and hence the stability of translation, may be under genetic control allows ready dismissal of a seeming paradox of the error theory, namely that if present-day cells really are unable to maintain the accuracy of protein synthesis, it is hard to explain how primitive cells, presumably with less accurate translation, were ever able to reproduce. It has been suggested (Orgel, 1963; Orgel, 1973; Holliday, 1974) that cellular selection could maintain healthy growth within a population of cells, but this requires that at least some of the population can maintain a stable error level from one cell generation to the next. The notion that cells may normally operate at a stable error level, but that they may be subject to a continual risk of randomly crossing a threshold error level onto an essentially irreversible pathway towards error catastrophe is consistent with the HKH model if $q_{threshold}$ is not too far below q_{stable} (Kirkwood & Holliday, 1975; Kirkwood,

1977). It is also possible that error propagation may be largely confined to the somatic cells of multicellular organisms and that special mechanisms may exist to enhance the accuracy of protein synthesis in the germ line (Orgel, 1973).

Elsewhere (Kirkwood, 1977), I have shown that the concept of controlling stability by modifying R and S can be used to relate the error theory to the evolution of ageing in higher animals. It is supposed that high accuracy protein synthesis requires energy (Hopfield, 1974) and that in multi-cellular organisms an energy-saving strategy of reduced error regulation in somatic cells, but not in the germ-line, allows the organism to reproduce more effectively though at the cost of ultimately experiencing senescence due to propagation of macromolecular errors in somatic tissue. This principle would not apply to unicellular organisms which may be expected to maintain relatively high accuracy in protein synthesis and should, therefore, be fairly stable against error catastrophe.

8. Experimental Test of the Error Theory

To test the error theory decisively it will be necessary to show whether or not some cells do indeed succumb to lethal feedback of error in information transfer. This presents considerable difficulties, particularly since error catastrophe is only loosely defined and further work is necessary to determine what would, in practice, constitute a lethal breakdown in the accuracy of information transfer. Certainly, the change in accuracy would not need to be large to cause a significant reduction in the proportion of error-free proteins synthesized: for example, if the error level $(1-q)$ were to increase from 10^{-4}, say, to 10^{-3}, the proportion of error-free proteins containing only 100 amino acid residues would drop from 99% to 90%, while for larger proteins of 1000 residues the drop would be from 90% to 37%. Thus a relatively small increase in error level could seriously threaten a cell's continued viability. It should also be noted, in this context, that the first effect of a drop in accuracy of macromolecular synthesis may be to prevent a cell from undergoing further mitosis, rather than to cause its immediate death; this is particularly significant in experimental systems, such as fibroblast culture, where the measure of senescence is loss of cell proliferative capacity, rather than cell death itself.

Nevertheless, two types of experiment are possible which should greatly aid in determining the theory's credibility. Firstly, naturally senescing cells may be studied for evidence of a general build-up of errors. Experiments of this kind, some of which were cited in section 1, are limited in that they can only demonstrate the existence of errors, not their propagation, and in that the isolation and characterization of heterogeneous and potentially labile erroneous proteins may be difficult. Secondly, experiments involving deliberate *in vivo* modification of the accuracy of transcription and translation should allow the stability of translation apparatus to be measured. If the error theory is correct, these experiments may accelerate naturally occurring senescence or induce senescence in normally immortal cell lines. Theoretical models of the translation apparatus can play an important role in designing and interpreting this second kind of experiment, a few of which have already been performed.

Holliday & Tarrant (1972) grew human fibroblasts in medium containing a low concentration of the base analogue 5 fluorouracil, which is incorporated into RNA and may be expected to raise the error level. They observed that, after addition of the analogue, their cultures initially grew normally but that the treated cultures showed the characteristic morphological symptoms of senescence much earlier than control cells and that they accumulated erroneous protein more rapidly. Also, to try to induce error catastrophe in cells which normally multiply indefinitely, several workers have cultured *Escherichia coli* in medium containing aminoglycoside antibiotics which increase the rate of ribosomal mistranslation. Conflicting results have been reported. Branscomb & Galas (1975) grew *E.coli* in medium containing low levels of streptomycin and reported that after addition of the drug they found a progressively increasing level of erroneous protein together with a delayed progressive decrease in growth rate and increase in the number of visibly sick cells. They concluded that error feedback in protein synthesis does seem to occur in *E.coli* and that these cells can be induced to experience error catastrophe. These observations have recently been confirmed and extended by Rosenberger and Holliday (1979). However, Edelmann & Gallant (1977) performed similar experiments and found that an initial rise in error frequency after addition of streptomycin subsequently declined and that the error frequency reached a limiting value. Despite the fact that this limiting error frequency was about 40 times higher than normal, the cells did not show signs of further progress towards error catastrophe and reverted in a few generations to normal when the streptomycin was removed. Edelmann and Gallant interpreted their results as evidence against the possibility of uncontrolled error feedback and estimated that, in terms of Orgel's (1970) model, the feedback parameter, α was approximately 0·8 (this, being less than 1, corresponds to stability). They also considered their results in terms of the HKH model and commented that, since translation was apparently stable, $R < R^* < \lambda/n(\lambda - 1)$.

Consideration of the HKH model shows that these results are not fundamentally incompatible. The effect of adding streptomycin to the culture medium is roughly equivalent to decreasing the specificity, S, while leaving the other parameters of the model unchanged. *E.coli* may be assumed normally to be stable, but there may be variation among strains in the degree of stability and in the sensitivity of S to streptomycin. Adding streptomycin will cause a downward displacement of the curve in the q-diagram and, depending on dose and the cell strain used, may or may not result in instability (Fig. 3). If it does, a progressive increase in error level will occur, leading to catastrophe. In some cases, adaptation within the cells or selection of streptomycin-resistant variants within the population may prevent the culture from dying out completely, and apparent recovery may occur. This was discussed by Branscomb & Galas (1975). Alternatively, if

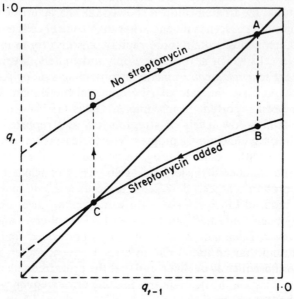

FIG. 3. Effect of streptomycin on the accuracy of translation in a stable organism (e.g. *E.coli*) according to the HKH model. Without streptomycin, the accuracy remains at the stable point, A on the upper curve. Adding streptomycin decreases specificity, S, and displaces the curve DA down to the curve CB. Accuracy falls to lower stable point, C. On removing streptomycin, curve CB is restored to DA and accuracy returns to the stable point A. In practice, there would be a time lag while streptomycin was taken up by the cells and diluted out from them, so the change in accuracy between A and C, and back again, would not exactly follow the indicated pathway but would be more gradual. If curve CB had no stable point (not shown) accuracy would fall progressively after addition of streptomycin and would only return to A on removal of streptomycin if it had not already fallen below $q_{threshold}$ for curve DA. Note that the Figure shows only the upper right corner of the q-diagram and that the error level, c_t, is equal to $1 - q_t$.

the addition of streptomycin does not result in instability, it will, nevertheless, decrease the stable level of accuracy so that the error level will increase to a new limiting value which may be considerably in excess of normal. On removing streptomycin the error level would return to normal, as observed by Edelmann & Gallant (1977). (It is worthy of note that, although Edelmann and Gallant estimated α to be less than 1, their figure of $\alpha \simeq 0 \cdot 8$ does suggest that a considerable amount of error feedback was occurring in their cultures). In this way, judicious experimentation using aminoglycoside antibiotics with different cell strains, including mutants known to have high levels of mistranslation, may provide a direct means of testing the qualitative predictions of the theoretical models and of estimating values for their parameters.

9. Conclusions and Discussion

As highlighted by correspondence following a recent debate by Hill & Franks (1977) and Holliday (1977), there is considerable controversy as to whether cellular senescence is more likely to be due to error feedback or to some sort of genetic programming. However the final answer to this turns out, Orgel's (1963, 1973) error theory has an intrinsic logic which makes it potentially relevant to every living cell. Errors in information transfer are known to occur and indeed, at the basic level of quantum uncertainty in chemical bonds, are inevitable. Erroneous information handling enzymes are also known to exist and it is therefore certain that there is real potential for error feedback in any information transfer loop. What we do not yet know is whether this theoretical possibility is so remote as to be irrelevant to ageing, whether it may actually be the primary mechanism of cellular senescence, or whether the truth lies somewhere between.

Since experimental evidence to date is sufficiently ambiguous as not to provide a conclusive answer either way, there is scope for theoretical models of information transfer systems to make a significant contribution to realistic discussion of this question. As with all biological problems, the difficulty is to define models which are both sufficiently realistic to be practically useful and sufficiently simple to be readily understood. Inevitably, these two requirements tend to be in opposition. It has been argued that Orgel's (1970) model is too simple to generate useful predictions of experimental results, while flaws have been identified in the models of Hoffman (1974) and Goel & Ycas (1975). This leaves only the Kirkwood & Holliday (1975) modification of Hoffman's model. As noted previously, this is not by any means the ideal model, since it suffers from a number of obvious limitations, and it is to be hoped that progress will be made towards new and better alternatives.

In the meantime, the HKH model does provide a useful tool for elucidating the general characteristics to be expected of the translation apparatus and, for example, forms a convenient basis for relating the error theory to the evolution of ageing (Kirkwood, 1977). In particular, the identification of the two controlling parameters, R (proportion of normal activity retained by erroneous adaptors) and S (specificity of normal adaptors), while clearly an oversimplification, should prove useful in relating experimental results to the error theory's prediction.

Future models will need (i) to replace the all-or-none specificity by a more gradual progression between states, (ii) to take account both of the stochastic and of the continuous time nature of the process, (iii) to permit consideration of individual information processing units rather than averages, (iv) to allow for multiple feedback loops at different levels of information transfer. It is to be hoped that concurrent advances in experimental research will generate realistic development of the theoretical work. Even if the error theory turns out not to play a major part in cellular senescence this research is likely, as Holliday (1969) remarked, to throw a good deal of new light on the mechanism of information transfer from DNA to protein, as well as on some other basic cellular functions.

I thank E. W. Branscomb, D. J. Galas, J. A. Gallant, J. Ninio, L. E. Orgel, R. Rosenberger and, in particular, R. Holliday for helpful discussions.

REFERENCES

BRANSCOMB, E. W. & GALAS, D. J. (1975). *Nature, Lond.* **254**, 161.
EDELMANN, P. & GALLANT, J. (1977). *Proc. natn. Acad. Sci. U.S.A.* **74**, 3396.
EIGEN, M. (1971). *Naturwissenschaften.* **58**, 465.
EVANS, C. H. (1977). *Exp. Geront.* **12**, 169.
FULDER, S. J. & HOLLIDAY, R. (1975). *Cell.* **6**, 67.
GERSHON, H. & GERSHON, D. (1970). *Nature, Lond.* **227**, 1214.
GERSHON, H. & GERSHON, D. (1973). *Proc. natn. Acad. Sci. U.S.A.* **70**, 909.
GOEL, N. S. & YCAS, M. (1975). *J. theor. Biol.* **54**, 245.
GOEL, N. S. & YCAS, M. (1976). *J. math. Biol.* **3**, 121.
GOEL, N. S. & ISLAM, S. (1977). *J. theor. Biol.* **68**, 167.
HAYFLICK, L. & MOORHEAD, P. S. (1961). *Exp. Cell. Res.* **25**, 585.
HILL, B. & FRANKS, L. M. (1977). *Trends biochem. Sci.* **2**, N80.
HOFFMAN, G. W. (1974). *J. mol. Biol.* **86**, 349.
HOLLIDAY, R. (1969). *New Scientist* **42**, 532.
HOLLIDAY, R. (1974). *Fed. Proc. Fed. Am. Soc. exp. Biol.* **34**, 51.
HOLLIDAY, R. (1977). *Trends biochem. Sci.* **2**, N80.
HOLLIDAY, R. & TARRANT, G. M. (1972). *Nature, Lond.* **238**, 26.
HOPFIELD, J. J. (1974). *Proc. natn. Acad. Sci. U.S.A.* **71**, 4135.
KIRKWOOD, T. B. L. (1977). *Nature, Lond.* **270**, 301.
KIRKWOOD, T. B. L. & HOLLIDAY, R. (1975). *J. mol. Biol.* **97**, 257.
LEWIS, C. M. & HOLLIDAY, R. (1970). *Nature, Lond.* **228**, 877.

LINN, S. KAIRIS, M. & HOLLIDAY, R. (1976). *Proc. natn. Acad. Sci. U.S.A.* **73**, 2818.
LOFTFIELD, R. B. (1963). *Biochem. J.* **89**, 82.
NINIO, J. (1975). In: *Ecole de Roscoff 1974—l'Evolution des Macromolécules Biologiques* (C. Sadran, ed.) pp. 51–68. Paris: Editions du C.N.R.S.
ORGEL, L. E. (1963). *Proc. natn. Acad. Sci. U.S.A.* **49**, 517.
ORGEL, L. E. (1970). *Proc. natn. Acad. Sci. U.S.A.* **67**, 1476.
ORGEL, L. E. (1973). *Nature, Lond.* **243**, 441.
ROSENBERGER, R. & HOLLIDAY, R. (1979). In preparation.
SCOTT, D. B. & TIMS, S. R. (1966). *Mathematical Analysis: An Introduction* Cambridge: University Press.

4

MISTRANSLATION AND AGEING IN *NEUROSPORA*

C. M. Lewis and R. Holliday

ORGEL[1] has proposed that cellular senescence could be attributed to an accumulation of errors in the protein synthesizing machinery of cells. Slight alterations in the structure of enzyme proteins involved in biosynthetic pathways might not have any major effect and might, at worst, be expected to reduce the metabolic efficiency by a fixed amount[2,3]. But slight alterations in other enzymes concerned with transcription and translation might be expected to increase the frequency of errors in all proteins of the cell and thus give rise to the ever decreasing metabolic efficiency which is so characteristic of many ageing tissues.

Two mutants of *Neurospora crassa* have characteristics of senescence. One is the *leu-5* auxotroph, the ageing of which is sensitive to temperature. At low temperatures, 20°–25° C, this mutant grows indefinitely, but when mycelium is transferred to high temperatures, about 37° C, growth continues apparently quite normally for 3 to 4 days and then suddenly ceases. Printz and Gross[4] have provided evidence that the mutant has an abnormal leucyl-tRNA synthetase which seems to charge leucyl-tRNA with other amino-acids. The characteristics of all the other enzymes that Printz and Gross examined and which were unrelated to the synthetase changed in a way which suggested that the mistranslation extended to all the proteins of the cell.

Another mutant of *Neurospora* which senesces is *nd*, natural death, which was isolated by Sheng[5]. This mutant grows at a constant rate until, quite suddenly, the culture dies. It is also temperature sensitive to some extent, for it grows for a much longer time at low temperatures than at high temperatures. Holliday[6] has shown that in an *ad-3 nd* strain the leakiness of the missense *ad-3* mutation increased with age of the *nd* and this suppressive effect suggests that there is a concomitant increase in error frequency which may be also responsible for the ultimate death of the mutant.

In both cases, if Orgel's theory is to be supported, it is important to know whether the cessation of growth and death of the fungus result from an exponentially increasing frequency of errors or whether a constant reduction in metabolic efficiency brought about by a fixed level of altered proteins will have the same effect.

Although experiments with *Drosophila* and fungi[6,7] suggest that senescence results from a breakdown in the maintenance of accurate protein synthesis, there is little

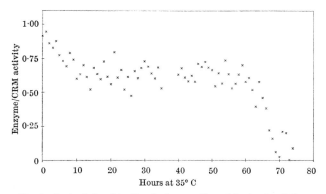

Fig. 1. Ageing induced by transferring mycelium of the *leu-5* mutation growing in Vogel's minimal medium supplemented with 150 μg/ml. of leucine from 25° to 35° C. The graph shows the change with time in the ratio of glutamic dehydrogenase to enzyme cross-reacting material (CRM), one unit of cross reacting material being that which inhibits one unit of wild type enzyme activity. Antisera and assays for CRM were based on the methods of Roberts and Pateman[9] and Roberts[10].

biochemical evidence to support the data. It would be pertinent, therefore, to demonstrate that during the senescent period there is a progressively increasing frequency of errors in the structure of individual proteins which might reflect a generalized imprecision in protein biosynthesis.

We have investigated this question by two approaches the results of which are shown in Figs. 1–6. Figs. 1, 2 and

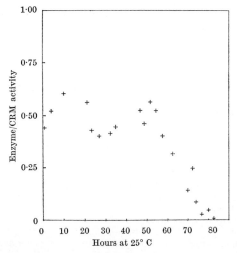

Fig. 2. The change with time in the ratio of specific activity of glutamic dehydrogenase to CRM with the *nd* mutant. The initial age of the culture was 6 days.

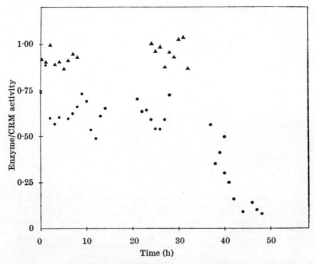

Fig. 3. The change with time of the ratio of specific activity of glutamic dehydrogenase to CRM with mycelium grown in ethionine (●), 0·1 mg/ml., and mycelium maintained in saline (▲).

3 show the results of experiments designed to determine whether the decrease in active enzyme which is observed during senescence is the result of a reduction in synthesis or whether it is the result of synthesis of inactive enzyme protein. If mistranslation increases exponentially, the ratio of active to inactive enzyme might be expected to change exponentially too. Glutamic dehydrogenase was chosen for its convenience of assay[8], and crude extracts prepared from mycelium of different ages were assayed both for active enzyme and for the quantity of glutamic dehydrogenase cross-reacting material (CRM)[9,10].

The time course of the changes in ratio in mycelium of the *leu-5* auxotroph can be determined from the onset of ageing, that is, when mycelium is transferred from 25° C to 35° C. The earliest stages of the *nd* mutation, however, are more difficult to assay. This mutant is maintained in a heterokaryon and in isolating an *nd* monocaryon some 4 to 6 days' growth has taken place before there is sufficient mycelium to assay. It is not therefore known whether the amount of enzyme activity during the early stages of growth of this mutant ever approaches that in the wild type.

Amino-acid analogues have been used to induce alterations in proteins[6,7]. Harrison and Holliday proposed that these increased the frequency of mistranslation, and they did prove effective in shortening the life span. That this was the result of an accumulation of errors and

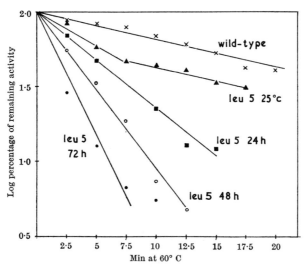

Fig. 4. The effect on glutamic dehydrogenase of heating at 60° C crude extracts from *leu-5* of different ages.

not merely the result of toxic effects of the anti-metabolites is suggested by the results shown in Fig. 3. The ratio of enzyme to enzyme cross-reacting material from mycelium grown in ethionine undergoes changes which are very similar to those of the two senescent mutants.

The extent of the defects in proteins is reflected by the subsequent rate of growth of samples taken from the culture growing in ethionine. Although all samples ultimately recovered, presumably as a result of heterogeneity in the mycelium, the growth of the later samples was severely inhibited for some 48 h.

Fig. 3 also shows the values for wild type mycelium maintained in saline for 2 days. The fact that the ratio of enzyme to cross-reacting material remains constant in these conditions when mycelium stops growing would seem to contradict any interpretation that the changes in ratio observed with the mutants is simply the result of instability of enzyme as growth ceases.

Although these results provide evidence that there is a progressive increase in inactive enzyme protein during the senescent period, they do not reveal much about the exact alterations. It may be that the *leu-5* and *nd* mutants both cause mistranslation in some form and that death occurs when, in the case of the *leu-5*, a certain percentage of all the leucine residues in all the proteins of the cell become saturated with an incorrect amino-acid. Fig. 1, however, suggests that this is not entirely true. Initially, when mycelium is grown at 25° C the error frequency

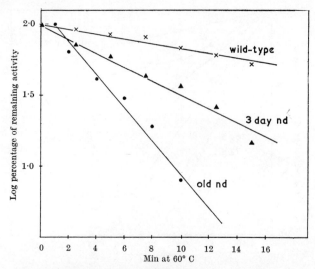

Fig. 5. The lability of glutamic dehydrogenase at 60° C in crude extracts of the *nd* mycelium after growth at 25° C for 3 days (7 days after isolation) and when old (14 days after isolation).

seems to be very low, for the enzyme to CRM ratio approaches that of wild type. Once mycelium is transferred to 35° C there is an initial rapid loss of enzyme activity, the enzyme to CRM ratio then remaining constant for some time. Studies with cycloheximide have shown that the half life of glutamic dehydrogenase *in vivo* is of the order of 4–5 h, so that the initial decrease and ensuing steady state are possibly the result of insertion of an alternative amino-acid in place of the leucine residues as new, defective enzyme is synthesized. The rather constant enzyme to CRM ratio which is maintained for some time might be best explained as resulting from a constant level of mistranslation at the point of insertion of leucine. The important point in all cases, however, is that the ultimate death of the *nd* and *leu-5* and the cessation of growth of mycelium in ethionine is preceded by a sharp decline in enzyme activity which is not accompanied by a decrease in the amount of cross-reacting material. The explanation of this observation is that this is the point where the inaccuracy in protein biosynthesis has extended to other enzymes concerned with transcription and translation.

Further evidence that the incidence of alterations does not remain constant but increases progressively as these mutants age comes from Figs. 4–6. Glutamic dehydrogenase in crude extracts of mycelium of different ages, 24, 48 and 72 h in the case of *leu-5*, and 3 days and

62

Table 1. REVERSION RATES OF CONIDIA ISOLATED FROM THE *leu-5* MUTANT
AT 20° AND 35° C

Mutant strain:	Conidia from culture grown at 20° C	Conidia from culture grown at 35° C
leu-5	$2 \cdot 3 \times 10^{-8}$	$1 \cdot 54 \times 10^{-5}$
ad-3B strains with *leu-5*		
2–17–12	$2 \cdot 0 \times 10^{-8}$	$3 \cdot 0 \times 10^{-8}$
2–17–181	$4 \cdot 0 \times 10^{-7}$	$6 \cdot 0 \times 10^{-7}$
2–17–16	$2 \cdot 0 \times 10^{-8}$	$2 \cdot 5 \times 10^{-8}$
2–17–158	$8 \cdot 2 \times 10^{-6}$	$3 \cdot 35 \times 10^{-5}$
2–17–165	$7 \cdot 3 \times 10^{-6}$	$1 \cdot 25 \times 10^{-4}$
2–17–167	$2 \cdot 4 \times 10^{-6}$	$4 \cdot 6 \times 10^{-6}$
ad-3B strains alone		
2–17–12	Not determined	$2 \cdot 0 \times 10^{-8}$
2–17–181	Not determined	$1 \cdot 9 \times 10^{-7}$
2–17–16	Not determined	$2 \cdot 5 \times 10^{-8}$
2–17–158	Not determined	$2 \cdot 0 \times 10^{-8}$
2–17–165	Not determined	$1 \cdot 7 \times 10^{-7}$
2–17–167	Not determined	$1 \cdot 5 \times 10^{-7}$

Conidia were washed from the surface of mycelium grown in 250 ml. flasks on a layer of medium supplemented with 150 μg/ml. of leucine. Three flasks were used at each temperature and the result is the average of two experiments. Conidia were plated on Vogel's minimal medium with sorbose. For reversion rates of the *ad-3B* mutants the mycelium was grown on the same medium supplemented with 10 μg/ml. of adenine and the conidia were plated on Vogel's sorbose minimal medium supplemented with 150 μg/ml. of leucine. For each determination at least 5×10^9 conidia were plated and the result again represents the average of two experiments.

10 days in the case of *nd*, shows increasing thermolability at 60° C with increasing age. Many mutant forms of β-galactosidase have altered thermal stability[11,12]. We propose that the same is true here and, moreover, that the degree of thermolability is to some extent proportional to the number of alterations in the enzyme protein. Fig. 6 shows the thermolability of glutamic dehydrogenase is also increased in extracts of mycelium grown in medium containing amino-acid analogues.

If the alterations in proteins ultimately become widespread in these mutants, one might predict that all enzymes and structural proteins would be involved. In fact, enzymes concerned with nuclear function and DNA replication might be expected to contain alterations, and it has been shown that alterations in DNA polymerase can confer mutator activity[13]. It is also known that in cultural conditions which induce a high level of mistranslation in bacteria there is a corresponding increase in mutation frequency[13,14]. It seemed possible therefore that the *leu-5* auxotroph might have mutator activity when grown at the high temperature which induces mistranslation. This was found to be the case. Table 1 shows the frequency of reversion to prototrophy of the *leu-5* mutation at a low temperature, 20° C, and at the "ageing" temperature of 35° C. The reversion rates of six *ad-3B* mutants[15] alone and when combined with the *leu-5* mutation are also given. The reversion rates of three of the *ad-3B* mutations in the *ad-3B*, *leu-5* strains and of the *leu-5* mutation itself are very much greater at the higher temperature. Not all *ad-3B* mutants under-

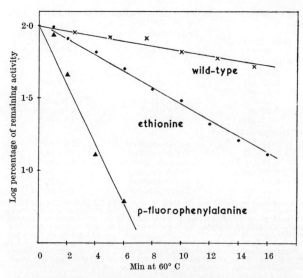

Fig. 6. The lability of glutamic dehydrogenase at 60° C in crude extracts of wild type grown in Vogel's minimal medium with 0·1 mg/ml. of ethionine and 0·1 mg/ml. of para-fluorophenylalanine.

went a higher frequency of reversions when combined with the *leu-5* and grown at the restrictive temperature; this might be a property of the original mutation. Those which did undergo increased reversion, 2–17–158, 2–17–167 and 2–17–165, are all mutants which are readily phenotypically revertible by 5-fluorouracil and 8-azaguanine (personal communication from F. J. de Serres), that is, they will grow in the absence of adenine if the medium contains these analogues. They are base substitution mutants which could either revert by back mutation or possibly as a result of mutation at a suppressor locus. The other mutants do not respond to the analogues. One is of the addition deletion type and the others are probably nonsense base substitution mutations[15]. It is not yet known why *leu-5* does not affect the reversion frequencies of the latter.

Two facts contribute to the belief that the reversions represent nuclear mutations and are not merely the result of continued phenotypic reversion through mistranslation by *leu-5*. For one thing, all the reversions, whether *leu-5* or *ad-3B*, can be subcultured indefinitely at 20° C, when the level of mistranslation is low, without regaining any requirement for either leucine or adenine. The other piece of evidence comes from the results of backcrossing 10 of the *leu-5* revertants with wild type; ninety-six progeny were examined in each case. Two of the mutations causing reversion were either very closely

Table 2. FORWARD MUTATIONS ISOLATED FROM CONIDIA OF *leu-5* MYCELIUM
GROWN FOR 3 DAYS AT 35° C

	No. of isolates	No. of presumptive mutants (many subsequently died)	Actual mutants	Auxotrophic requirements
Experiment 1	2,380	172	9	2 vitamin 4 nucleic acid base 3 amino-acid
Experiment 2	1,720	183	19	1 vitamin 6 nucleic acid base 12 amino-acid

linked to the original *leu-5* or were mutant in the same gene, for no *leu-5* auxotrophs were recovered: of the other eight, the frequency of recovery of *leu-5* was in all cases close to 25 per cent, a frequency which would be expected if the reversion were due to an unlinked suppressor mutation[16]. It is unfortunately not possible to rule out selection pressures which may obtain during culture of the mutants at the high temperature, so that these figures cannot be taken as an accurate measure of the potency of *leu-5* as a mutator.

Table 2 shows the results of experiments to assess the frequency of forward mutations to auxotrophy in a culture maintained at 35° C. Many apparent mutants died after isolation; presumably in these cases the level of mistranslation had reached the irreparable point of no return, but a minority of mutants could be maintained and proved to be quite stable at low temperatures. Tatum *et al.*[17] found only one spontaneous auxotroph among 3,000 tested microconidia; *leu-5* increases this rate to approximately 150-fold.

We tried to determine the rate of reversion of the *nd* mutation, but out of some 4,000 isolated, viable conidia from young and old mycelium, no reversions were found. Reversions do occur, however, and can be isolated readily as outgrowths when the front of mycelium stops growing. Reversion rates of other loci and forward mutations were not investigated because of the inability, with this strain, to "switch off" the ageing process.

Because amino-acid analogues can alter the properties of proteins and may cause mistranslation, they might also be effective mutagens. Of several that have been tested, some have proved to be mutagenic. These results will be reported elsewhere.

Although amino-acid analyses of a protein from cultures of different ages may be the only direct way of proving that mistranslation extends to many amino-acids, the biochemical results we have presented suggest that, at least as far as fungi are concerned, once the degree of mistranslation reaches a certain level, it becomes more extensive and finally results in the death of the organism. Orgel's theory therefore seems to be true for the *leu-5* and *nd* mutations. Holliday[6] has discussed senescence and vegetative death in wild type strains of other species of fungi, and although it is possible that *nd* and *leu-5* are exceptional cases it would be surprising, because the phenotypic effects are so similar, to find that ageing in these species did not occur in the same way.

It is interesting that the *leu-5* auxotroph has mutator activity, for it has been suggested many times that ageing could be the result of deleterious effects which might build up if spontaneous mutations accumulated in both dividing and non-dividing cells[18,19]. It has been shown that the spontaneous mutation rate in the micronucleus of *Paramecium* increases with age[20]. Damage accumulating in the cell nucleus, which is ultimately required to exercise control over the whole cell, might be expected to lead to cellular inefficiency. These results with *leu-5* suggest that Orgel's theory and this theory of somatic mutation may not be as diametrically opposed as they seem.

We thank Mrs G. Tarrant for help with enzyme assays, and Dr F. J. de Serres for *ad-3B Neurospora* stocks.

Received June 10, 1970.

[1] Orgel, L. E., *Proc. US Nat. Acad. Sci.*, **49**, 517 (1963).
[2] Crestfield, A. M., Stein, W. H., and Moore, S., *J. Biol. Chem.*, **238**, 2413 (1963).
[3] Richmond, M. H., *J. Mol. Biol.*, **6**, 284 (1963).
[4] Printz, D. B., and Gross, S. R., *Genetics*, **55**, 451 (1967).
[5] Sheng, T. C., *Genetics*, **36**, 199 (1951).
[6] Holliday, R., *Nature*, **221**, 1224 (1969).
[7] Harrison, B. J., and Holliday, R., *Nature*, **213**, 990 (1967).
[8] Fincham, J. R. S., *J. Mol. Biol.*, **4**, 257 (1962).
[9] Roberts, D. B., and Pateman, J. A., *J. Gen. Microbiol.*, **34**, 295 (1964).
[10] Roberts, D. B., *J. Bact.*, **91**, 1888 (1966).
[11] Langridge, J., *J. Bact.*, **96**, 1711 (1968).
[12] Langridge, J., *Molec. Gen. Genet.*, **103**, 116 (1968).
[13] Drake, J. W., Preparata, R. M., Allen, E. F., Forsberg, S. A., and Greening, E. O., *Nature*, **221**, 1128 (1969).
[14] Speyer, J. F., Karam, J. D., and Lenney, A. B., *Cold Spring Harbor Symp. Quant. Biol.*, **31**, 693 (1966).
[15] Malling, H. V., and de Serres, F. J., *Mut. Res.*, **5**, 359 (1968).
[16] Yanofsky, C., *Bact. Rev.*, **24**, 221 (1960).
[17] Tatum, E. L., Barratt, R. W., Fries, N., and Bonner, D. M., *Amer. J. Bot.*, **37**, 38 (1950).
[18] Szilard, L., *Proc. US Nat. Acad. Sci.*, **45**, 30 (1959).
[19] Curtis, H. J., *Science*, **141**, 686 (1963).
[20] Sonneborn, T. M., and Schneller, M., *J. Protozool.*, **2**, Suppl. 6 (1955).

5

Reprinted by permission from *Nature* **254**:161–163 (1975)

PROGRESSIVE DECREASE IN PROTEIN SYNTHESIS ACCURACY INDUCED BY STREPTOMYCIN IN *ESCHERICHIA COLI*

E. W. Branscomb and D. J. Galas

THE passage of genetic information from gene to polypeptide does not proceed with perfect fidelity[1-3], but it is not known what short and long-term cellular responses may be induced by decreases in the accuracy of protein synthesis. Errors in gene expression may tend to increase their own rate of production[4], and thus could constitute a limiting instability of cellular life. The synthesis of inaccurate RNA, RNA polymerases, tRNAs, tRNA modifying enzymes, tRNA charging enzymes and ribosomal proteins are potential pathways for the perpetuation or amplification of errors. Such errors may be inducible *in vivo* by external chemical agents and may constitute an indirect pathway for mutagenesis[5-8]. An enhanced rate of such errors may be the mechanism by which certain deleterious mutations exert their effects. Here we report the results of experiments with

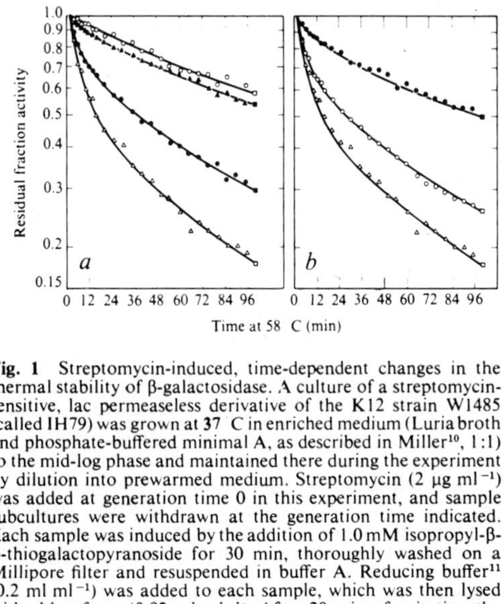

Fig. 1 Streptomycin-induced, time-dependent changes in the thermal stability of β-galactosidase. A culture of a streptomycin-sensitive, lac permeaseless derivative of the K12 strain W1485 (called IH79) was grown at 37 °C in enriched medium (Luria broth and phosphate-buffered minimal A, as described in Miller[10], 1:1) to the mid-log phase and maintained there during the experiment by dilution into prewarmed medium. Streptomycin (2 µg ml⁻¹) was added at generation 0 in this experiment, and sample subcultures were withdrawn at the generation time indicated. Each sample was induced by the addition of 1.0 mM isopropyl-β-D-thiogalactopyranoside for 30 min in this experiment, and sample subcultures were withdrawn at the generation time indicated. Each sample was induced by the addition of 1.0 mM isopropyl-β-D-thiogalactopyranoside for 30 min, thoroughly washed on a Millipore filter and resuspended in buffer A. Reducing buffer[11] (0.2 ml ml⁻¹) was added to each sample, which was then lysed with chloroform (0.02 ml ml⁻¹). After 20 min of agitation the samples were centrifuged at 4,000 r.p.m. for 20 min, and the remaining chloroform was removed from the supernatant by evaporation with agitation at 37 °C. (Samples lysed in a pressure cell gave identical kinetics of β-gal thermal denaturation.) Portions (1 ml) of each sample were incubated in plastic tubes at 58 °C (stable to within ± 0.05 °C) for various times determined by varying the starting time. The incubation was stopped by agitating all the tubes in 19 °C water for 1 min. The tubes were then allowed to stand at 37 °C for 90 min before assay for β-gal activity by hydrolysis of orthonitrophenol-β-D-galactopyranoside at 37 °C (ref. 12). The residual fraction of enzymatic activity is plotted as a function of the time at 58 °C. The zero and 100-min points are each the mean of five samples. The generation time for each curve is the mid-point of the sample induction period calculated from the history of absorbance readings (650 nm) taken from the main culture. Repeated runs with improved resolution showed that, with this procedure, the decay kinetics of β-gal from untreated cultures is first order to within 2–3% for 200 min after the first 5 min. *a*, Curves for samples out to 7.5 generation, *b*, the two subsequent samples with the 7.5 generation curve repeated. Samples taken at the indicated number of generations (doublings): *a*, ○, 0; ▲, 1; ●, 4.1; △, 7.5. *b*, △, 7.5; ○, 11; ●, 16.

Escherichia coli designed to detect evidence of self amplification of errors in protein synthesis. If such error feedback is quantitatively significant, a slight increase in the pre-existing error rate should be followed by a progressive decrease in the accuracy of protein synthesis. While this process could lead to a new stable level and thus need not precipitate a catastrophic breakdown, some late effects should be observable if the mechanisms are at all as hypothesised.

We have manipulated the error rate in protein synthesis by exposing the cells to low levels of streptomycin, an aminoglycoside antibiotic which increases the rate of ribosome-mediated misincorporation of amino acids[9]. The fidelity of protein synthesis in a culture at a given time was assayed by measuring the thermal decay kinetics of the enzyme β-galactosidase (β-gal) produced by a freshly divided and induced subculture. The thousandfold inducibility of this enzyme permits a reasonably sharp definition of the time at which the accuracy of protein synthesis is measured.

In the basic experiment in this series, a culture was sampled just before and at various times after addition of streptomycin at concentrations in the range 0.5 to 5 µg ml⁻¹. Typical results are shown in Fig. 1. Before addition of streptomycin the β-gal thermal denaturation curve had reasonably first-order characteristics, indicating a correspondingly homogeneous enzyme population. After addition of the drug a progressively increasing fraction of the newly synthesised β-galactosidase molecules showed enhanced thermolability.

Simultaneously with these changes in the β-gal thermolability pattern, the cultures themselves showed a delayed, progressively decreasing growth rate and a progressive increase in the production of multinucleate, filamentous cells. Recovery of the culture, as measured by thermal decay of β-gal, (Fig. 1*b*) was also accompanied by increased growth rate and disappearance of elongated cells. These observations support the idea that thermal decay assay detects a specific consequence of a general cell-wide effect.

A series of experiments similar to that depicted in Fig. 1 lead us to several additional observations and tentative conclusions concerning this effect. (1) Both the maximum extent and speed of development of the effect showed nonlinear, direct dependence on the concentration of streptomycin. (2) Maximum extent of the effect was correlated with the speed of its development. The smaller the effect, the longer it takes to develop; it was maximal in 8 to 10 generations for high drug concentration (2–5 µg ml⁻¹) and in 14 to 17 generations for low concentration (about 1 µg ml⁻¹). These observations seem inconsistent with the idea that the simple replacement of some cell component by its post-treatment equivalent (without feedback) is responsible for the progressive nature of the effect. (3) The effect was sensitive to the growth medium and growth history of the culture. (4) Within a fourfold range of growth rates (produced by varying the growth medium), generation time and not the actual time seemed to determine the kinetics of development. (5) With treatments mild enough to allow growth for a few generations after addition of streptomycin, the cultures ultimately recovered. Recovery to normal growth rate and cell morphology was always correlated with decreased levels of thermolabile β-gal. We have, however, never seen complete recovery in as many as 25 generations after addition of the drug.

We do not know whether recovery was due to adaptation within the cells or to selection of some subtly resistant variants. If a favoured phenotype was being selected, however, it must have been present initially at a concentration of about 1 in 10^3. The recovered populations are not resistant to streptomycin at levels of 20 µg ml^{-1} or higher; hence, classical streptomycin-resistant alleles of the *strA* locus are not involved.

Since we observed increased heterogeneity of the thermal decay rates for β-gal, it seems unlikely that the change was due to some other factor produced by the treated cells, which persisted through our washing, resuspension and lysis procedures and enhanced the thermolability of the enzyme. To check this possibility, however, we mixed lysates from induced but untreated cells in various ratios with extracts from late, drug-treated cultures that contained only basal levels of β-gal. The thermolability of the enzyme was unaffected even when the extract from treated cells was in fivefold excess and when parallel induced samples of the treated cultures showed the expected enhanced thermolability.

A convincing test of the same question comes from an experiment designed to show whether the increased thermolability is due to modification of the enzyme after synthesis. Since enzyme synthesised before addition of streptomycin remained unaffected (Fig. 2), we tentatively attribute the observed alteration of the protein to faulty protein synthesis.

Another experiment showed that the growth of late streptomycin-treated cultures was much more sensitive to temperature than that of untreated cultures (Fig. 3), implying that thermolabile, but otherwise functional, protein required for growth is being produced in the treated cells.

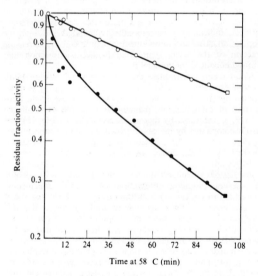

Fig. 2 Thermolability seems to be conferred at the time of protein synthesis. Induced cells grown overnight to produce maximum levels of β-gal were deinduced by washing on a Millipore filter, and resuspended in medium containing streptomycin (2 µg ml^{-1}). A control sample was taken immediately (but not further induced) and the culture was allowed to grow for about 5.5 generations. It was then divided into two samples (○ and ●), one of which (●) was induced for 30 min. The procedures and conditions were otherwise as described in Fig. 1. This figure shows the thermal decay curves resulting from these sample cultures. The decay kinetics of the control sample were characteristic of untreated cultures and indistinguishable from the zero generation curve in Fig. 1 and from curve ○ in this figure.

If we were observing a progressive breakdown in the accuracy of protein synthesis, it could have been due to causes other than error feedback: a slow increase in the uptake of streptomycin, possibly the result of a feedback cycle directly affecting permeation, or of some process involving the metabolism or binding of the drug. Indirect evidence that such effects are not implicated came from our examination of the immediate effect of high doses of streptomycin on the production of thermolabile β-gal (data not shown). Substantial fractions of thermolabile enzyme were produced as an immediate response

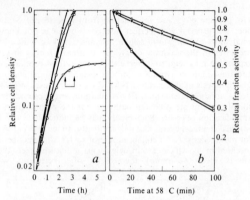

Fig. 3 Thermal sensitivity of the growth of treated cultures. Two freshly divided cultures of IH79, one with and one without streptomycin (2 µg ml^{-1}) were grown at 37 °C for four generations in exponential growth medium. At time zero, each culture was diluted into two flasks of the same medium, one held at 37 °C and the other at 42.5 °C. During subsequent incubation with agitation at these temperatures the growth of the cultures was measured by reading the absorbance at 650 nm. At the time indicated by the first arrow (*a*), samples were removed and induced at 37 °C. Induction was stopped at the time of the second arrow and the thermal degradation kinetics of the β-gal in each sample determined (*b*). Procedures and conditions were otherwise as described in the legend to Fig. 1. The strain IH79 normally grows well following a temperature upshift from 37 °C to 43 °C, but stops abruptly if shifted to 45 °C. ○, 42.5 °C, +streptomycin (2 µg ml^{-1}); △, 42.5 °C, −streptomycin; □, 37 °C, +streptomycin (2 µg ml^{-1}); ●, 37 °C, +streptomycin.

to streptomycin and limitations of permeability of the drug (revealed by pretreatment with EDTA–Tris[13]) played some role. The maximum immediate effect was, however, significantly less than the maximum late effect caused by streptomycin at 2 µg ml^{-1}, and the high concentrations required ($\gtrsim$ 30 µg ml^{-1}) caused rapid cell killing. It therefore seems unlikely that the observed progressive effects of streptomycin resulted from changes in the intracellular concentration of active drug.

Mutations that confer resistance to high concentrations of streptomycin are mapped in the gene (*strA*) coding for a 30S subunit ribosomal protein (S12) and depress both the *in vitro* and *in vivo* error-producing effects of streptomycin[9]. Two resistant derivatives of IH79 and two independent resistant (*strA*) strains were found to have normal β-gal thermolability at concentrations of drug below 100 µg ml^{-1} but to produce a small amount of thermolabile enzyme (20% or less) when treated with 200 µg ml^{-1}. We conclude that the *strA* locus mediates the production by streptomycin of the progressive effect.

The results reported here suggest that error feedback in gene expression exists in *E. coli*, and constitutes a basis for further investigation of the stability of information transfer in gene expression.

This work was performed under the auspices of the US Atomic Energy Commission.

Received December 2, 1974; revised January 13, 1975.

1 Loftfield, R. B., *Biochem. J.*, **89**, 82–92 (1963).
2 Loftfield, R. B., and Vanderjagt, D., *Biochem. J.*, **128**, 1353–1356 (1972).
3 Davies, J., *Prog. molec. subcell. Biol.*, **1**, 47–81 (1969).
4 Orgel, L. E., *Proc. natn. Acad. Sci. U.S.A.*, **49**, 517–521 (1963); **67**, 1476 (1970).
5 Speyer, J. F., Karam, J. D., and Lenny, A. B., *Cold Spring Harb. Symp. quant. Biol.*, **31**, 693–697 (1966).
6 Lewis, C. M., and Holliday, R., *Nature*, **228**, 877–880 (1970).
7 Lewis, C. M., and Tarrant, G. M., *Mutation Res.*, **12**, 349–356 (1971).
8 Talmud, P., and Lewis, D., *Nature*, **249**, 563–564 (1974).
9 Gorini, L., and Davies, J., *Current Topics Microbiol. Immun.*, **44**, 100–122 (1968).
10 Miller, J. H., in *Experiments in Molecular Genetics*, 432–433 (Cold Spring Harbor Laboratory, 1972).
11 Revel, H. R., Luria, S. E., and Rotman, B., *Proc. natn. Acad. Sci. U.S.A.*, **47**, 1956–1967 (1961).
12 Kepes, A., and Bequin, S., *Biochim. biophys. Acta*, **123**, 546–560 (1966).
13 Leive, L., *Proc. natn. Acad. Sci. U.S.A.*, **53**, 745–750 (1965).

69

6

Reprinted from *IRCS Med. Sci.* **10:**874–875 (1982)

STREPTOMYCIN-INDUCED PROTEIN ERROR PROPAGATION APPEARS TO LEAD TO CELL DEATH IN *ESCHERICHIA COLI*

R.F. Rosenberger

Genetics Division, National Institute for Medical Research, Mill Hill, London NW7 1AA

Paper received: 16th September, 1982; amended 7th October, 1982

Hydrogen bonding between complementary bases cannot by itself explain the great accuracy of protein synthesis (1). The stereochemical specificity of enzymes is largely responsible for the precision observed and proteins are thus responsible for the accuracy of their own synthesis (1). Accidental errors in the structure of proteins involved in maintaining accuracy could make them less specific and create further protein errors (2). This kind of positive feedback (error propagation) has been suggested to be a cause underlying ageing (3, 4). If this were true, a potentially immortal organism like *Escherichia coli* would differ from mortal ones by preventing or eliminating errors in protein synthesis more efficiently. Conversely, increasing the protein error rate in *E. coli* artificially should give it a limited lifespan and this is the experiment I have attempted.

For such experiments, it is necessary to increase protein errors in a highly specific way. The antibiotic streptomycin is known to have this effect (8). Point mutations in a single ribosomal protein, rpsL, make *E. coli* strains resistant to a thousand times the normal lethal concentration; the resistance is due to lack of streptomycin binding by the mutant ribosomes and not to failure of the antibiotic to enter cells. Ribosome function is thus the only target for streptomycin at concentrations of 500 μg/ml or less and on the ribosome it is known to decrease the accuracy of mRNA translation (8).

Materials and methods: *E. coli*, strain 4515-9 (5), was grown with shaking at 37 °C in Luria broth (6) containing 10^{-3} M isopropyl ß-D thiogalactoside, an inducer of ß-galactosidase (7). Streptomycin was added in various concentrations and the cultures diluted every 1.5 generations with pre-warmed medium containing streptomycin in order to keep cell numbers between 5×10^7 and 2×10^8 cells/ml. Total cell numbers were determined by particle counting in a Coulter Counter, Model ZBI, fitted with a 30 μm orifice and by optical density measurements. Viable counts were obtained by plating dilutions in triplicate on Luria broth solidified with agar. ß-galactosidase activity was measured as described in (5).

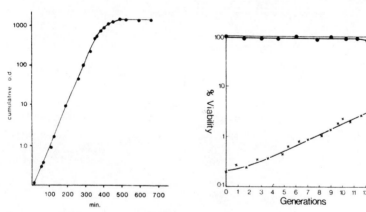

Growth curve (left-hand graph) and viability (○) and β-galactosidase activity (×) (right-hand graph) of 4515-9 growing with 0.5 μg streptomycin/ml. β-galactosidase specific activity is nmol/min/mg protein.

Results and discussion: Streptomycin binds to bacterial ribosomes and increases the error rate in protein synthesis (8). Various low concentrations of streptomycin were screened to find one that would raise protein errors, allow cultures to grow for more than 10 generations and finally lead to a loss of viability. Continuous growth with 0.5 μg streptomycin/ml consistently gave this result, as shown in the figure. Errors in protein synthesis were measured by the misreading of the nonsense mutation in the ß-galactosidase gene of 4515-9 (5). Only erroneous processing of the mutant codon can produce active enzyme and the rise in ß-galactosidase activity shows that protein errors increased continuously after streptomycin addition. Although erroneous proteins were accumulating, cultures grew with a constant doubling time of 27 min for the first 10 generations. Growth then slowed and extensive cell death occurred after approximately 14 generations.

It is very unlikely that the equilibration of external and internal streptomycin would take anything like this time. Thus, the prediction that erroneous proteins can accumulate under constant culture conditions and during a period of apparently normal growth appears to be justified. In keeping with these findings, *E. coli* growing with low streptomycin concentrations accumulates abnormally heat-labile proteins (9) and shows marked changes in its protein turnover (Carrier, M. and Hipkiss, A., personal communication).

In the above experiments, cells died after extended growth with streptomycin. Gallant and Palmer (10) showed that sublethal streptomycin concentrations can increase protein errors but still allow cells to grow indefinitely. I have obtained similar results with concentrations < 0.5 μg/ml. It is important to realise that these findings do not contradict the concept of error propagation. Erroneous proteins will only accumulate above a critical error rate (2, 3) and this may not have been reached with < 0.5 μg/ml or the streptomycin concentrations used in (10).

Definitive proof of an error feedback will require isolation of the protein synthesising machinery and demonstration of its progressive deterioration. The above experiments can thus be only indicative. However, given the specificity of streptomycin action and the long interval before viability was affected, they suggest that the *E. coli* translational apparatus may be operating quite close to the zone of instability postulated in (2).

1. Fersht, A.R. (1981) *Proc. Roy. Soc. Lond. B.*, **212**, 351-379
2. Kirkwood, T.B.L. and Holliday, R. (1975) *J. Mol. Biol.*, **97**, 257-265
3. Orgel, L.E. (1973) *Nature, Lond.*, **243**, 441-445
4. Holliday, R. (1977) *Trends Biochem. Sci.*, **2**, N80-N82
5. Rosenberger, R.F. and Foskett, G. (1981) *Mol. Gen. Genet.*, **183**, 561-563
6. Miller, J.H. (1972) *Experiments in Molecular Genetics*, p. 433, Cold Spring Harbor Laboratory, New York
7. Miller, J.H. (1982) *Experiments in Molecular Genetics*, pp. 356-359, Cold Spring Harbor Laboratory, New York
8. Gorini, L. (1974) in *Ribosomes*, (Nomura, M., Tissieres, A. and Lengyel, P., eds.), pp. 791-803, Cold Spring Harbor Laboratory, New York
9. Branscomb, E.W. and Galas, D.J. (1975) *Nature, Lond.*, **254**, 161-163
10. Gallant, J. and Palmer, L. (1979) *Mech. Ageing Dev.*, **10**, 27-38

7

Reprinted from *Proc. Natl. Acad. Sci. (U.S.A.)* **74:**3396–3398 (1977)

On the translational error theory of aging

(error propagation/translational fidelity/clonal senescence)

PATRICIA EDELMANN* AND JONATHAN GALLANT

Department of Genetics, University of Washington, Seattle, Washington 98195

Communicated by Herschel L. Roman, April 15, 1977

ABSTRACT   Theoretical treatments of error feedback in translation have revealed that two different modes of behavior are possible, depending on the values of certain parameters. In mode I, the error frequency will rise steadily toward randomness, inevitably reaching whatever value is catastrophic for cell survival; the "error catastrophe" theory of aging implicitly assumes this mode of behavior. In mode II, the error frequency will converge to a stable value, which may or may not have toxic consequences. We have performed an experimental test of the behavior of the translation system in *Escherichia coli* cells: we altered the system's intrinsic fidelity by means of the error-promoting drug streptomycin, and monitored the kinetics of change in error frequency by means of a specific assay of one kind of mistranslation (incorporation of cysteine into flagellin). We find that the system behaves according to mode II. Moreover, *E. coli* cells in which the error frequency has stabilized at a value as high as 50 times greater than normal continue to proliferate, albeit abnormally slowly, and their viability is not detectably reduced. Earlier results by Gorini and his associates point in the same direction. These observations diminish the plausibility of the error catastrophe theory of aging.

Orgel has pointed out that the fidelity of translation contains an element of positive feedback: errors in protein components of the translation apparatus itself (e.g., ribosome proteins or activating enzymes) would increase the frequency of subsequent translational errors (1). At first thought, this seems to imply that the fidelity of translation must deteriorate progressively, leading to an inevitable "error catastrophe." The seeming inevitability of the error catastrophe has offered an attractive explanation of the equally inevitable death of normal mammalian cell lines or of organisms that undergo clonal senescence and, by implication, has offered an explanation of mortality itself (2–7).

However, this implication may be in error. In a significant amendment to the original theory, Orgel (8) noted that the error frequency might not increase indefinitely, but could approach a limiting value. This issue Orgel posed is as follows. Let C_n be the error frequency in "generation" n, assuming a simplified model in which successive generations of the protein synthetic apparatus are discrete and distinguishable. This error frequency can be partitioned between an intrinsic error frequency that occurs independently of prior errors, and a further contribution assumed to be linearly dependent on the preexisting error frequency in the synthetic apparatus. Let R be the intrinsic error frequency per generation, and let α be the proportionality constant between errors already present in the synthetic apparatus and new errors in freshly synthesized protein. Then:

$$C_n = R + \alpha C_{n-1}, \text{ and } C_{n-1} = R + \alpha C_{n-2}, \text{ etc.}$$

Because $C_1 = R$ by definition, $C_n = R + R\alpha + R\alpha^2 + R\alpha^3$

$\ldots + R\alpha^{n-1}$. If $\alpha \geq 1$, then the error frequency must increase without limit, eventually no doubt reaching catastrophic levels. But if $\alpha < 1$, then the geometric series converges and C_n approaches $R/(1 - \alpha)$.

The problem, then, is to determine whether or not α is less than one. Recently, Hoffman (9) has developed a more detailed analysis of error feedback, which has been generalized further by Kirkwood and Holliday (10) and by Goel and Ycas (11). In effect, these models make explicit a number of parameters which go into Orgel's R and α, and define more complex recursion relationships between translational fidelity in a given "generation" and the preceding one. The general issue, however, remains the same as that pointed out by Orgel in 1970: for certain values of the parameters, the fidelity of translation declines steadily toward randomness, but for other values of the parameters, a stable solution is possible wherein the error frequency converges to a value which remains constant from one generation to the next.

In the most general terms, then, the problem is to determine whether stable states of translational accuracy can be attained. We have approached the problem by effecting large changes in the intrinsic error frequency, in *Escherichia coli* cells, through the addition or removal of streptomycin, a drug known to induce translational errors (12–14). We then followed the kinetics of change in the frequency of a particular type of translational error, which ought to be proportional to the aggregate error frequency C_n. The type of error we measured was the illegitimate incorporation of cysteine into flagellin, a protein that normally contains no cysteine residues. We have described elsewhere our method for detecting cysteine incorporation into flagellin, and have presented evidence that it reflects misreading of the two arginine codons CGU and CGC for the cysteine codons UGU and UGC (15).

METHODS AND MATERIALS

E. coli strain C92 (HfrC, $phoS^- relA^- spoT^-$) was employed in these studies. Conditions of growth, and the method for detecting cysteine incorporation into flagellin, were exactly as described previously (15). Briefly, the cells were double-labeled with [^{35}S]sulfate and [^{3}H]alanine under conditions where incorporation of sulfate into methionine is blocked altogether; flagella were highly purified and disassociated, and the flagellin monomer was resolved by sodium dodecyl sulfate/polyacrylamide gel electrophoresis; the ^{35}S/^{3}H ratio in the flagellin peak was recorded by liquid scintillation counting. The left-hand ordinate of Fig. 1 presents these ^{35}S/^{3}H ratios. In order to convert these ratios into absolute misreading frequencies, the pmol of cysteine incorporated per pmol flagellin made during the labeling period was determined, as described previously (15), in four experiments, two in uninhibited cells and two in

* Present address: Department of Biological Sciences, California State University, Chico, Chico, CA 95929.

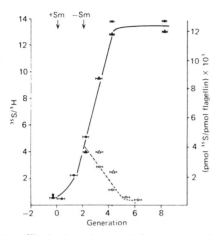

FIG. 1.   Kinetics of error frequency in the presence and absence of streptomycin (Sm). A C92 culture was cultivated in the absence of streptomycin and the basal error frequency was measured. At the time indicated by the arrow, 5 μg/ml of streptomycin was added and the error frequency was monitored over the next eight generations. After two generations of growth in the presence of streptomycin, an aliquot of the culture was removed, washed twice with the minimal medium minus supplements to remove streptomycin, and resuspended in minimal medium plus supplements. The error frequency was again monitored. Cultures were labeled with [³H]alanine (53 Ci/mmol) and [³⁵S]sulfate (2 μCi/nmol) and flagellin was isolated as described previously (15). — And filled symbols, error frequency in cells grown in the presence of streptomycin. - - - And open symbols, error frequency in cells after streptomycin removal. The symbols ●○ and ▲△ indicate data obtained from two experiments. The length of the horizontal bars indicates the fraction of the generation during which labeling occurred.

cells growing in streptomycin at 2 μg/ml; these measurements showed that a $^{35}S/^{3}H$ ratio of 1.0 corresponded on average to 0.95 ± 0.14 (SEM) $\times 10^{-4}$ pmol of cysteine per pmol of flagellin produced. The $^{35}S/^{3}H$ values were accordingly scaled to this value on the right-hand ordinate of Fig. 1.

For the streptomycin studies, cells were grown in medium containing streptomycin at 2, 5, or 8 μg/ml. In the presence of these low concentrations of the drug, growth inhibition developed gradually, as other workers have observed (16), reaching maximal levels after a few generations; maximal growth inhibition by streptomycin at 2, 5, and 8 μg/ml was 15%, 20%, and 40%, respectively.

Total cell number was determined by cell count in a Petroff–Hauser chamber. Viable cell number was determined by plating on LB agar (21).

RESULTS AND DISCUSSION

Fig. 1 presents the results of two types of experiment. In the first, we simply added streptomycin and monitored the error frequency at intervals thereafter (filled symbols). It can be seen that the error frequency rose gradually, in agreement with Orgel's original notion of positive feedback in error, as well as with observations on the gradual increase in nonsense suppression (17) and the production of thermolabile β-galactosidase (16) in similarly treated cells.

However, the error frequency did not continue to increase progressively, but rather reached a limiting value after a few generations of growth in streptomycin. In other experiments we found that the error frequency did not change between 8

Table 1.   Error frequency and cell viability in cells grown in streptomycin

Streptomycin, μg/ml	Doublings	Relative error frequency	(Cells/ml) × 10⁻⁸ Total	(Cells/ml) × 10⁻⁸ Viable	Proportion viable cells
0	—	0.5	4.8	4.4	0.92
2	8	3.86	2.8	3.0	1.07
2	16	3.45	—	—	—
5	8	13.8	4.5	4.8	1.07
5	18	—	4.8	4.8	1.0
8	8	26.6	3.2	3.25	1.02

Cells were cultivated as described, with dilution every few doublings, in the indicated streptomycin concentrations. Relative error frequencies are $^{35}S/^{3}H$ ratios in purified flagellin as in the left-hand ordinate of Fig. 1; the values without streptomycin, and after eight generations of growth in streptomycin at 2 μg/ml, are averaged from four and six replicate determinations, respectively, and were presented previously in table 1 of ref. 15. The other values were determined once each, with simultaneous measurement of cell viability. (Closely similar misreading frequencies were also detected at each of the streptomycin concentrations with the related strain C91.)

and 16 generations of growth in streptomycin. This implies that a stable state of translational accuracy can be attained, even at an error frequency more than 20 times greater than normal, as in Fig. 1. In terms of Orgel's 1970 model, this implies that α is indeed less than one.

A more decisive test of this point is provided by the experiment illustrated in the open symbols of Fig. 1. In this experiment, we removed streptomycin after two generations of growth in its presence, thus reducing the intrinsic error frequency—Orgel's R—back to its low, normal value. If α were greater than one, then the first equation predicts that the error frequency should continue to rise. On the contrary, the error frequency dropped steadily after removal of streptomycin. This observation is inconsistent with any value of $\alpha \geq 1$; instead, the data indicate a value of α in the vicinity of 0.8.

In more general terms, our results show that a stable state of translational accuracy is maintained even at an error frequency much higher than normal. Moreover, a substantial increase in the error frequency, as seen after two generations in streptomycin, does not propagate itself into a further escalation of error toward catastrophe after removal of the drug. Both of these observations suggest that, whatever model one entertains, the parameters are such that translational accuracy converges to a stable value.

It follows that the immortality of vegetatively growing *E. coli* cells need not be ascribed to cellular selection, but may simply reflect the fact that the normal limiting error frequency is tolerable. In agreement with this view, we find that much higher limiting error frequencies are tolerable. Table 1 shows the error frequency and cell viability determined simultaneously in cultures grown in three concentrations of streptomycin for eight generations—long enough to reach a limiting error frequency as judged by Fig. 1. Even with the antibiotic at 8 μg/ml, cells that manifested an error frequency 53 times greater than normal were essentially all viable. Microscopic examination of these cells revealed no gross abnormalities. The cells did grow slower than normally, which is roughly consistent with the expected increase in the proportion of inactive enzymes produced through mistranslation. We conclude that a large increase in the error frequency is neither lethal in itself nor capable of triggering an escalating error catastrophe.

A considerable literature points to the occurrence of abnormal proteins in senescent cells, suggesting the possibility of

Proc. Natl. Acad. Sci. USA 74 (1977)

increased translational error (3–5, 18–20). The least we can say from our data is that a large increase in error frequency is not necessarily lethal for the type of cell we have studied. Therefore, the mere demonstration of increased translational error in senescent cells is clearly insufficient for arguing a causal relationship between translational error and cell death. What is required, our experiments emphasize, is an independent calibration of the error level that cells can or cannot tolerate. If other types of cells are as tolerant of error as *E. coli* appears to be, then evidence in support of the error catastrophe theory of aging will have to demonstrate very high error frequencies indeed in senescent cells.

Garvin *et al.* have measured the misreading of poly(U) *in vitro* by crude S_{30} preparations derived at different times from cells growing in dihydrostreptomycin (table 2 of ref. 17). They detected a sizeable increase in poly(U) misreading, but this increase reached a constant value within a few generations of growth in the antibiotic. Our measurements of misreading *in vivo* (Fig. 1) show the same type of convergence. Both observations imply that the parameters of error feedback are such that stable states of translational accuracy are indeed attained, in agreement with theoretical considerations outlined by Goel and Ycas (11). It follows that error catastrophe is not an inevitable property of error feedback in protein synthesis.

This work was supported by National Institutes of Health Grants GM 13626 to J.G. and AG 05043 to P.E.

1. Orgel, L. E. (1963) *Proc. Natl. Acad. Sci. USA* **49**; 517–521.
2. Holliday, R. (1969) *Nature* **221**, 1224–1228.
3. Lewis, C. M. & Holliday, R. (1970) *Nature* **238**, 877–880.
4. Holliday, R. & Tarrant, G. M. (1972) *Nature* **238**, 26–30.
5. Lewis, C. M. & Tarrant, G. M. (1972) *Nature* **239**, 316–318.
6. Orgel, L. E. (1973) *Nature* **243**, 441–445.
7. Holliday, R. (1974) *Fed. Proc.* **34**, 51–55.
8. Orgel, L. E. (1970) *Proc. Natl. Acad. Sci. USA* **67**, 1476.
9. Hoffman, G. H. (1974) *J. Mol. Biol.* **86**, 349–362.
10. Kirkwood, T. B. L. & Holliday, R. (1975) *J. Mol. Biol.* **97**, 257–265.
11. Goel, N. S. & Ycas, M. (1975) *J. Theor. Biol.* **55**, 245–282.
12. Davies, J., Gilbert, W. & Gorini, L. (1964) *Proc. Natl. Acad. Sci. USA* **51**, 883–889.
13. Davies, J., Gorini, L. & Davis, B. (1965) *J. Mol. Pharmacol.* **1**, 93–106.
14. Davies, J., Jones, D. S. & Khorana, H. G. (1966) *J. Mol. Biol.* **18**, 48–57.
15. Edelmann, P. & Gallant, J. (1977) *Cell* **10**, 131–137.
16. Branscomb, E. W. & Galas, G. (1975) *Nature* **254**, 161–163.
17. Garvin, R. T., Rosset, R. & Gorini, L. (1973) *Proc. Natl. Acad. Sci. USA* **70**, 2762–2766.
18. Gershon, H. & Gershon, D. (1970) *Nature* **227**, 1214–1217.
19. Gershon, H. & Gershon, D. (1973) *Proc. Natl. Acad. Sci. USA* **70**, 909–913.
20. Zeelon, P., Gershon, H. & Gershon, D. (1973) *Biochemistry* **12**, 1743–1750.
21. Miller, J. H. (1972) *Experiments in Molecular Genetics* (Cold Spring Harbor Laboratory, Cold Spring Harbor, NY), p. 433.

Part II

CELLULAR AGING

Editor's Comments
on Papers 8 Through 14

Following the development of cell culture techniques, it was widely accepted that populations of avian or mammalian cells were capable of indefinite growth. Carrell claimed that chick fibroblasts had been serially propagated in his laboratory for 34 years—longer than the lifetime of the donor animal. Subsequently, it was discovered that cell populations could sometimes be obtained from human or rodent tumors, which could be subcultured indefinitely (permanent lines). Many workers failed to confirm Carrell's results, but this was almost always attributed to faulty technique (Witkowski, 1980). It was also shown, initially by Swim and Parker (1957), that human diploid fibroblasts could not be grown for long periods in culture. This very important observation was more thoroughly documented by Hayflick and Moorhead (1961) and by Hayflick in Paper 8. They realized that the limit to fibroblast growth was not due to inadequate techniques but was intrinsic to the cells. Many experiments were described that effectively ruled out other possible explanations of the limited in vitro growth (defective serum or medium, presence of infective agents, and so on).

The major features of in vitro growth were defined as follows: Phase I, the establishment of primary cultures from tissue explants; Phase II, the long period of proliferation (40 to 50 population doublings), during which the morphological features of the cells remain unchanged; and Phase III, the period during which the rate of growth and yield per bottle or flask declines and the cells become progressively abnormal in appearance. These Phase III cultures invariably die out, without the emergence of permanent lines. The transition from Phase II to Phase III was shown to be related to the number of passages, or population doublings, achieved rather than to the actual chronological time the cells were kept in culture. It was also shown in several laboratories that chick fibroblast cultures have an even shorter growth potential than human cells, which discredited Carrell's claims for their immortality.

In Paper 8, Hayflick makes the explicit claim that the limited growth of human fibroblasts is due to aging at the cellular level, and he concludes that they provide appropriate material for fundamental experimental investigations of the aging process. In this connection, he made the important discovery that cells frozen in liquid nitrogen stayed at the same in vitro age, which meant that cultures of a well-characterized strain, such as WI-38 or MRC-5, could be reconstituted for innumerable experimental studies over a long time. In the same paper, he makes clear the distinction between diploid cells, which have normal histological features, maintain their karyotypic stability, and have finite growth, and transformed cells, which have the abnormal morphology of cancer cells, are karyotypically unstable, and grow indefinitely in culture.

In support of his view that the aging of cultured cells could be related to the aging of the organism, Hayflick (Paper 8) provided evidence that the growth potential of populations of fibroblasts was related to the age of the donor. Whereas fetal lung fibroblasts grew for an average of 48 population doublings, lung fibroblasts obtained at death from individuals aged 58 to 87 years grew for an average of 20 population doublings. This was more fully documented in Paper 9. In a study of skin fibroblasts from a large number of individuals in the age range 10 to 90 years, a statistically significant decline in growth potential was seen. For any one age group, however, a considerable range of population doublings was recorded. The longevity of only one culture from each individual was measured, but it is now clear that there are important stochastic processes that influence the longevity of clones of cells or whole populations (Papers 11 and 12). In view of this point, it is improbable that the growth of a single culture can be accepted as a quantitative measure of in vitro longevity, and some of the conclusions may not be valid—for example, that the actual site of the skin biopsy influences longevity. As well as the documentation of the relationship between in vivo and in vitro aging with normal individuals, the important discovery was made that cells from patients with Werner's syndrome have very limited growth potential in culture. This recessive autosomal condition appears to accelerate many of the normal features of aging, and the expectation of life of affected individuals is severely reduced (for reviews, see Epstein et al., 1966; Salk, Fujiwara, and Martin, 1985). Correlations between accelerated in vivo and in vitro aging add considerable support to Hayflick's contention that fibroblasts provide a means for studying the nature of the intrinsic processes of aging. It is important to realize, however, that neither he nor others has stated that the aging of fibroblasts in vivo is important in the aging of the whole organism.

In Paper 10 Orgel followed up his earlier theoretical work on protein error propagation with a wide-ranging discussion of the aging of cultured mammalian cells. He discusses several possible causes of cellular aging, including molecular clocks and programmed aging, the deterioration of mitochondria, the dilution of essential cellular components, as well as protein or general error theories. He makes clear the distinction between the aging of individual cells, cell lineages, and cell populations, particularly in relation to the immortality of transformed cultures. Paper 10 is often cited as an updated account of the relevance of the error theories to fibroblast aging, but its major contribution is to provide a clear and succinct discussion of the overall problem of cellular aging and how it can best be approached experimentally. If it had been more widely read in this light, a variety of ambiguous, confusing, or contradictory assertions in the literature need never have been published.

With regard to human diploid fibroblasts, there are three fundamental observations that need explanation. First, all members of the population become committed to senescence, since permanent lines never emerge spontaneously. Second, parallel populations derived from very early passage cultures, or individual clones, vary considerably in their growth potential. Third, the immortalization of diploid cells can occur under given circumstances (for example, after SV40 infection or in tumor tissue) to produce heteroploid permanent lines. The commitment theory of cellular aging (Paper 11; Holliday, 1975; Kirkwood and Holliday, 1975; Holliday, Huschtscha, and Kirkwood,

1981) attempts to provide an explanation for these observations, but it should also be noted that it was in part derived from the experimental demonstration of the commitment to senescence in *Podospora anserina* (Marcou, 1961). In cell populations of this fungus, a random event occurs that has no effect on growth rate or appearance of the culture but is followed by a constant incubation period, at the end of which the culture dies. This model was applied to cells and cell lineages of human fibroblasts (Kirkwood and Holliday, 1975). It was proposed that an initial founder population of cells in uncommitted and capable of indefinite growth. These cells give rise during division to committed cells, with a probability P, that are irreversibly destined to become senescent and die after a given period of proliferation. The number of divisions from commitment to cell death is referred to as the incubation period, *M*. (To simplify the mathematical treatment, it is assumed that the rate of cell division is constant during the incubation period.) It was shown that if *P* is of the order of .25 and *M* is at least 55 divisions; then the proportion of uncommitted cells progressively declines during Phase II, and these will eventually be lost before any cells have reached the end of the incubation period. The whole population is then committed to die at some later stage. The model showed that cell population could have finite or infinite growth, depending on the value of *P* and *M,* and also that the population size, *N,* would affect longevity. For a population with finite growth, it was predicted that the rate of population doubling would decline when the earliest committed cells reached the end of their incubation period and thereafter stay constant until a few population doublings before the population died. This was verified for fetal lung strains MRC-5 and MG-4 (Kirkwood and Holliday, 1975; Holliday, Huschtscha, and Kirkwood, 1981).

The stochastic form of the model was developed in Paper 11, which took account of the fact that dilution and loss of uncommitted cells would be subject to random fluctuation, or drift, as uncommitted cell numbers became small. (The mathematical details of this model have not been published.) Computer simulations showed that the parallel populations established from a very early-passage culture will vary considerably in their growth potential, as indeed is the case for MRC-5, WI-38, and TIG-1 (Paper 11; Ohashi et al., 1980). Simulations were also used to predict the effect of a transient, drastic reduction in population size ("bottlenecks") at different population doubling levels. Bottleneck experiments with MRC-5 (Paper 11) and MG-4 (Holliday, Huschtscha, and Kirkwood, 1981) strikingly confirmed the computer predictions. Another strong prediction of the theory is that any very large population of fibroblasts should grow indefinitely, since uncommitted cells would not be diluted out. Such an observation would demonstrate that the Hayflick limit is, in a sense, an artifact of the way cells are serially cultured. Unfortunately, the experiment is not practical, since the population size would have to be approximately 10^{10} cells. Finally, the model predicts that mixtures of fibroblasts from individuals with two genetically distinguishable types (e.g. female cells heterozygous for the X-linked markers, G6PD A and B) will often produce Phase III populations that consist entirely of only one phenotype, even in the absence of cellular selection for this cell type. This prediction has been confirmed in an independent study by Zavala, Fialkow, and Herner (1978).

Since the model has only three simple parameters, one of which is usually set by the experimenter (population size, N), it would be surprising if it was compatible with all features of growth and death of whole populations or single clones of human fibroblasts. A series of elegant studies has been carried out by Smith and his associates (Paper 12; Smith and Braunschweiger, 1979; Smith, Pereira-Smith, and Good 1977; Smith, Pereira-Smith, and Schneider, 1978). A major finding is the extraordinary variability in growth potential of these individual clones. In general, a subset of cells has very limited growth potential (1 to 8 divisions), while the remainder, although very heterogeneous in their growth, achieve on average the number of population doublings observed with the parental mass population. Paper 12 illustrates the technique used and the results obtained. It also showed that sister cells often have major differences in growth potential, which clearly establishes the stochastic nature of the aging process. In other studies, the relationship between donor age and fibroblast longevity has been confirmed by measuring the growth potential of individual clones (Smith, Pereira-Smith, and Schneider, 1978).

Although the variability in the longevity of individual clones is compatible with the major features of the commitment theory of fibroblast aging, other observations on clonal growth show that the theory has to be refined or modified. The experiements clearly demonstrate, for instance, that the incubation period could not be the same for all lineages within a clone arising from a single committed cell. Prothero and Gallant (1981) and Angello and Prothero (1985) have attempted to modify the theory to account for some of these observations. It must be borne in mind, however, that cells in the environment of a mass culture may behave differently from cells that are isolated from each other. For example, cloning efficiencies of early-passage cells vary enormously from laboratory to laboratory, whereas it is easy to demonstrate by ^{3}H thymidine autoradiography that almost all such cells are dividing in mass cultures.

Many experiments have been done in which young and senescent cells were fused to form hybrids or in which cells were reconstituted from young or senescent nuclei and cytoplasm (cybrids). In principle, such experiments should establish whether the senescent condition is dominant or recessive and whether the primary events are nuclear or cytoplasmic. In practice, the experiments are very difficult to carry out and the results have been rather equivocal (for a review, see Muggleton Harris, 1979). One recent study, again from Smith's laboratory (Periera-Smith and Smith, 1982) gives particularly clear-cut results, however. When senescent cells were fused with young cells, the growth potential of the hybrid clones is much closer to that of the senescent parent than to that of the young parent. Thus, senescence is dominant.

In other studies, in which normal fibroblasts were fused with transformed cells, such as HeLa cells, it was found that most hybrids have limited growth potential, showing that the indefinite growth of transformed cells is a recessive characteristic (Bunn and Tarrant, 1980). This research has been followed by more detailed experiments, including those in which two different transformed cell types have been fused with each other (Paper 13). The conclusion is that immortalization—that is, the ability of a line to grow indefinitely—is a recessive characteristic, since the fusion of different lines can produce a hybrid with finite growth. The implications of this are far-reaching, since it implies that the activity of specific genes is necessary for cell aging, or mortalization, and that with modern techniques such genes could be isolated and studied.

This section has dealt with aging of human cells, but it is clear from many studies that the behavior of rodent cells (mouse, rat, and hamster) is very different. For example, they frequently transform spontaneously to permanent lines, and this appears to be a necessary prerequisite for full transformation. With human cells, immortalization is a rare event, which has so far been demonstrated only in cells infected with SV40 (for a review, see Huschtscha and Holliday, 1983). The recently discovered relationships between oncogenes, transformation, and immortalization in rodent cells (Land, Parada, and Weinberg, 1983; Newbold and Overell, 1983) have highlighted the importance of experimental studies of cellular aging.

With the exception of Paper 10, Part II has covered the major features of human cell aging rather than the underlying mechanisms. The commitment theory, for example, is a formal one, compatible with a program or clock for aging, error accumulation, or simply prolonged, unbalanced growth (Holliday and Kirkwood, 1975). Paper 14 is a review and is included because it forms a link between studies of cellular aging and the experiments documented in Parts III and IV, which attempt to test in one way or another the possibility of error accumulation in senescent cells. The title is taken from the authorative review, "An Unsolved Problem in Biology," by Medawar (1952). In the intervening period innumerable studies of cell aging have been carried out, but the mechanism of the process remains as elusive as ever. However, the final section of Paper 14 attempts to unify current knowledge about cell aging, with information about error accumulation and the evolution of aging in higher organisms. A more general presentation, covering similar ground, is also published in the same monograph (Kirkwood, 1984), and another review, covering 100 years of "cytogerontology," takes as its starting point the fundamental realization by Weismann that the aging of the organism must be understood at the cellular level (Kirkwood and Cremer, 1982).

REFERENCES

Angello, J. G., and J. W. Prothero, 1985, Clonal Attenuation in Chick Embryo Fibrobalsts, *Cell Tissue Kinet.* **18:**27–43.

Bunn, C. L., and G. M. Tarrant, 1980, Limited Lifespan in Somatic Cell Hybrids and Cybrids, *Exp. Cell Res.* **127:**385–396.

Epstein, C. J., G. M. Martin, A. L. Schultz, and A. G. Motulsky, 1966, Werners Syndrome. A Review of Its Symptomology, Natural History, Pathologic Features, Genetics and Relationships to the Natural Ageing Process, *Medicine (Baltimore)* **45:**177–221.

Hayflick, L., and P. S. Moorhead, 1961, The Serial Cultivation of Human Cell Strains, *Exp. Cell Res.* **25:**585–621.

Holliday, R., 1975, The Growth and Death of Diploid and Transformed Human Fibroblasts, *Fed. Proc.* **34:**51–55.

Holliday, R., L. I. Huschtscha, and T. B. L. Kirkwood, 1981, Further Evidence for the Commitment Theory of Cellular Ageing, *Science* **213:**1505–1508.

Huschtscha, L. E., and R. Holliday, 1983, The Limited and Unlimited Growth of SV40 Transformed Cells from Human Diploid MRC-5 Fibroblasts, *J. Cell Sci.* **63:**77–79.

Kirkwood, T. B. L., 1984, Towards a Unified Theory of Cellular Ageing, *Monogr. Dev. Biol.* **17:**9–20. Karger, Basel.

Kirkwood, T. B. L., and R. Holliday, 1975, Commitment to Senescence: A Model for the Finite and Infinite Growth of Diploid and Transformed Human Fibroblasts in Culture, *J. Theor. Biol.* **53:**481–496.

Kirkwood, T. B. L., and T. Cremer, 1982, Cytogerontology since 1881: A Reappraisal of August Weismann and a Review of Modern Progress, *Hum. Genet.* **60:**101–121.

Land, H., L. F. Parada, and R. A. Weinberg, 1983, Tumorigenic Conversion of Primary Embryo Fibroblasts Requires at Least Two Co-operating Oncogenes, *Nature* **304:**596–602.

Marcou, D., 1961, Notion de longévité et nature cytoplasmique du déterminant de la sénescence chez quelque champignons, *Annu. Sci. Natl. Bot.* **12:**653–764.

Medawar, P. B., 1952, *An Unsolved Problem in Biology,* Lewis, London. (Reprinted in *The Uniqueness of the Individual,* 1957, Methuen, London.)

Muggleton Harris, A. L., 1979, Re-assembly of Cellular Components for the Study of Ageing and Finite Life Span, *Int. Rev. Cytol.* **9**(Suppl):279–301.

Newbold, R. F., and R. W. Overell, 1983, Fibroblast Immortality is a Prerequisite for Transformation by EJc-Ha-*ras* Oncogene, *Nature* **304**:648–651.

Ohashi, M., S., Aizawa, H. Ooka, T. Ohsawa, K. Kaji, H. Kondo, T. Kobayashi, T. Noumura, M. Matsuo, Y. Mitsui, S. Murota, K. Yamamoto, H. Ito, H. Shimada, and T. Utakoji, 1980, A New Human Cell Strain, TIG-1, for the Research on Cellular Ageing, *Exp. Gerontol.* **15**:121–133.

Pereira-Smith, O. M., and J. R. Smith, 1982, The Phenotype of Low Proliferative Potential is Dominant in Hybrids of Normal Human Fibroblasts, *Somatic Cell Genet.* **8**:731–742.

Prothero, J., and J. A. Gallant, 1981, A Model of Clonal Attenuation *Proc. Natl. Acad. Sci. (U.S.A.)* **78**:333–337.

Salk, D., Y. Fujiwara, G. M. Martin, eds., 1985, *Werner's Syndrome and Ageing,* Plenum Press, New York.

Smith, J. R., and K. I. Braunschweiger, 1979, Growth of Human Embryonic Fibroblasts at Clonal Density: Concordance with Results from Mass Cultures, *J. Cell Physiol.* **98**:577–602.

Smith, J. R., and L. Hayflick, 1974, Variation in the Life-Span of Clones Derived from Human Diploid Cell Strains, *J. Cell. Biol.* **62**:48–53.

Smith, J. R., O. M. Pereira-Smith, and P. I. Good, 1977, Colony Size Distribution as a Measure of Age in Cultured Human Cells. A Brief Note, *Mech. Ageing and Dev.* **6**:283–286.

Smith, J. R., O. M. Pereira-Smith, and E. L. Schneider, 1978, Colony Size Distribution as a Measure of *in vivo* and *in vitro* Aging, *Proc. Natl. Acad. Sci. (U.S.A.)* **75**:1353–1356.

Swim, H. A., and R. F. Parker, 1957, Culture Characteristics of Human Fibroblasts Propagated Serially, *Am. J. Hyg.* **66**:235–243.

Witkowski, J. A., 1980, Dr. Carrel's Immortal Cells, *Med. Hist.* **24**:129–142.

Zavala, C., P. J. Fialkow, and G. Herner, 1978, Evidence for Selection in Cultured Diploid Fibroblast Strains, *Exp. Cell Res.* **117**:37–144.

8

Reprinted from *Exp. Cell Res.* **37**:614–636 (1965)

THE LIMITED *IN VITRO* LIFETIME OF HUMAN DIPLOID CELL STRAINS[1,2]

L. HAYFLICK

The Wistar Institute of Anatomy and Biology, Philadelphia, Pa., U.S.A.

Received May 4, 1964

Previous reports from this laboratory have emphasized the fact that serially cultured human diploid cell strains have a finite lifetime *in vitro* [2, 12, 14, 15]. After a period of active multiplication, generally less than one year, these cells demonstrate an increased generation time, gradual cessation of mitotic activity, accumulation of cellular debris and, ultimately, total degeneration of the culture. The limited *in vitro* multiplication of many kinds of cultured cells has been a common observation of cell culturists. Until recently [14], technical difficulties were invoked as an explanation for this event. This phenomenon in the course of *in vitro* cultivation of human fetal diploid cell strains, which we refer to as Phase III, has been shown to occur after 50 ± 10 serial passages *in vitro* using a 2:1 subcultivation ratio [14]. This event has now been confirmed in other laboratories [36, 41] and appears to be causally unrelated to conditions of cell culture, the media composition used, presence of mycoplasma or latent viruses, or the depletion of some non-replicating intracellular metabolic pool [14]. Consequently, we advanced the hypothesis that the finite lifetime of diploid cell strains *in vitro* may be an expression of aging or senescence at the cellular level. Experiments to be described extend the studies of this phenomenon and have further bearing on an interpretation based on a theory of senescence.

MATERIALS AND METHODS

Medium.—The medium used was Eagle's Basal Medium [7] supplemented with 10 per cent calf serum. Sufficient $NaHCO_3$ was added so that the medium, upon equilibration to 37°C, reached a pH of 7.4. For prevention of microbial (including

[1] This investigation was supported (in part) by USPHS Career Development Award 5-K3-CA-5938, Research Grant CA-04534 and Contract PH-43-62-157 from the National Cancer Institute.

[2] Part of this work was presented at a symposium on "Cytogenetics of Cells in Culture, Including Radiation Studies" sponsored by the International Society for Cell Biology, Pasadena, Calif., Oct. 14–17, 1963.

mycoplasma) contamination, 50 µg/ml of Aureomycin (Lederle product no. 4691–96, intravenous) was used. The material, packaged in 500 mg amounts, was reconstituted, with agitation, in 50 ml of sterile distilled water at 37°C. Five ml aliquots of this stock concentrate were stored at −20°C. Each liter of warm medium was supplemented before use with 5 ml of stock concentrate. Mycoplasma determinations made over a three-year period on approximately 2000 cell cultures have revealed a total absence of mycoplasma where Aureomycin has been used. Subsequent growth of treated cultures in antibiotic-free medium for extended periods of time has also been proven negative for the presence of these microorganisms. Trypsin was prepared as previously described [14], pre-warmed to 37°C, and raised to pH 7.5 before use.

Human diploid cell strains.—Strains WI-26, WI-38 and WI-44 were used. WI-26 was derived from male fetal human lung and WI-38 and WI-44 from female human fetal lung. All embryos were obtained from surgical abortions and were of approximately three months' gestation.

Subcultivation of confluent cultures.—The method of subcultivation was a modification of that previously described [14]. The medium from confluent cell sheets was removed and pre-warmed (37°C) trypsin solution (Difco 1:250) was added to each culture for 1 min. All except 1 or 2 ml of the trypsin was then decanted and the bottle culture allowed to stand at room temperature for about 30 min. A small amount of Eagle's medium was added and splashed over the loosened cell sheet. The suspension was then vigorously aspirated with a narrow-bore 5 or 10 ml pipette to obtain discrete single cells. Sufficient additional medium was added for the total volume of the suspension to cover twice the surface area from which it was obtained. This is referred to as a 2:1 split ratio. Cultures were incubated at 37°C.

Initiation of the new strains, chromosome analysis, preservation in liquid nitrogen and reconstitution were performed as previously described [12, 14, 40]. These strains have characteristics similar to others previously reported [14]. Chromosome analyses have shown the human diploid cell strains WI-26, WI-38 and WI-44 to be normal or classic diploid [30]. Preliminary studies on the human cell strains of adult lung origin also indicate classic diploidy [24].

EXPERIMENTAL RESULTS

Reconstitution of frozen cells.—As previously shown, human fetal diploid cell strains, preserved at sub-zero temperatures and subsequently reconstituted, enter Phase III at a total number of passages (2:1 split ratio) of 50 ± 10 [12, 14, 15]. This compares favorably with the passage level at which Phase III occurs in the original passage series of the strain which had never been frozen [12, 14, 15]. Further experiments with human diploid cell strains WI-26 and WI-38 have confirmed and extended these results. As indicated in Fig. 1, the average passage level at which mitoses ceased (Phase III) in 20 ampules of WI-26, reconstituted from various passage levels and preserved for periods of time up to 19 months, was 47 passages. The

85

range was 38–60 passages. This compares favorably with 50 passages for the original unfrozen culture.

Seven hundred and fifty ampules of WI-38 were preserved at the eighth passage level. The average passage level at which Phase III occurred in 18 of these

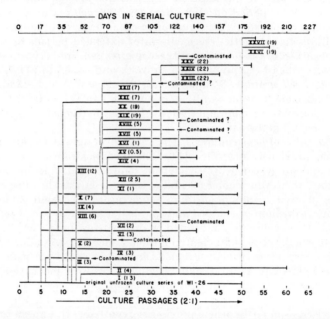

Fig. 1. Diagrammatic representation of the history of human diploid cell strain WI-26. Original unfrozen culture series represents the continuous subcultivation of the strain through 50 passages, during which time surplus cultures from each passage were stored in liquid nitrogen. Other series represented by the roman numerals were reconstituted at passage levels denoted by the origin of the vertical lines. The numbers in parentheses denote the number of months that the ampule giving rise to each series was stored. The average number of passages at which Phase III occurred is 47. The range is 38–60.

ampules preserved for periods of time up to 15 months was 47 passages. The range was 39 to 58 passages. These results are similar to the 48 passages obtained with the original unfrozen culture, as shown in Table I.

These results indicate that, regardless of the passage level at which a human fetal diploid cell strain is frozen, the total number of passages that can be expected at a 2:1 split ratio is about 50 ± 10, including those made prior to preservation. Therefore, it can be concluded that the onset of Phase III cannot be related to absolute calendar time but is related to the time during which the culture is actively proliferating.

This observation has been repeated with a number of different human diploid cell strains of embryonic origin and confirmed in other laboratories [36, 41]. Furthermore, none of the approximately 200 laboratories that have received cultures of strains WI-26 and WI-38 has reported success in subculturing them indefinitely. All recipients, successfully cultivating these

TABLE I. *Passage levels at which Phase III occurred in thawed ampules of the 8th passage of WI-38.*

Culture	Passage level reached prior to cessation of mitoses[a] (Phase III)	Weeks preserved in liquid nitrogen	Culture	Passage level reached prior to cessation of mitoses[a] (Phase III)	Weeks preserved in liquid nitrogen
Parental	48	—	X	44	36
(never frozen)			XI	45	40
I	48	3	XII	47	40
II	48	8	XIII	47	40
III	42	10	XIV	49	50
IV	50	12	XV	47	50
V	43	20	XVI	39	59
VI	47	24	XVII	42	65
VII	53	30	XVIII	45	68
VIII	58	32		Average 47	
IX	47	32		Range 39–58	

[a] All passages done at a 2:1 split ratio. Figures include 8 doublings prior to preservation, except the parental culture.

strains, have reported that Phase III occurred at "about the 50th passage" when a 2:1 split ratio was used.

Split ratio effect on Phase III.—The measurement of Phase III as a function of the number of subcultivations may be, in effect, equivalent to a measurement of accumulated generations or cell doublings. No one has reported the successful cultivation of any human diploid cell strain in suspension culture, where it is possible to keep a cell population in continuous logarithmic growth, thereby permitting a more accurate determination of cell doublings.

Numerous attempts by us to grow human diploid cell strains in agitated fluid suspension have consistently failed under conditions allowing for the luxuriant growth of heteroploid cell lines [14].

The mechanical or enzymatic methods which must be used to subcultivate cells grown as monolayers in static cultures repeatedly cause the cultures to pass through a "lag-log-lag" pattern. Thus, during the first 24 hr post-sub-cultivation little, if any, mitotic activity is observed. From 24 hr post-sub-cultivation (depending upon the inoculation density) there is a logarithmic

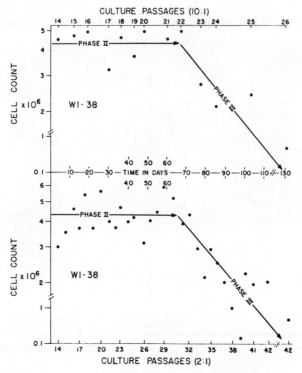

Fig. 2.—Cell counts determined at each passage of WI-38 for two different split ratios (10:1 and 2:1). Total cell counts are plotted as a function of time in days and actual passages. Although fewer actual passages occur when the culture is split 10:1, the total calendar time accruing is the same as that of the 2:1 split ratio in respect to commencement and termination of Phase III.

cell increase, followed by a lag associated with confluency of the monolayer culture.

In order to ascertain the correlation between commencement of Phase III and the total number of accrued doublings, rather than the total number of subcultivations (passages), sister cultures of strain WI-38, taken at the 14th passage, were subcultivated at two different split ratios, 2:1 and 10:1. At each subcultivation the cells in four to six parallel cultures were counted. Two persons manipulated these cultures independently so that the entire experiment was performed simultaneously and in duplicate. Each set of cultures was grown on medium from two different sources to circumvent microbial contamination at any point in the five-month period of this experiment.

The human diploid cell strain WI-38 was thawed at the eighth passage, then at passage 14; cells from four separate confluent cultures were counted. Two cultures were then serially passaged by each of the two individuals, one culture at a 2:1 split ratio and the other at a 10:1 split ratio. All cultures were subcultivated when confluency was reached, at which times cell counts were made and averaged to accumulate the data plotted in Fig. 2. Those cultures which were split 2:1 were subcultivated when they reached confluency every three and four days alternately during Phase II (the period of active cell proliferation) and at increasingly longer intervals during Phase III. Cultures split 10:1 were subcultivated as soon as they became confluent. As can be observed in Fig. 2, the total calendar time accruing until Phase III commenced was similar for both split ratios. Phase III can be defined as the terminal period, during which, time intervals between population doublings are progressively greater. The accrued calendar time at which all cell mitoses ceased and culture degeneration began was also similar for both split ratios. From Fig. 2, a computation of the number of generations accruing for each split ratio before Phase III gives a total of 17 for the 2:1 split ratio and 27 for the 10:1 split ratio. Thus, for the 18 passages of the 2:1 split ratio accruing prior to Phase III, there was an average of 0.96 cell doublings per passage, whereas the eight passages of the 10:1 split ratio accruing during the same period yield an average of 3.33 cell doublings per passage. The figure 0.96 is very close to 1.00, which would, theoretically, be expected to be the number of doublings per passage of cultures split 2:1. The figure 3.33 is also very close to 3.25, the theoretical expectation of cell doublings per passage in cultures split 10:1.

It was assumed that regardless of the split ratio, the total number of cell doublings would be identical during the entire series of passages until Phase III. Contrary to expectation, the split ratio appears to affect the actinal number of generations. The greatest effect was observed when cell cultures were permitted to reach confluency more often during cultivation (2:1 split ratio) and the smallest effect under more efficient conditions where cell confluency occured least often. Variations in these patterns of efficiency have yielded the data shown in Table II. Clones were isolated, transferred to bottles and allowed to reach a density of 4×10^6 before the first 2:1 subcultivation was made. It is evident from the data in Table II, therefore, that the total number of doublings to be expected from a human diploid cell strain is reduced in

proportion to the number of times the culture is permitted to achieve confluency (lag period). Since mitotic activity lessens once these cultures become
confluent [14], a greater, although finite, number of generations may be
expected from a human diploid cell strain which has been cultivated almost

TABLE II. *Doublings of WI-38 as a function of split ratio.*

Treatment	Actual number of splits until Phase III	Number of doublings[a] until Phase III
All 2:1 splits	42	40
2:1 splits until passage 14, then 10:1 splits	27	57
2:1 splits until passage 9, cloned, then 2:1 splits	37	57
2:1 splits until passage 8, cloned, then 2:1 splits to passage 15, recloned, then 2:1 splits	20	63

[a] Based on 0.96 doublings per 2:1 split; 3.33 doublings per 10:1 split; and 22 doublings for a single cell to reach a density of 4×10^6 (confluent culture) when it was then split 2:1.

constantly in the "log phase" of growth. A 2:1 split ratio permits a considerably shorter overall period of time in log phase than does a culture that has
been cloned, allowed to reach a maximum density and then recloned. This
effect has also been described by Todaro, Wolman and Green [36], who, using
different inoculation densities throughout the lifetime of a human diploid
cell strain, observed that, generally, the total number of doublings at the high
inoculation densities was reduced. This effect of realizing higher doubling
potential as a result of more efficient conditions of growth remains to be
clarified; but under the best conditions, eventual failure (Phase III) of the
culture ensues.

The shape of the curves in Fig. 2 is of considerable importance in attempting to understand the mechanism of Phase III. It was desirable, therefore, to
repeat the experiment with another human diploid cell strain and to accumulate cell counts from the earliest possible 2:1 passage until termination
of Phase III. This was done starting with the fourth passage of strain WI-44
and the results are given in Fig. 3. The higher average cell counts obtained

with WI-44 as compared with WI-38 during Phase II resulted from the use of a slightly larger culture vessel. It is apparent from Fig. 3 that the shape of this curve is similar to those in Fig. 2. Such curves are similar to those obtained with the multiple-hit or multiple-target inactivation theory and an

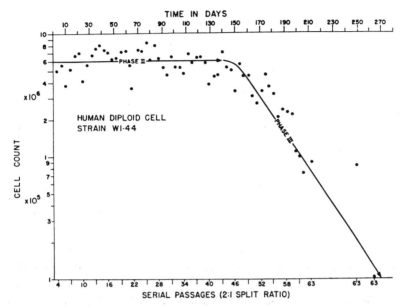

Fig. 3.—Cell counts determined at each passage of strain WI-44. This figure, like Fig. 2, results in a curve suggestive of multiple-hit or multiple-target inactivation phenomena as an explanation for the mechanism of the occurrence of Phase III. The initial plateau during Phase II, with no apparent loss of biological function as measured by constant doubling time, is followed by Phase III, where doubling time increases exponentially.

hypothesis of the mechanism of Phase III based on this phenomenon will be considered subsequently.

Cloning experiments.—Any satisfactory explanation of the finite lifetime of human diploid cell strains *in vitro* involves the question of whether each cell in the population is endowed with the "50 passage potential" or, alternatively, whether there exists a random distribution of passage potentials among individual cells composing the population which results in an average 50 passage potential for the entire population.

In order to investigate the possibility that cell progeny derived from different single cells of the same strain may show differences in the passage level at which they enter Phase III, three randomly selected clones were isolated from three Petri dish cultures of WI-38. This was done at passage level two of the original unfrozen series. The three clones, designated C1, C2 and C3, were each transferred to a milk dilution bottle, incubated at 37°C in a CO_2 incubator, and allowed to reach confluency. They were then subcultivated semi-weekly at a 2:1 split ratio and the total number of pas-

TABLE III. *Accrued doublings of three cloned populations of WI-38.*

Figures based on one doubling per 2:1 split.

Clone	Number of doublings before cloning		Calculated number of doublings for 1 cell to reach a density of 4×10^6		Number of doublings accruing after first 2:1 split, post-cloning		Total number of doublings
C1	2	+	22	+	30	=	54
C2	2	+	22	+	27	=	51
C3	2	+	22	+	26	=	50

sages was recorded starting with number one at this first bottle culture (first 2:1 split, post-cloning). As summarized in Table III, the three cloned cultures entered Phase III at passages 30, 27 and 26, respectively. These numbers do not include the two subcultures carried out before cloning and the cell *doublings* that occurred after cloning and prior to reaching confluency in the milk dilution bottle.

A more accurate appraisal of the results should be based on a compilation of numbers of generations or, more precisely, average doublings of the population. The number of cell doublings necessary to reach a density of 4×10^6 (average cell content of a confluent culture in a milk dilution bottle) is 22. Thus the results tabulated in Table III include these data.

Unless the choice of these three clones was fortuitous, it appears that each *clonable* cell within the population is endowed with the same "50 doubling potential". Although the populations had only undergone a total of 32, 29 and 28 serial 2:1 subcultivations before entering Phase III, the added number of doublings accumulated during the cloning manipulation, in which a single cell was raised to a population of 4×10^6 cells, resulted in the expected value of a total of 50 doublings.

The results of this experiment simply serve to underline the fact that the number of generations expected from a human diploid cell strain is not strictly a function of the number of 2:1 serial subcultivations but rather of the number of cell doublings, and that each clonable cell is endowed with the same doubling potential.

Furthermore, it would be predicted that it should not be possible to initiate a cloned population from, for example, cells of the 35th doubling or greater,

TABLE IV. *Occurrence of Phase III in mixed populations.*

Mixture	Passages of "oldest" component before mixing	Passages accruing after mixing		Total passages	Passages of "oldest" culture (unmixed control)
WI-26 VIII Passage 21 + WI-26 II Passage 45	45	+ 29	=	74[a]	60 (WI-26 II)
WI-26 VIII Passage 21 + WI-26 IX Passage 11	21	+ 42	=	63	46 (WI-26 VIII)
WI-26 IX Passage 11 + WI-26 II Passage 45	45	+ 38	=	83[a]	60 (WI-26 II)
		Average ...		73.3[a]	55.7

[a] Values greatly in excess of any ever observed.

and expect that population to increase to 4×10^6 cells. This failure would be expected since the 35 doublings preceding cloning must be added to the 22 doublings necessary to raise a single cell to a density of 4×10^6. This range of 57 doublings would be near the outside limit, at which time Phase III would be expected to occur. Such is the case. Seven attempts to clone a fetal human diploid population at doublings greater than 30 have consistently failed to yield populations giving rise to as many as 4×10^6 cells. The clones that do result reach Phase III before any subcultivation is necessary.

Mixing populations of cells with different doubling potentials.—We had suggested [14] that the Phase III phenomenon was intracellularly determined. In that previously described experiment the male human diploid cell strain WI-1 at the 49th passage (Phase III) containing many metabolically active, non-dividing cells was mixed with a suspension of actively dividing female cells of strain WI-25 at passage 13 (Phase II). Metaphases from the mixed population, examined at 17 passages post-mixing, were found to consist entirely of female cells. The female cells of the mixed culture ultimately entered Phase III at about the same passage level as the unmixed control culture of WI-25. Thus it was concluded that "old" cells had no detectable effect upon "young" cells or *vice versa*. This experiment also demonstrated the unlikelihood that latent micro-organisms (or media composition) could account for Phase III, since it is probable that the virus spectra of various human diploid cell strains are qualitatively similar [12, 14, 40]. Final proof

TABLE V. *Occurrence of Phase III in mixed populations.*

Mixture	Passages of "youngest" component before mixing		Passages accruing after mixing		Total passages	Passages of "youngest" culture (unmixed control)
WI-26 VIII Passage 21 + WI-26 II Passage 45	21	+	29	=	50	47 (WI-26 VIII)
WI-26 VIII Passage 21 + WI-26 IX Passage 11	11	+	42	=	53	55 (WI-26 IX)
WI-26 IX Passage 11 + WI-26 II Passage 45	11	+	38	=	49	55 (WI-26 IX)
				Average ...	50.7	52.3

of these conclusions would, however, depend upon the outcome of an experiment in which cells from a given strain, approaching Phase III, were mixed with cells from the same strain reconstituted from frozen stock at an earlier passage (Phase II).

Cell cultures of human diploid cell strain WI-26 were reconstituted from frozen stock on different calendar dates at passage levels two (series II, Fig. 1), five (series VIII, Fig. 1) and seven (series IX, Fig. 1). When each had reached passage levels 45, 21 and 11, respectively, by semi-weekly subcultivations at a 2:1 split ratio, two cultures of each were counted and split 2:1. Since all experiments were performed in duplicate, one culture of each served as a control and the other two were mixed in equal numbers while in suspension (according to Table IV), planted and subsequently split 2:1, semi-weekly, exactly as were the controls.

The purpose of this experiment, therefore, was to ascertain whether cells of the same strain from various widely separated passage levels would influence each other, as measured by the commencement of Phase III. If it were assumed that the "oldest" component of each mixed population survived (a possibility made implausible by the experiment previously described employing mixed male and female cells), Table IV would result. This explanation of replacement by the "older" component results in passage levels of 74 and 83; levels not in keeping with any values ever obtained. Furthermore, these figures do not compare favorably with the controls.

Based on the supposition in Table V that the passage level at which Phase

TABLE VI. *A comparison of the passage levels at which Phase III occurred in human diploid cell strains of adult and fetal origin.*

All strains cultivated at a 2:1 split ratio. Fetal strains derived from donors of 3–4 months' gestation obtained by surgical abortion. Adult and fetal strains derived from both male and female tissue.

Fetal lung		Adult lung			
Strain	Passage level at which Phase III occurred (cell doublings)	Strain	Passage level at which Phase III occurred (cell doublings)	Age of donor	Cause of death
WI-1	51	WI-1000	29	87	Heart failure
WI-3	35	WI-1001	18	80	Cerebral vascular accident
WI-11	57	WI-1002	21	69	Bronchial pneumonia
WI-16	44	WI-1003	24	67	Dissecting aneuryism
WI-18	53	WI-1004	22	61	Renal failure
WI-19	50	WI-1005	16	58	Rheumatoid arthritis
WI-23	55	WI-1006	14	58	Pulmonary embolus
WI-24	39	WI-1007	20	26	Auto accident
WI-25	41				
WI-26	50				
WI-27	41				
WI-38	48				
WI-44	63				
Average	48		20		
Range	35–63		14–29		

III occurred in the mixture was a function of the continuing multiplication of the "youngest" half of the mixed population after total loss of the "oldest" component, it is apparent that the observed values of total passages after mixing conform to expectations. The following conclusions can thus be drawn: Phase III in a human diploid cell population occurs at that time when the "youngest" cell component in a mixed population is expected to reach Phase III. The "older" cell component has no apparent effect upon the "younger" cells in such mixed populations. This experiment, incidentally, using a homogeneous cell system, again substantiates the previous conclusion [14] that Phase III cannot be explained by the presence of a latent virus, myco-plasma or media composition.

Occurrence of Phase III in adult human diploid cell strains.—If, as has been determined, human diploid cell strains of fetal lung origin reach Phase III

TABLE VII. *A comparison of the passage levels at which Phase III occurred in parental and frozen substrains of adult lung human diploid cell strains.*

All strains cultivated at a 2:1 split ratio.

Strain	Parental culture (never preserved) Passage level at which Phase III occurred (cell doublings)	Substrain reconstituted from liquid nitrogen		
		Preserved at passage	No. of weeks frozen	Passage level at which Phase III occurred (cell doublings)
WI-1000	29	6	6	23
WI-1004	22	6	5	20
WI-1003	24	8	2	20

at about 50 ± 10 passages, it would be of interest to compare the occurrence of Phase III in similar cultures derived from adult human lung. Eight diploid strains of adult human cells have been compared with our 13 fetal strains under identical cultivation conditions. A number of strains from both groups were cultivated simultaneously utilizing common sources of reagents and glassware. In addition, all strains were cultivated in duplicate by two different individuals.

All of the fetal lung strains were derived from fetuses of approximately three months' gestation. The adult strains were derived from lung tissue obtained at death at ages and from causes indicated in Table VI. The human adult diploid cell strains were morphologically indistinguishable from the fetal strains. Such properties as growth rate, nutritional requirements and others that were investigated indicated that no parameter other than the total number of doublings occurring during Phase II, could distinguish adult from fetal human diploid cell strains. A comparison of the number of doublings obtained prior to Phase III for the eight adult and 13 fetal human diploid cell strains is given in Table VI.

The average number of cell doublings accruing in adult lung strains prior to cessation of mitotic activity is 20 (range 14–29) which is significantly less than the average number of 48 doublings (range 35–63) obtained with fetal lung strains.

There appears to be no exact correlation between the age of the donor and the doubling potential of the derived strain. If such a relation does, in fact, exist, it cannot be detected by the present crude methods of determining doubling potential. It is clear, however, that there is a significant difference between the doubling potentials of human diploid cell strains when derived from lung tissue of either fetal or adult origin.

It was also of interest to determine whether the doubling potential of adult strains preserved in liquid nitrogen is similar to that of the original unfrozen culture. The results of such a comparison, utilizing three different adult strains and one thawed ampule of each, are given in Table VII. The interpretation of these results is identical to that obtained with the fetal strains, in that the onset of Phase III is unrelated to absolute calendar time but is related to the total time during which the culture is actively proliferating. Furthermore, the passage potential of frozen adult strains is similar to that of the original unfrozen cultures from which they were derived. In both cases the passage potential is substantially less than that obtained with unfrozen or frozen strains derived from human fetal lung.

DISCUSSION

The finite lifetime of human diploid cell strains *in vitro* has been quantitatively examined and found to be related only indirectly to numbers of subcultivations at a particular split ratio. The effect is more precisely related to a finite number of cell doublings. Cloning experiments have led to the conclusion that the doubling potential is the same for each clonable cell in the population.

This event is not influenced by the presence of cells in the culture with a reduced doubling potential, as demonstrated by an experiment in which cells of the same strain were mixed at three widely separated passage levels. In all cases the total doublings accrued by the mixed cultures before extinction was a function of the doubling potential of the "youngest" component of the population. This coincided with the passage level at which Phase III occurred in the unmixed controls. It is possible that the Phase III phenomenon of cultured human diploid cell strains may be related to senescence *in vivo*. In this regard four points are relevant.

The cellular theory of aging.—A cellular theory of aging is generally considered unacceptable because of the apparent "immortality" of cell cultures

[4, 22, 23, 26]. This general belief is based on the "immortality" of those cell cultures (cell lines) now known to share many, if not all, of the characteristics associated with malignant cells [12, 14]. During the development of cell culture techniques from the beginning of this century until the early 1930's, it was apparent that, regardless of the vertebrate tissue of origin, cell populations derived *in vitro* could be kept in an active state of multiplication for a varied but finite period of time. This finite period of cell proliferation could not, generally, be extended much over a year. Variations in media composition, cultivation techniques, incubation temperature, and other parameters investigated by early workers failed to change this course of events. In fact, it was concluded that the primitive methodology used for *in vitro* cell cultivation was reason enough for the short term cultivation of cells *in vitro*. It is our contention that the finite lifetime of unaltered or diploid cell strains is an innate characteristic of the cells, unrelated to known techniques for optimum cell cultivation. One possible exception to this generalization was the highly popularized development from Carrel's laboratory wherein it was claimed that a population of cells derived from embryonic chick heart tissue was kept in serial cultivation for 34 years [25]. Since, even with more modern and sophisticated cell culture techniques, chick fibroblast cultures do not survive more than a year, there is serious doubt that the common interpretation of Carrel's experiment is valid. An alternative explanation of Carrel's experiment is that the method of preparation of the chick embryo extract, used as a source of nutrient for his culture and prepared daily under conditions permitting cell survival, contributed new, viable, embryonic cells to the chick heart strain at each subcultivation or feeding [9]. A consideration of the details of this experiment [9] indicates that waves of mitotic activity in his cultures were coincidental with the periodic addition of chick embryo extract. In any event, Carrel's experiment has never been confirmed.

It remained for Gey [10] in 1936 and Earle [8] in 1943 to demonstrate that cell populations derived from a number of mammalian tissues, including human tissue, could unequivocally be kept in a state of rapid multiplication for apparently indefinite periods of time. Since this pioneer work, cell populations with the extraordinary capacity of being able to multiply *in vitro* indefinitely have been spontaneously derived from at least 225 mammalian tissues [13]. A consideration of the characteristics of these populations, referred to by us previously as "cell lines" [14], has led to the conclusion that such cell lines, regardless of whether the tissue of origin is normal or cancerous, share properties with cancer cells. First, they are heteroploid, as are all transplantable solid tumors. Second, when inoculated into suitable hosts, they form tumor masses. Third, less definitive tests, such as staining and microscopic examination, have indicated that cell lines share those properties that are usually descriptive of cancer cells. Conversely, cell strains are diploid and fail to exhibit such properties. The relationship that cell lines bear to cell strains is identical to the relationship that transplantable tumors bear to normal tissue. The former two systems are assumed to be *in vitro* expressions of the latter *in vivo* systems.

HETEROPLOID CELL LINES	:	TRANSPLANTABLE TUMORS	=	DIPLOID CELL STRAINS	:	NORMAL SOMATIC TISSUE
(*in vitro*)		(*in vivo*)		(*in vitro*)		(*in vivo*)

1. Heteroploid
2. Cancer cells (histological criteria)
3. Indefinite growth

1. Diploid
2. Normal cells (histological criteria)
3. Finite growth

Thus the phenomenon of the alteration of a cell strain to a cell line [14] is important because, in its simplest terms, it can be regarded as oncogenesis *in vitro*. Spontaneous alterations do occur in human cell cultures but at a very low frequency and only a few photographs of this event have ever been published [11]. A set of precise environmental conditions under which alterations of human cells could take place were, until recently, unknown. The spontaneous alterations described in the literature [13] have arisen under many different kinds of cultural conditions. Reproducible conditions for inducing such alterations would be a most powerful tool for the study of the *in vitro* conversion of normal human diploid cells to cancer cells. Recently, it was discovered [18, 19, 32, 36] that the infection of primary cultures or human diploid cell strains, with the virus S.V.$_{40}$ could provide these conditions.

Since heteroploid cell lines are known to possess qualities characteristic of cancer cells, the cellular theory of *in vivo* aging should be related to activities of *normal diploid cells in vitro* rather than to *heteroploid cell lines in vitro*. On this basis the cellular theory of aging must be reconsidered, since it has been shown [14, 36, 41] that normal human diploid cell strains *in vitro* are, in fact, "mortal". To our knowledge no one has thus far reported that cells having the karyotype of the tissue of origin have been able to multiply *in vitro* longer than the lifespan of the animal species from which the tissue was obtained. Cells which can be cultivated indefinitely *in vitro* (heteroploid cell lines) can only be compared with continuously cultivable cells *in vivo*, i.e., transplantable tumors. Likewise, diploid cells having a finite lifetime *in vitro* can only be compared with normal cells *in vivo*, i.e., normal somatic cells.

The finite lifetime of cells in vivo.—The above relationship had led us [14] to consider an experiment designed to test the question as to the length of time normal tissue could be grown when transplanted from animal to animal in an attempt to escape from the normal senescence of the host. Recently, two series of experiments have been performed which bear upon this question.

A series of transplantation experiments devised by Krohn [20] appears to demonstrate that there *is* a finite period of cultivation of normal mammalian somatic cells *in vivo*. Using skin transplants from inbred strains of mice, Krohn attempted to determine whether skin tissue has an indefinite lifespan when transplanted from one host to another. His studies revealed that grafts from young donors remained in satisfactory condition for about 650 to 1000 days and after two to five transplantations. However, the grafts began to decrease in size at that time and between 850 and 1750 days the transplants had become "minute areas of skin which were unsuitable for further transplantation". In comparison with the longest recorded lifespan of $3\frac{1}{2}$ years for any mouse [28], the maximum lifespan of skin transplants ranged from $4\frac{1}{2}$ to 5 years. What is most important is that the skin transplants *did* demonstrate a finite period of cultivation *in vivo*, as do normal diploid mouse fibroblasts *in vitro* [16, 29, 37]. Transplanted normal mouse tissue does not exhibit the kind of immortality characteristic of transplantable mouse tumors, a number of which have been passaged for decades *in vivo* [38], as have heteroploid mouse cell lines *in vitro*.

Krohn [20] investigated time-chimeras by studying the viability of aged skin grafted onto young animals and observed that after initial growth the old grafts failed at an overall age far short of the normal lifespan of the host mouse. Thus, the results of these *in vivo* transplantation experiments with mouse tissue parallel directly the results reported here for human diploid cell strains *in vitro* derived from fetal and adult lung tissue. That is, the passage potential or expected number of cell doublings is generally related to the age of the donor. Any successes with indefinitely cultivable mouse cells that have been reported have always been the result of the diploid cell population altering to a heteroploid or to a near-diploid cell line. This event always occurs *in vitro* in far less time than the average life expectancy of the mouse.

A similar series of experiments performed by De Ome and his associates with transplanted mouse mammary tissue has resulted in similar findings [6]. Normal mammary tissue from adult C3Hf/Crgl ♀ mice was transplanted into a group of three-week-old (virgin) C3Hf/Crgl ♀♀ mice who, when five months old served as donors for the next transplant generation. This process was repeated for 40 to 45 months when the outgrowths could no longer be found in the living animals and thus could not be further transplanted. This transplantation procedure has been repeated and the C3Hf normal tissue maintained by serial transplantation in animals less than five months of age was not recoverable after 40 to 45 months of transplantation. In addition, in two trials with normal mammary tissue from an adult C3H/Crgl ♀, which was serially transplanted using exactly the same method described above for the C3Hf/Crgl tissue, the normal tissue could no longer be recovered after 20 to 30 months. Finally, when normal mammary tissues from BALB/c/Crgl ♀♀ were used in a similar experiment, they could not be recovered after a period of only nine months. Similar serial transplantation experiments carried out with hyperplastic alveolar tissue rather than normal mammary tissue yielded, however, quite different results. Three such tissues which were serially transplanted for more than five years through more than 20 transplant generations have to date shown no reduction in growth rate and tumors derived from these three strains have been carried for many years without apparent loss of viability [6].

The implication of these two series of experiments may be, therefore, that the acquisition of potential for unlimited cellular division or the escape from senescent-like changes by mammalian somatic cells, even *in vivo*, can only be achieved by cells which have altered and assumed properties of cancer cells. This applies equally well to normal mammalian somatic cells growing *in vivo* or *in vitro*.

Chromosome anomalies associated with "old" cells in vivo and in vitro.—A number of studies [3, 14, 21, 34, 35] have shown that the karyotype of human fibroblast cells in long-term culture is very stable. More recently two reports have demonstrated that some aneuploidy and other aberrations do occur, but only in Phase III of the *in vitro* life of such cell populations [30, 41]. In the report of Saksela and Moorhead [30], aneuploidy was first detected at about the 40th passage in a number of sub-strains of both WI-26 and WI-38 as well as in very late passages of two other strains. In the work of Yoshida and Makino [41] there was no karyotype variation between the first and the 41st subcultures, but cells from the 45th to the 47th passages showed striking chromosome aberrations and their strain could not be subcultivated beyond the 48th passage. These metaphase studies only confirm an earlier study by Sax and Passano [31] in which it was shown that anaphase anomalies increased with age *in vitro* over a period of six months' subcultivation. Abnormally large interphase nuclei and bizarre nuclear shapes were also described by us in late passaged human diploid cell strains [14].

This direct correlation between age *in vitro* and the appearance of chromosome aberrations suggests that the chromosome damage itself may be responsible for the failure of the culture. Such an explanation would be similar to somatic mutation theories of aging already offered. It is equally plausible, however, that loss of proliferative ability and chromosome damage occur independently.

In this connection a relationship between spontaneous somatic cell aberrations *in vivo* and *natural* aging has been demonstrated by scoring anaphase anomalies in regenerating liver tissue of mice [5]. Strains of mice with different life expectancies revealed corresponding differences in the incidence of aberrations observed. Also, *within* each strain there was an age-correlated increase in anaphase and telophase aberrations scored following partial hepatectomy. Of even greater interest is the recent observation of Jacobs *et al.* [17] who have found in man that increased hypodiploid counts in peripheral blood leucocytes are correlated with the chronologic age of the donor. There exists, therefore, some *in vivo* evidence of age-associated chromosomal anomalies that may also be involved in the limitation upon the proliferation of human diploid cells *in vitro*.

Occurrence of Phase III as a function of donor age.—The experiments described which demonstrate a significantly decreased doubling potential for strains derived from human adult lung tissue appear to parallel the results of Krohn [20]; wherein, it was observed that the growth potential of skin transplants from older mouse donors was far less than that of skin transplants derived from young donors. The implications of these results are that normal tissue, when cultivated *in vitro* or transplanted *in vivo*, has a finite period of multiplication and, furthermore, that the age of the donor of such tissue, under either condition of cultivation, is directly related to the expected growth potential. In this connection, it was exactly 50 years ago that Carrel [1] observed that "fragments of connective tissue taken from an embryo, or from young adult and old animals and placed in normal adult chicken plasma grew at different rates". He concluded that "the velocity of the growth always varied in inverse ratio to the age of the animal from which the tissue had been extirpated".

MECHANISM: The mechanism of the Phase III phenomenon in cultured human diploid cell strains remains to be elucidated. When cell counts are made after each serial subcultivation of such strains and are plotted against time, the curves described in Figs. 2 and 3 are obtained.

The shape of these curves is similar to multiple-target or multiple-hit curves. Such survival curves are commonly obtained, for example, by plotting effects of irradiation on *E. coli* [39] or on human cells [27]. An initial threshold dose is required before an exponential form of the curve is established. Although it is not known whether background irradiation contributes to the Phase III phenomenon, the survival curves obtained (Figs. 2 and 3) are similar to such "dose-effect" curves and allow for hypotheses concerning mechanism:

(a) Each cell contains *n* targets, each of which must be inactivated. This is the multi-target concept.

(b) Each cell contains a single target which must sustain *n* hits before the target is inactivated. This is the multi-hit concept.

Ordinarily the extrapolation of the exponential portion of a multiple-target or multiple-hit curve to the ordinate gives a value for *n* which is the average number of targets struck per cell or the number of "hits" required to inactivate a single target. The crude method by which the exponential portion of the curves in Figs. 2 and 3 is obtained does not allow for an accurate extrapolation, since even slight inaccuracies in the slope of the exponential portion of the curve will result in a large error. However, it is apparent that the number of targets or number of hits on a single target would be large.

Such interpretations follow from the curves in Figs. 2 and 3 which describe an initial plateau with no apparent loss of potential to multiply as measured by the constant doubling time during Phase II, followed by an exponential loss of this function (exponential increase of doubling time in Phase III). Similar curves for other human diploid cell strains have also been described [36]. The plateau indicates that loss of function may require an accumulation of damage caused either by mechanisms (a) or (b) above.

Thus whatever the cell component(s) involved may be, the inactivation of which results in the Phase III phenomenon, the ultimate accumulation of non-dividing cells could be the result of accumulated heritable damage to some sensitive intracellular target(s). This accumulated damage may further result in what has now been described cytologically as an accumulation of aneuploid cells [30, 41] at Phase III.

We propose, therefore, that the basic step in the Phase III phenomenon is an accumulation of "hits" or errors in DNA replication which inactivates part of the genome. It is further assumed that the hits are random and that per unit period of time the probability that a part of the genome suffers such a hit is constant.

The mathematical expression of the type of multi-hit curve in Figs. 2 and 3 in which an initial threshold must be reached before an exponential form of the curve is established is similar to the mathematical model of *in vivo* aging postulated by Szilard [33] in which death occurs when the amount of genetic damage reaches a threshold level. The reduced doubling potential or earlier occurrence of Phase III in human adult diploid lung strains may imply that these cell populations grown *in vitro* had already acquired, *in vivo*, a significant number of hits. This may be analogous to the variation in inherited "fault loads" postulated by Szilard [33] to account for individual variations in time of death as a result of senescence.

Any satisfactory theory of senescence at the cellular level, including the somatic mutation theory, must include, as a corollary, an explanation for the apparent lack of senescent-like changes in transplantable tumors *in vivo* and in heteroploid cell lines *in vitro*. Lacking any evidence on this point, it could be argued that escape from the inevitability of aging by normal cells *in vivo* and diploid cell strains *in vitro* is only possible when such cells acquire, respectively, properties of transplantable tumors or heteroploid cell lines. One of the common denominators of these latter two systems is heteroploidy (usually modally distributed) which, when acquired by the cell may be the mechanism needed for it to circumvent the inevitability of death and thus escape from senescence.

SUMMARY

The time at which human diploid cell strains can be expected to cease dividing *in vitro* (Phase III) is not a function of the number of subcultivations but rather of the number of potential cell doublings. Each clonable cell within the population is endowed with the same doubling potential (50 ± 10). Cells of the same strain, but with different "doubling potentials", were mixed. Phase III in such mixed populations occurs at that time when the "youngest" cell component is expected to reach Phase III. The "older" component has no effect on the time at which Phase III is expected to take place in the "younger" component. An ancillary conclusion that Phase III cannot be explained by the presence of a latent virus, mycoplasma or media composition is confirmed. Human diploid cell strains derived from adult lung have a significantly lower doubling potential *in vitro* than do fetal strains. The Phase III phenomenon may be related to senescence *in vivo*. The cellular theory of aging must be related to normal cells *in vitro* and not to heteroploid cell lines. The former have a finite period of multiplication; the latter are indefinitely cultivable. *In vivo* experiments also indicate that transplanted normal tissue has a finite lifetime. Chromosome anomalies occurring in Phase III may be related to such anomalies occurring in the cells of older animals, including man. The survival curves obtained with human diploid cell strains are comparable to "multiple-hit" or "multiple-target" curves obtained with other biological systems where an initial threshold dose is required before an exponential form of the curve is established. Whatever cell component(s) may be involved in the finite lifetime of human diploid cell strains, the ultimate accumulation of nondividing cells could be the result of accumulated damage to a single cellular target or to inactivation of many targets.

The author gratefully acknowledges his indebtedness to Dr. Jack L. Titus of the Section of Pathologic Anatomy, Mayo Clinic, Rochester, Minnesota who provided the adult human lung tissue, to Dr. Sven Gard of the Karolinska Institutet Medical School, Stockholm, Sweden, and to Dr. Alvaro Macieira-Coelho and Mrs. Ruth Casper for much valuable assistance.

REFERENCES

1. CARREL, A., *J. Exptl Med.* **18**, 287 (1913).
2. CHU, E. H. Y., *Natl Cancer Inst. Monograph* **7**, 55 (1962).
3. CHU, E. H. Y. and GILES, N. H., *Am. J. Human Genet.* **11**, 63 (1959).
4. COWDRY, E. V., *in* LANSING, A. I. (ed.), Problems of Aging, p. 60. Williams and Wilkins Co., Baltimore, Md., 1952.
5. CROWLEY, C. and CURTIS, H. J., *Proc. Natl Acad. Sci. USA* **49**, 626 (1963).
6. DE OME, K. B., Personal communication.
7. EAGLE, H., *J. Exptl Med.* **102**, 595 (1955).
8. EARLE, W. R., *J. Natl Cancer Inst.* **4**, 165 (1943).
9. EBELING, A. H., *J. Exptl Med.* **17**, 273 (1913).
10. GEY, G. O. and GEY, M. K., *Am. J. Cancer* **27**, 45 (1936).
11. HAYFLICK, L., *Exptl Cell Res.* **23**, 14 (1961).
12. —— *in* POLLARD, M. (ed.) Perspectives in Virology. Vol. **3**, p. 213. Hoeber Medical Division, Harper and Row, New York, 1963.
13. HAYFLICK, L. and MOORHEAD, P. S., Handbook on Growth. Fed. Am. Assoc. Exptl Biol., Washington, D. C., 1962.
14. —— *Exptl Cell Res.* **25**, 585 (1961).
15. HAYFLICK, L., PLOTKIN, S. A., NORTON, T. W. and KOPROWSKI, H., *Am. J. Hyg.* **75**, 240 (1962).
16. HSU, T. C., *Intern. Rev. Cytol.* **12**, 69 (1961).
17. JACOBS, P. A., COURT BROWN, W. M. and DOLL, R., *Nature* (Lond.) **191**, 1178 (1961).
18. JENSEN, F., KOPROWSKI, H. and PONTÉN, J., *Proc. Natl Acad. Sci. USA* **50**, 343 (1963).
19. KOPROWSKI, H., PONTÉN, J. A., JENSEN, F., RAVDIN, R. G., MOORHEAD, P. S. and SAKSELA, E., *J. Cell. Comp. Physiol.* **59**, 281 (1962).
20. KROHN, P. L., *Proc. Roy. Soc. B* **157**, 128 (1962).
21. MAKINO, S., KIKUCHI, Y., SASAKI, M. S., SASAKI, M. and YOSHIDA, M., *Chromosoma* (Berl.) **13**, 148 (1962).
22. MAYNARD SMITH, J., *Proc. Roy. Soc. B* **157**, 115 (1962).
23. MEDAWAR, P. B., The Uniqueness of the Individual. Basic Books, Inc., New York, 1958.
24. MOORHEAD, P. S., Personal communication.
25. PARKER, R. C., Methods of Tissue Culture. Hoeber Medical Division, Harper and Row, New York, 1961.
26. PEARL, R., The Biology of Death. J. B. Lippincott Co., Philadelphia, 1922.
27. PUCK, T. T. and MARCUS, P. I., *J. Exptl Med.* **103**, 653 (1956).
28. ROBERTS, R. C., *Heredity* **16**, 369 (1961).

29. ROTHFELS, K. H., KUPELWIESER, E. B. and PARKER, R. C., *in* BEGG, R. W. (ed.). Canadian Cancer Conference, Vol. **5**. Academic Press Inc., New York, 1963.
30. SAKSELA, E. and MOORHEAD, P. S., *Proc. Natl Acad. Sci. USA* **50**, 390 (1963).
31. SAX, H. J. and PASSANO, K. N., *Am. Naturalist* **95**, 97 (1961).
32. SHEIN, H. M. and ENDERS, J. F., *Proc. Natl Acad. Sci. USA* **48**, 1164 (1962).
33. SZILARD, L., *Proc. Natl Acad. Sci. USA* **45**, 30 (1959).
34. TJIO, J. H. and PUCK, T. T., *Proc. Natl Acad. Sci. USA* **44**, 1229 (1958).
35. TJIO, J. H. and PUCK, T. T., *J. Exptl Med.* **108**, 259 (1958).
36. TODARO, G. J., WOLMAN, S. R. and GREEN, H., *J. Cell Comp. Physiol.* **62**, 257 (1963).
37. TODARO, G. J. and GREEN, H., *J. Cell Biol.* **17**, 299 (1963).
38. UNITED STATES ARMED FORCES INSTITUTE OF PATHOLOGY, Transplantable and Transmissable Tumors of Animals. U.S. Government Printing Office, Washington, D.C., 1959.
39. WITKIN, E. M., *Proc. Natl Acad. Sci. USA* **32**, 59 (1946).
40. WORLD HEALTH ORGANIZATION, Scientific Group on the Human Diploid Cell, Report to the Director General. WHO, Geneva (WHO/PA/140.62), 1962.
41. YOSHIDA, M. C. and MAKINO, S., *Jap. J. Human Genet.* **8**, 39 (1963).

9

Reprinted by permission from *Lab. Invest.* **23**:86–92 (1970)

Replicative Life-Span of Cultivated Human Cells

Effects of Donor's Age, Tissue, and Genotype

George M. Martin, M.D., Curtis A. Sprague, B.S., and Charles J. Epstein, M.D.

*Department of Pathology, University of Washington School of Medicine, Seattle, Washington 98105, and
Department of Pediatrics, University of California School of Medicine, San Francisco, California 94122*

In vitro studies of the longevity of fibroblast-like human diploid somatic cells in over 100 mass cultures and 200 clones from a variety of donors have provided evidence for a limited replicative life-span, a negative regression of growth potential on age of donor from the 1st to the 9th decades, and a variable growth potential as a function of the tissue of origin. The results lend support to the hypothesis that the limited growth potential of cultured somatic cells is a manifestation of senescence at the cellular level. A striking diminution of the growth potential of cultures from patients with Werner's syndrome, a hereditary disease manifested by early and widespread degenerative changes, was also observed.

Additional key words: Aging, Diabetes mellitus, Fibroblast cultures, Progeria, Rothmund's syndrome, Werner's syndrome.

Hayflick[10] described three stages in the natural history of cultivated human diploid "fibroblast" cultures derived from various somatic tissues. Stage I is the establishment of the cultures; stage II consists of reproducible cycles of growth throughout successive passages; and stage III comprises progressive increases in population doubling times, with eventual failure of cell replication. Hayflick suggested that this restricted growth potential is a manifestation of senescence at the cellular level. We now report experiments which lend support to this interpretation. The longevity of skin fibroblast cultures has been found to be inversely correlated with the age of the donor, confirming and extending the results of Hayflick[9-11] and of Goldstein, Littlefield, and Soeldner.[7] Furthermore, a specific autosomal recessive mutation—that responsible for the extensive degenerative pathology associated with Werner's syndrome—results in a sharply limited *in vitro* replicative life-span, confirming and extending earlier work from this laboratory.[3, 18] Finally, we emphasize the importance of standardizing the site of biopsy for *in vitro* studies on cell senescence because the replicative life-span of the cultures has also been found to be a function of the tissue of origin, and was first suggested by the work of Swim and Parker.[30]

MATERIALS AND METHODS

Tissue specimens were diced into 1-mm. cubes, and replicate cultures consisting of two such cubes per Leighton tube (Bellco Glass, Inc., Vineland, New Jersey) were established. The cubes of tissue were sandwiched between a glass slide (cut to fit the Leighton tube) and the bottom of the tube. Two milliliters of Waymouth's medium prepared in this laboratory with 9.0 per cent heat-inactivated newborn calf serum and

penicillin (50 units per ml.) were added, and the tubes were incubated at 37° C. in an atmosphere of 5 per cent CO_2 in air. The medium was changed once during the 1st week, then twice weekly until approximately 40 per cent of the glass surface was covered with fibroblasts. At that time they were trypsinized,[15] and the 30,000 to 100,000 cells harvested were transferred to a 4-ounce prescription bottle. Upon confluency, the cells were again trypsinized and were transferred to 6-ounce prescription bottles. Repeated trypsinizations were conducted when the monolayers became partially confluent, usually every 6 to 8 days during stage II, at which time cell counts ranged from 1×10^6 to 3.5×10^6. At each passage, the bottles were reseeded with 400,000 cells. Hemocytometer counts of unattached cells from randomly selected cultures of skin (16 to 20 hours after plating) indicated that 75 to 90 per cent of the plated cells attached to the glass, with somewhat fewer cells attaching during later passages than earlier passages. Cell counts performed with an electronic particle counter (Coulter Electronics Company, Hialeah, Florida) were deceptive in that increasing amounts of cytoplasmic debris, often in clumps, were encountered with later passages, yielding erroneously high "cell counts" and proportionately larger numbers of "unattached cells." The data to be presented are calculated on the basis of 100 per cent cell attachment and the assumption that all cells participate in the population doubling. Therefore, the total number of cell doublings calculated for any given culture is a minimal estimate. For example, in a typical passage during stage II, 7 days after an inoculum of 400,000 cells, 2.4×10^6 cells could be harvested by trypsinization. On the basis of the assumptions that all 400,000 cells survived the trypsinization and attached to the glass and that all of these cells participated in the growth of the culture, we would calculate an expenditure of 2.5 cell doublings for that particular passage. However, if only 75 per cent of the plated cells attached to the glass and if only two-thirds of these cells subsequently underwent mitoses, an average of 3.5 cell doublings would have been required to achieve a population of 2.4×10^6 cells. A culture was terminated when, after 1 month, the total yield of cells in a given passage was less than 400,000. The cumulative number of cell doublings from all previous passages (beginning with the transfer from the Leighton tube to the 4-ounce prescription bottle) was then determined and utilized in the construction of Figures 1 and 2. Statistical analysis of the data was performed through the courtesy of Dr. Julian I. E. Hoffman; regression lines were calculated by the method of least squares.[29] Mycoplasma or other microorganisms could not be cultured from randomly selected monolayers and supernatant media obtained throughout the course of the study. The cultures for mycoplasma were performed aerobically and anaerobically by Dr. George E. Kenny with the use of a soy peptone dialysate broth medium.[12]

Subjects were selected at random from laboratory personnel and from inpatients and available autopsies with a wide variety of congenital anomalies, metabolic disorders, and cardiovascular, degenerative, and neoplastic disease. Subjects known or suspected of having an infectious disease were never used. Analysis of the data failed to reveal any relationship between the cumulative number of cell doublings achieved by a culture and the sex of the patient, or, in the case of autopsy specimens, the time after death at which a biopsy was obtained. Except for a rare instance of microbial contamination, cultures could be regularly established up to at least 72 hours postmortem (in the present study, however, nearly all of the postmortem biopsies were obtained within 24 hours after death). No difference in growth potential could be discerned between groups of antemortem and postmortem specimens.

Summaries of Clinical Histories

Werner's syndrome is an autosomal recessively inherited disease characterized by a wide variety of striking degenerative features which we have referred to as a "caricature" of normal aging.[3] The earliest manifestation of the disease is a symmetrical retardation of growth, followed by graying of the hair, atrophy of subcutaneous fat and skin, hyperkeratosis, generalized loss of hair, alterations in the voice, cataracts, skin ulcers, and, in approximately one-half of the cases, diabetes mellitus. Severe arteriosclerosis, osteoporosis, skeletal muscle atrophy, and severe testicular atrophy are typical. At least 10 per cent of the patients develop serious neoplasms during their average life-span of 47 years. Progeria (the Hutchinson-Gilford syndrome), possibly also autosomal recessive in inheritance, results in even more severe pathology, with signs of "senility" appearing during the 1st decade of life, death sometimes resulting from complications of coronary artery atherosclerosis within the first 2 decades of life.[3] Rothmund's syndrome is another recessively inherited disease which has sometimes been believed to be related to Werner's syndrome.[3] Cultures were established from three patients with Werner's syndrome, from a 4-year-old girl with progeria, and from a 2-year-old boy with Rothmund's syndrome. Abstracts of their case histories follow.

H. Mc. G. (University of Washington Hospital, 077-54). An extensive report of clinical, laboratory, and pathologic findings of this Japanese-American housewife appears as case 1 in Reference 3. She displayed typical signs and symptoms of Werner's syndrome, including growth retardation, cataracts (diagnosed at age 22), graying and loss of hair, glycosuria, skin atrophy, ulcers, osteoporosis, and voice changes. The cultures reported in the present study were established when the patient was age 49.

P. N. (National Institutes of Health, 06-16-12). This Caucasian man with Werner's syndrome is cited in the addendum of Reference 3. He first developed difficulty at age 36 when hip pain led to the discovery of a calcific deposit. At age 37 he developed recurrent ulcerations of both feet, primarily at pressure points; before this, however, the patient had observed that his skin was quite dry and easily bruised. Bilateral cataracts were diagnosed at age 38 and were surgically removed at age 41. He had had a high pitched voice throughout his life. Because of frequent sore throats and difficulties in swallowing, he was operated on at age 42 and was told that a papilloma was removed; at that time a diagnosis of diabetes mellitus was made. The patient was the father of two normal children; however, since approximately age 45, there was loss of libido and impotence. Of seven siblings, one sister had diabetes and cataracts, but none of the other manifestations of Werner's syndrome. The parents of the patient were not related. Physical examination at age 48 revealed a thin, balding man who appeared considerably older than his stated age. There was generalized loss of body hair and slender extremities. The skin was dry and thin and markedly atrophic over both legs and feet, the latter areas showing both hyper- and hypopigmentation. Large purulent ulcers were present on the posterior aspects of both heels, and several small punched-out ulcers were present on the toes and other bony prominences of the feet. There was a grade 2 blowing systolic murmur along the left sternal border. The prostate was diffusely enlarged, the testis descended and firm, and the penis small. Laboratory studies revealed a moderate iron deficiency anemia, an elevated fasting blood sugar (122 mg. per 100 ml.), and normal values for calcium, phosphorus, and alkaline phosphatase. The serum γ-globulin was elevated to 1.9 gm. per 100 ml. An electrocardiogram revealed diffuse osteoporosis of the pelvis, lumbar spine, legs, and feet; tendinous and soft tissue calcification was present in several areas of both feet. Intravenous pyelogram revealed bilateral hydronephrosis with hydroureters and bilateral distal ureteral obstruction. A perineal needle biopsy revealed undifferentiated carcinoma of the prostate. The patient was treated with bilateral orchiectomy followed by stilbestrol; sections of the testes revealed spermatogenic arrest and moderate interstitial sclerosis. Cultures were established from both skin and testis from this patient at age 48.

W. L. (*National Institutes of Health, 07-00-63*). This unmarried Caucasian man with Werner's syndrome was well until age 13 when his voice was noted to be scratchy and hoarse; at that time, he was also treated with horse gonadotrophins because of a small phallus. At age 14 he developed pain in the bottom of both feet whenever he put weight on them, and calluses were noted to be present. At age 24 he developed an ulcer on the lateral aspect of his left heel; shortly thereafter bilateral cataracts were diagnosed. After developing a second heel ulcer at age 28, a diagnosis of Werner's syndrome was made. At age 31 he was found to have diabetes mellitus. No other members of his family had any stigmata of Werner's syndrome; however, it is of interest that the parents of the patient were first cousins. On physical examination at age 37, he appeared 20 years older. The hair was gray, the extremities thin and wasted, and there were flexion contractures of both knees. A grade 2 systolic murmur was noted along the left sternal border. The penis was small and the testes soft, measuring approximately 2 by 2 by 2.5 cm. Skin of the extremities was dry and taut, with several small ulcers on the feet, and large ulcers exposing the gastrocnemius tendons on the posterior aspects of both feet. There were hyperkeratoses over both elbows. Pubic and axillary hair was sparce, but a grayish beard was noted. X-rays revealed diffuse vascular calcifications and diffuse osteoporosis especially in the legs and feet, and soft tissue calcifications of the knees, feet, and ankles. The 2-hour postprandial blood sugar was elevated and was followed by hypoglycemia at 5 hours. Skin cultures were established at age 37.

M. M. (*University of Washington Hospital, 509-86*). This Caucasian girl was referred because of growth failure, noted only after the 1st year of life. Although at birth she weighed 8 pounds 6 ounces and was 21 inches long, she weighed only 19 pounds 11 ounces and was only 30¾ inches long at 28 months. At approximately age 2, her hair began to fall out, her nails were noted to be thin and fragile, strabismus developed, and the veins of the scalp appeared to be quite prominent. Except for weakness and some fatigue, the patient seemed to be otherwise in good health and only rarely suffered from infections. Her parents (unrelated) were age 27 and of normal stature, as were the maternal and paternal grandparents; three female siblings were living and well. When examined at age 38 months, she weighed 21.9 pounds and was 33.1 inches tall. On physical examination, she had the appearance of a thin and tiny old lady. The skin was very thin, and there was striking atrophy of subcutaneous fat—especially over the scalp. The nose appeared "beaked." The nails were thin and furrowed. Scalp hair was very blonde and extremely sparse. There were no cataracts, skin ulcers, or hyperkeratoses, and there was no evidence of diabetes. The serum cholesterol varied from 168 to 213 mg. per 100 ml. and the serum triglycerides, from 82

to 252 mg. per 100 ml. A diagnosis of progeria was made by Professor David W. Smith and his colleagues, to whom we are grateful for the opportunity of evaluating this patient. Skin fibroblast cultures were established when the patient was 58 months old.

G. A. (*University of Washington Hospital, 747-08*). This Caucasian boy was seen at age 34 months, with a chief complaint of peculiar patterns of skin pigmentation with areas of scarring. Except for a scaly red rash over the forehead, the child's skin had been clear (but pale) until age 6 to 7 months, at which time a reticular macular facial hyperpigmentation was noted, gradually also involving arms, legs, and buttocks, with areas of hypopigmentation, hyperpigmentation, flakiness, drying, and scarring. The unrelated parents were age 25, and there was a normal male sibling; no similar pathology had been observed in former generations of the pedigree. On physical examination, the small but symmetrically developed child was 85 cm. tall and weighed 10.66 kg. There was a broad (1 to 10 mm.) reticular network of hyperpigmentation over the face, neck, arms, legs, and buttocks, with circular areas of depigmentation of the thighs and buttocks. The skin of the extremities was dry, scaly, and focally hyperkeratotic. Scattered small telangiectases were noted on the cheeks. There were no eyebrows and only two eyelashes; scalp hair was blond, dry, abundant, and fine. Subcutaneous tissue was present over all areas of the body. There was no tightening or ulceration of the skin. An ophthalmologist had prescribed glasses for a refractive error, but cataracts were not reported. The right thumb (including the first metacarpal) was congenitally absent, and the left thumb was hypoplastic. There was a proximal fusion of the radius and ulna bilaterally; the secondary ossification centers of L4 had not formed. The Dermatology, Medical Genetics, Pediatric, and Radiology Services agreed on a diagnosis of Rothmund's syndrome. Skin fibroblast cultures were established at age 34 months.

RESULTS

From 100 subjects, 102 mass cultures of skin (epidermis and dermis of the mesial aspect of the mid-upper arm) were studied. The data shown in Figure 1 indicate that the longevity of a culture is a function of the age of the skin donor. Statistical analysis of this data (excluding cultures from fetuses and from patients with Werner's syndrome, Rothmund's syndrome, and progeria) indicates a regression of growth potential of the age of the donor from the 1st to the 9th decades. For the 1st to the 9th decades, the regression coefficient is -0.20 cell doublings per year, with a standard deviation of 0.05 and a correlation coefficient of -0.50. This regression coefficient is significantly different from 0 ($p < 0.01$). The average growth potential of approximately 44 cell doublings (Fig. 1), realized with cultures from fetal skin (from donors ranging in fetal age from 65 to 168 days), appears to be in good agreement with the results of Hayflick with the use of the fetal lung.[9] However, for various reasons noted in the Materials and Methods and Discussion sections of this paper, these data are not strictly comparable, and in any case represent only minimal estimates of the growth potential of such cells.

When a single skin biopsy from a normal 35-year-old man was divided and used to establish duplicate mass cultures, the results were nearly identical (34.0 and 35.0 cell doublings). However, when two independent biopsies, one from the left arm and another from the right arm, were established from the same individual (a 6-month-old female with Down's syndrome and an unbalanced G/D translocation), somewhat different results were obtained (37.6 and 30.7 cell doublings). On the other hand, when the numbers of cumulative cell doublings were calculated for 59 clones independently derived from an established culture of human newborn foreskin fibroblasts, the results were remarkably consistent, with a mean of 63.1 cell doublings and a standard deviation of 7.6.[20]

Several biopsies were also obtained at necropsy from skeletal muscle (psoas) and lumbar vertebral bone marrow spicules, and in several cases the cumulative number of cell doublings was compared with that obtained with skin explants from the same individual. Cultures derived from the skin achieved the greatest number of cell doublings, and bone marrow spicules yielded the least, with skeletal muscle giving intermediate results (Fig. 2).

In addition, established mass fibroblast culture from

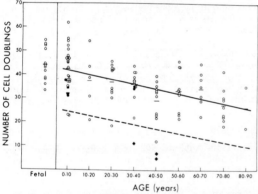

Fig. 1. The cumulative number of cell doublings achieved by human skin fibroblast cultures plotted as a function of the age of the donor. *Open circles*, Control cultures (see text); *closed diamonds*, Werner's syndrome homozygotes; *closed triangle*, Hutchinson-Gilford syndrome homozygote; *open triangle*, Hutchinson-Gilford syndrome heterozygote; *closed square*, Rothmund's syndrome homozygote; *closed circles*, parallel cultures from the same donor (left and right arm biopsies from a 6-month-old girl and bisected single biopsy from a 35-year-old man). The mean for each group is indicated by a *horizontal bar*. The calculated linear regression line (*solid line*) for the control group is drawn between the 1st and 9th decades and has a regression coefficient of −0.20 ± 0.05 standard deviation cell doublings per year with correlation coefficient of −0.50. The *dashed line* is the lower 95 per cent confidence limit for the regression line. In addition to Werner's syndrome, diagnoses for patients whose cultures fell below this line were: congenital heart disease with anomalous pulmonary venous drainage (age 2 weeks), cystic fibrosis (age 2), diabetes mellitus (age 17), acute metabolic encephalopathy ("Darvon" toxicity) (age 27), and carcinoma of the colon (age 44).

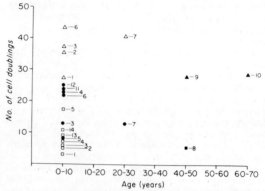

Fig. 2. The cumulative number of cell doublings achieved by human fibroblast cultures plotted as a function of the tissue of origin. *Open triangles*, skin (mesial aspect of midupper arm); *closed circles*, skeletal muscle (psoas); *open squares*, bone marrow spicules (lumbar vertebral); *closed triangles*, testis (controls); *closed square*, testis (Werner's syndrome). The numbers to the *right* of each culture identify the patients.

various regions of skin, subcutaneous fat, liver, kidney, or spleen of 40 other individuals and 150 skin fibroblast clones (including 68 secondary clones) have undergone "senescence" in our laboratory although the numbers of cell doublings achieved by these cultures are not known or were calculated on the basis of varying culture conditions. Thus, a finite replicative life-span appears to be an invariable characteristic of all cultures.

The original purpose of obtaining data for skin biopsies from individuals of various ages was to confirm our earlier report[3, 18] that a marked deficiency of *in vitro* growth potential is characteristic of fibroblasts derived from patients with Werner's syndrome. The present study (Fig. 1) demonstrates that the numbers of cell doublings for the skin fibroblast cultures from the two older patients with Werner's syndrome were more than two standard deviations below the mean of the distribution of control cultures for the 5th decade; the number of cell doublings was more than three standard deviations below the mean of the distribution of control cultures for the 4th decade in the case of our 37-year-old patient. The culture from the patient with progeria is less readily differentiated from the control cultures; however, it ranks number 23 of 26 cultures in this age group. Similarly, the culture from the patient with Rothmund's syndrome ranks 22 of 26 in that group. A fibroblast culture derived from the testis biopsy of the patient with Werner's syndrome and carcinoma of the prostate also had a sharply restricted life-span in comparison with control testis cultures derived from other patients with carcinoma of the prostate (Fig. 2).

In addition to the three cultures from patients with Werner's syndrome (which were not included in the regression analysis), five skin fibroblast cultures were found to fall below the lower 95 per cent confidence limit for the regression line (Fig. 1); the age and diagnosis of each of these patients is given in the legend to Figure 1.

DISCUSSION

Four major conclusions emerge from the present study.

Without exception, mass cultures and clones of somatic cells from a wide variety of human genomes and tissues eventually cease to replicate. Similar observations or observations interpreted as evidence of a progressive decline in growth rate have been made in other laboratories by using a wide variety of media and sera.[1, 4, 6-11, 14, 18, 21, 27, 28, 30, 33, 37] Therefore, it appears that one may now make the generalization that all cultures of normal somatic cells (normal in the sense that they carry the genome of the original host organism) have a finite replicative life-span *in vitro*. One of us[17] has referred to such cultures as "hyperplastoid" because they may serve as *in vitro* models of hyperplasia; they have also been referred to as "homonuclear" cell lines.[13] The prototype of the hyperplastoid cell line is the skin fibroblast culture—representing proliferating cells with a restricted growth potential, presumably a sampling of cells which undergo hyperplasia in a healing wound of the skin.

In contrast, cultures which appear to have an infinite growth potential have almost invariably been shown to carry genomes which differ from those of the original host organisms—even with crude cytogenetic techniques revealing only gross differences in chromosome number and karyotype.[13] One of us[17] has referred to such cultures as "neoplastoid" because they may serve as models for certain aspects of the neoplastic process; they have also been referred to as "heteronuclear."[13]

There are possible exceptions to the above generalization,[13] notably certain apparently immortal cell lines derived from the peripheral blood of chromosomally marked donors in which exogenous contamination by established neoplastoid cell lines[19] has been ruled out.[23] However, the techniques utilized in establishing such cultures, involving extremely large cell inoculums and prolonged latent periods,[22] suggest that these cultures derive from spontaneously or virally transformed cells.

The replicative life-span of cultured human somatic cells is a function of the tissue of origin (Fig. 2). Even in the case of explants obtained from the same individual, there is a far greater replicative life-span of cells derived from skin than from bone marrow, with cells derived from skeletal muscle yielding approximately intermediate results. The reasons for these differences are unknown, but presumably reflect differences in the previous *in vivo* history of the cells, such as their patterns of differentiation and their previous replicative history. It is therefore clear that interstrain comparisons must be based upon a standard site of biopsy. Thus, we have chosen the mesial aspect of the skin of the upper arm as our standard tissue because of (1) its convenience as a biopsy site, and (2) the relatively greater growth potential achieved by skin fibroblasts, permitting estimation of a wide range of cumulative cell doublings exhibited by various strains and providing greater numbers of cells for correlative biochemical and morphologic studies (none of which, however, are reported in the present communication).

The replicative life-span of cultured human skin fibroblasts is inversely related to the age of the donor (Fig. 1). The present series of cases, the largest so far published, confirms a similar conclusion made by others[7, 9-11] and leads to a calculated regression coefficient of -0.2 ± 0.05 standard deviation cell doublings per year from the 1st to the 9th decades, the first such data to be reported.

The replicative life-span of cultured human skin fibroblasts is a function of the genotype of the donor (Fig. 1). By using an independent biopsy and a greater number of controls, the present study confirms earlier work from this laboratory,[3, 18] indicating a striking decrease in growth potential of fibroblasts from a patient (H. Mc. G.) with Werner's syndrome. In addition, comparable results with biopsies from two other patients from different pedigrees now permit one to conclude that homozygosity for a single gene mutation can have a striking effect on the longevity of cultured somatic cells.

Goldstein[6] reported that skin fibroblasts from a 9-year-old boy with typical progeria could be subcultured only twice, whereas cultures from age-matched controls could be subcultured 20 to 30 times. Fibroblasts from our patient with progeria did not show such a striking impairment of growth potential; however, the fact that this culture ranked 23 of 26 in that age group suggests the possibility that this mutation may also result in diminished growth potential. Investigation of several additional patients is necessary to establish this point; it is possible, as is becoming increasingly evident for various other heritable disorders, that there are different types of progeria. The culture from the single patient with Rothmund's syndrome, the other recessively inherited disease which is sometimes confused with Werner's syndrome, ranks 23 of 26 cultures in that age group, and therefore this disorder may also be characterized by some degree of impairment of growth potential *in vitro*; nevertheless, the behavior of these cells *in vitro* is clearly different from that of skin fibroblasts derived from patients with Werner's syndrome. As regards the five "control" cultures which were found to fall below the lower 95 per cent confidence limit for the regression line (Fig. 1), the patient with diabetes mellitus may be of special interest in view of the recent publication of Goldstein *et al.*[7] Although these workers concluded that the life-spans of mass cultures from normals could not be differentiated from the life-spans of mass cultures from the progeny of conjugal diabetics, statistical analysis of their data (Fig. 1 of that paper, excluding their atypical case 6) by the *t*-test[29] indicates a significant difference ($p = 0.02$) between the means of the life-spans of cultures from prediabetic individuals over age 30 (mean = 49.1 ± 7.5 standard deviation cell doublings) and those from nondiabetics over age 30 (mean =

59.5 ± 5.2 standard deviation cell doublings). Their important discovery of a diminished cloning efficiency of fibroblasts from prediabetics, comparable to that which they observe in aging control cultures, suggests that the diabetic genotype may result in accelerated senescence, either *in vivo* or *in vitro*. Vracko and Benditt have evidence[34-36] that endothelial cells of diabetics may undergo increased numbers of replication *in vivo*. Studies with the alkaline phosphatase marker have suggested that a proportion of cells which migrate from a skin explant and participate in the establishment of the culture may in fact be endothelial cells.[24] It is possible that fibroblasts from diabetics also have a previous *in vivo* history of comparatively increased mitotic activity. In view of the finite growth potential of somatic cells, explants from diabetics, especially older diabetics, might then be expected to yield cells which have already utilized a significant proportion of their "allotted" number of cell doublings and hence display a more restricted growth potential *in vitro*. The same interpretation could of course be made in the case of Werner's syndrome; all of our patients gave evidence of diabetes, a characteristic feature of the disease.[3] However, the simplest explanation of our results with Werner's syndrome is that there is an intrinsic deficiency of growth potential as a result of homozygosity for the mutant gene.

Little can be made of the unusually low growth potential of the single case of cystic fibrosis (Fig. 1); additional cultures must be investigated with suitable controls. It is interesting, however, that cultured fibroblasts from patients with cystic fibrosis (as well as from a variety of other disorders) are characterized by abnormal accumulations of mucopolysaccharides.[2]

It should be emphasized that the absolute numbers of cell doublings reported in this and other[6-11, 32, 33] studies are merely estimates and are likely to vary from laboratory to laboratory depending upon the precise methods employed. First of all, there is no quantitative information with regard to the number of cells which migrate from a given explant and the number of cell doublings required to establish a culture. In the calculation of their data, Goldstein *et al.*[7] arbitrarily set at 10 the number of cell generations used up by the primary explants in attaining the first confluent monolayers; in the present study, this unknown number of cell doublings was not included in the calculations, and therefore comparatively lower figures for cumulative cell doublings are reported. Secondly, we have little information concerning interstrain variation in regard to the heterogeneity of the population of cells which do migrate from the explant and the extent to which they contribute to the population of mass cultures during the life histories of such cultures.[5, 16, 24-26] Furthermore, although we have indicated a plating efficiency (percentage of cells plated which stick to glass) of 75 to 90 per cent from randomly selected mass cultures of skin, it is possible that more substantial variations of plating efficiency occurred with cultures from tissues other than skin or from certain strains of skin fibroblasts. Finally, it is clear from the results of Todaro

and Green[32] that media composition can influence the longevity of cultured somatic cells. Additional studies are required to determine the extent to which such factors may contribute to strain variations, particularly as a function of the tissue of origin, inasmuch as such factors as cellular heterogeneity, plating efficiency, and optimal media requirements might be expected to vary significantly from tissue to tissue.

In conclusion, we believe that our results, particularly with the skin fibroblast cultures, strengthen the hypothesis that human hyperplastoid cell lines[17] may serve as models for the study of senescence at the cellular level,[9-11] or more specifically, that aspect of senescence manifested by the loss of the mitotic potential of somatic cells. It is the challenge for the future to determine whether this senescence is specifically programmed by the genome or, alternatively, can be attributable to stochastic events (such as somatic cell mutation) which are merely modulated by the genome.

Note Added in Proof. By using the methods described in this paper, a fourth patient with Werner's syndrome (case 3 of Reference 3; biopsy of skin of distal ventral forearm obtained at age 43) yielded only 4.5 cell doublings. G. S. Merz and J. D. Ross. (J. Cell Physiol. 74:219, 1969).

Accepted for publication February 27, 1970.

This work was partially supported by Research Grants AM 04826, GM 13543, and GM 00100 from the National Institutes of Health.

REFERENCES

1. Cristofalo, V. J., Kabakjian, J. R., and Kritchevsky, D. Enzyme activities of some cultured human cells. *Proc. Soc. Exp. Biol. Med. 126:* 273, 1967.
2. Danes, B. S., and Bearn, A. G. A genetic cell marker in cystic fibrosis of the pancreas. *Lancet 1:* 1061, 1968.
3. Epstein, C. J., Martin, G. M., Schultz, A. L., and Motulsky, A. G. Werner's syndrome. A review of its symptomatology, natural history, pathologic features, genetics and relationship to the natural aging process. *Medicine (Balto.) 45:* 177, 1966.
4. Fraccaro, M., and Mannini, A. Persistence of the isochromosome in long term cultures from an XO/X-isochromosome X mosaic. *Atti. Assoc. Genet. It. (Pavia) 11:* 403, 1966.
5. Franks, D. Antigenic heterogeneity in cultures of mammalian cells. *In Vitro 2:* 74, 1966.
6. Goldstein, S. Life-span of cultured cells in progeria. *Lancet 1:* 424, 1969.
7. Goldstein, S., Littlefield, J. W., and Soeldner, J. S. Diabetes mellitus and aging. Diminished plating efficiency of cultured human fibroblasts. *Proc. Nat. Acad. Sci. U. S. A. 64:* 155, 1969.
8. Hay, R. J., and Strehler, B. L. The limited growth span of cell strains isolated from the chick embryo. *Exp. Geront. 2:* 123, 1967.
9. Hayflick, L. The limited *in vitro* lifetime of human diploid cell strains. *Exp. Cell Res. 37:* 614, 1965.
10. Hayflick, L. Cell culture and the ageing phenomenon. In *Topics in the Biology of Aging*, pp. 83–100. New York, Interscience Publishers, Inc., 1966.
11. Hayflick, L. Human cells and aging. *Sci. Amer. 218*, No. 3: 32, 1968.
12. Kenny, G. E. Serological comparison of ten glycolytic mycoplasma species. *J. Bact. 98:* 1044, 1969.
13. Krooth, R. S., Darlington, G. A., and Velazquez, A. D. The genetics of cultured mammalian cells. *Ann. Rev. Genet. 2:* 141, 1968.
14. Macieira-Coelho, A., Pontén, J., and Philipson, L. The division cycle and RNA synthesis in diploid human cells at different passage levels *in vitro*. *Exp. Cell Res. 42:* 673, 1966.

15. Martin, G. M. Use of tris (hydroxymethyl) aminomethane buffers in cultures of diploid human fibroblasts. *Proc. Soc. Exp. Biol. Med. 116:* 167, 1964.

16. Martin, G. M. Clonal variation of derepressed phosphatase in chromosomally mosaic cell cultures from a child with Down's syndrome. *Exp. Cell Res. 44:* 341, 1966.

17. Martin, G. M. Mitotic recombination in cultured human somatic cells. Discussion of paper by F. K. Zimmermann. In *Proceedings of the Second International Conference on Biochemical Pathology. Biochem. Pharmacol.*, in press.

18. Martin, G. M., Gartler, S. M., Epstein, C. J., and Motulsky, A. G. Diminished lifespan of cultured cells in Werner's syndrome. *Fed. Proc. 24:* 678, 1965.

19. Martin, G. M., Sprague, C., and Dunham, W. B. Chromosomal analysis of "leukocyte" cell lines. *Lab. Invest. 15:* 692, 1966.

20. Martin, G. M., and Tuan, A. A definitive cloning technique for human fibroblast cultures. *Proc. Soc. Exp. Biol. Med. 123:* 138, 1966.

21. Miles, C. P. Prolonged culture of diploid human cells. *Cancer Res. 24:* 1070, 1964.

22. Moore, G. E., and Minowada, J. Human hematopoietic cell lines. A progress report. *In vitro 4:* 100, 1969.

23. Moore, G. E., Porter, I. H., and Huang, C. C. Lymphocytoid lines from persons with sex chromosome anomalies. *Science 163:* 1453, 1969.

24. Papayannopoulou, T. G., and Martin, G. M. Alkaline phosphatase "constitutive" clones. Evidence for de-novo heterogeneity of established human skin fibroblast strains. *Exp. Cell Res. 45:* 72, 1966.

25. Pious, D. A. Antigenic variation in a rabbit fibroblast strain. *Genetics 56:* 601, 1967.

26. Pious, D. A., Hamburger, R. N., and Mills, S. E. Clonal growth of primary human cell cultures. *Exp. Cell Res. 33:* 495, 1964.

27. Saksela, E., and Moorhead, P. S. Aneuploidy in the degenerative phase of serial cultivation of human cell strains. *Proc. Nat. Acad. Sci. U. S. A. 50:* 390, 1963.

28. Sax, H. J., and Passano, K. N. Spontaneous chromosome aberrations in human tissue culture cells. *Amer. Naturalist 95:* 97, 1961.

29. Snedecor, G. W., and Cochran, W. G. *Statistical Methods*, Ed. 6. Ames, Iowa, Iowa State University Press, 1967.

30. Swim, H. E., and Parker, R. F. Culture characteristics of human fibroblasts propagated serially. *Amer. J. Hyg. 66:* 235, 1957.

31. Therkelsen, A. J. "Sandwich" technique for the establishment of cultures of human skin for chromosome investigation. *Acta Path. Microbiol. Scand. 61:* 317, 1964.

32. Todaro, G. J., and Green, H. Serum albumin supplemented medium for long term cultivation of mammalian fibroblast strains. *Proc. Soc. Exp. Biol. Med. 116:* 688, 1964.

33. Todaro, G. J., Wolman, S. R., and Green, H. Rapid transformation of human fibroblasts with low growth potential into established cell lines by SV_{40}. *J. Cell Physiol. 62:* 257, 1963.

34. Vracko, R. Skeletal muscle capillaries in diabetics. A quantitative analysis. *Circulation 41:* 271, 1970.

35. Vracko, R. Skeletal muscle capillaries in non-diabetics: A quantitative analysis. *Circulation 41:* 285, 1970.

36. Vracko, R., and Benditt, E. P. Capillary basal lamina thickening: Its relationship to endothelial cell death and replacement. *J. Cell Biol.*, in press.

37. Yoshida, M. C., and Makino, S. A chromosome study of non-treated and an irradiated human *in vitro* cell line. *Jap. J. Hum. Genet. 5:* 39, 1963.

38. Castor, C. W., Prince, R. K., and Dorstewitz, E. L. Characteristics of human "fibroblasts" cultivated *in vitro* from different anatomical sites. *Lab. Invest. 11:* 703, 1962.

39. Hakami, N., and Pious, D. A. Mitochondrial enzyme activity in "senescent" and virus-transformed human fibroblasts. *Exp. Cell Res. 53:* 135, 1968.

40. Merz, G. S., and Ross, J. D. Viability of human diploid cells as a function of in vitro age. *J. Cell Physiol. 74:* 219, 1969.

10

Copyright © 1973 by Macmillan Journals Limited
Reprinted by permission from *Nature* **243**:441–445 (1973)

Ageing of Clones of Mammalian Cells

L. E. ORGEL

The Salk Institute for Biological Studies, San Diego, California 92112

HAYFLICK has emphasized a possible connexion between the senescence of clones of mammalian cells maintained in tissue culture, and the ageing of members of the corresponding species[1,2]. He showed that the history of a tissue culture of human fibroblasts can be divided into three phases. His phase I corresponds to the establishment of a primary culture, his phase II to a period when vigorous growth follows each successive subculture, and his phase III to a period in which growth capacity declines until it is impossible to maintain further subcultures. These findings have been confirmed in a number of laboratories.

Hayflick reported that when fibroblast cultures are established from embryonic lung, the cells divide about fifty times before the cultures die out, but that cultures derived from adult lung die after about twenty divisions on average. There was a wide experimental scatter in these experiments, but more recent work has put Hayflick's principal conclusion on a sound statistical basis; the average number of divisions obtained declines by 0.20 for each year of the donor's life[3].

Hayflick's experiments and his tentative conclusions have been criticized on a number of grounds. It has been asserted that the limited capacity of human fibroblasts to proliferate in tissue culture reflects deficiencies of the medium rather than any intrinsic property of the cells. Supplementation of the medium, for example, with hydrocortisone[4,5] or with tyrosine[6] leads to a substantial increase in the number of doublings that occur before cultures enter phase III. Further, the Hayflick phenomenon is said not to be completely general. Chick fibroblasts, like human fibroblasts, cannot be maintained in culture indefinitely[7]. But it is claimed that a number of diploid cell lines derived from rats have been maintained in culture for well over a hundred doublings, without deterioration or obvious evidence of transformation[9]. This conclusion is disputed[10].

The dependence of the longevity of cell lines on the donor species, the cell type and the culture conditions is important. It warns us not to draw far-reaching conclusions from very limited experimental data. Hayflick's experiments do not show that humans age because their fibroblasts lose the capacity to proliferate after about fifty divisions. Hayflick's own important conclusion, that fibroblasts from adult human donors are reproducibly different from cells from embryos with respect to their proliferative capacity under a wide variety of culture conditions, is, however, not in doubt. More recently it has been claimed that size distribution of human fibroblasts in culture also varies with the age of the donor; phase II cultures derived from adult tissue are said to resemble phase III cultures derived from embryonic tissue[11].

The techniques of tissue transplantation and cell transfer permit the study of the survival and duplication of animal cells over periods longer than the life span of a single animal. The results that have been obtained in this way do not fit any single pattern. In many cases proliferation has been shown to be limited, often to a clonal life span similar to the life span of the animal. In others, tissues have continued to survive in an apparently normal state for much longer than a single life span. The nature of the tissue, the donor species, and the technique of transfer or transplantation all influence the outcome of the experiments[12].

The work of Williamson and Askonas provides a particularly clear-cut example of cellular ageing *in vivo*[13]. They used bone marrow from mice that had been immunized with a DNP-protein conjugate to repopulate acceptor mice whose immune system had been destroyed by heavy irradiation. The technique of focusing electrophoresis was then used to show that, in many cases, the acceptor mice were producing a unique antibody to DNP, presumably because a single clone of DNP-specific cells had become established.

Williamson and Askonas next attempted to carry out serial transfers of bone marrow to one generation after another of irradiated acceptors. They found that the marked clone could be transferred efficiently up to four times. After that antibody production fell off with each successive transfer and the original clone soon died out. It is estimated that the cells of the marked clone went through about ninety doublings in all.

Numerous studies on tissues taken from whole animals have shown that the total activities of many enzymes change with age. These studies, however, have not led to any important generalizations—the activities of some enzymes go down with age, but many enzyme activities are unaffected and some increase[14,15].

Studies of the specific activity and thermolability of enzymes, quantities which reflect the quality of the proteins, have given more promising results. When glucose-6-phosphate dehydrogenase (G6PD) or 6-phosphogluconate dehydrogenase (6PGD) from young cultures of MRC-5 fibroblasts is heated, enzyme activity decays exponentially, showing that the corresponding enzyme is homogeneous. The activities of these enzymes when isolated from cells in phase III, on the contrary, decay in a way that reveals the presence of up to 25% of abnormally thermolabile protein[16]. In related experiments it has been shown that the specific activity of lactic dehydrogenase isolated from cultures of MRC-5 cells in phase III is lower than that of the same enzyme isolated from cultures in phase II[17]. Gershon and Gershon have found similar alterations with age in the properties of the aldolase isolated directly from rat liver[18] and of isocitrate lyase from a nematode[19]. They interpret their results to show that the materials obtained from old animals consist of mixtures of an unaltered protein and a unique altered species.

In their experiments on human fibroblast cultures Holliday and Tarrant found[16] (1) that the proportion of thermolabile protein increases with the age of a culture (Fig. 1); (2) that the ratio of the proportion of thermolabile G6PD to the proportion of thermolabile 6PGD remains constant as cultures age; (3) that treatments which arrest the growth of cells, for example deprivation of serum, do not cause the

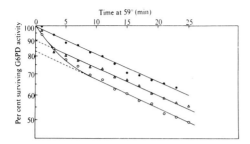

Fig. 1 Inactivation at 59° C of glucose-6-phosphate dehydrogenase obtained from young, middle aged and senescent cells. ●, Passage 22; △, passage 48; ○, passage 61. Reproduced with permission from Holliday and Tarrant[16].

accumulation of thermolabile protein; (4) that cells grown in the presence of low levels of 5-fluorouracil, a base analogue known to induce errors of protein synthesis, age prematurely. The ageing cells accumulate thermolabile protein after a relatively few subcultures; (5) that tissue cultures derived from a donor with Werner's syndrome already contain substantial amounts of thermolabile protein when they enter phase III after less than ten subcultures (R. Holliday, personal communication).

Holliday's results taken together with those of Lewis and Tarrant make it very probable that MRC-5 cells in phase III contain substantial amounts of protein that differ from the standard proteins present in cells in phase II. It seems unlikely that thermolabile isozymes of two independent enzymes are produced in old cultures, so aberrant protein is presumably present. These experiments, however, do not establish that protein synthesis is inaccurate. An alternative explanation is that post-synthetically modified proteins accumulate in old cells.

It is theoretically possible that the thermolabile proteins in old cells are derived entirely, or in part, from non-dividing (dead) cells and represent a protein fraction that is partially denatured or partially degraded. The enzymatic deamination of asparagine and glutamine residues, if it occurred at random positions in proteins, could well produce the observed results. A mechanism of this kind is not excluded by control (3) because serum-deprived cells, unlike old cells, might not contain the relevant degradative enzymes in an activated form. The experiment with 5-FU is not conclusive because the induction of thermolabile protein by 5-FU might be an indirect consequence of cell death rather than a direct consequence of errors of protein synthesis induced by the analogue. Recent experiments by Holliday (unpublished) argue against this explanation. They suggest that the proportion of thermolabile protein in moribund cultures is less than in cultures in phase III.

If it turns out that, in old cultures, newly synthesized proteins do have incorrect sequences, it will then be necessary to decide whether inaccurate protein synthesis is responsible, or whether the cultures have accumulated a wide variety of independent mutations in the G6PD and 6PGD genes. Holliday's results on the rate of appearance of thermolabile enzymes are not easily explained in terms of somatic mutations, but they do not completely exclude such an explanation. The distinction between mutations and errors of protein synthesis could be made by studying clones derived from single cells, by studying the components of the protein synthetic apparatus in vitro, or by examining virus-specified proteins.

It seems likely, on theoretical grounds, that mutations accumulate in tissue cultures as they age. The evidence on this point, however, is restricted to preliminary observations

by Fulder[20], who has shown that glucose-6-phosphate dehydrogenase mutants are more common in old than in young human fibroblast cultures. There is a good deal of evidence that the frequency of gross abnormalities in chromosome structure increases with age in cells taken from whole animals[21].

Old tissues differ from young ones in another way that implies a difference in their macromolecular composition. There are many indications that the induction of enzymes in old tissues is abnormal. The induction of tyrosine aminotransferase in old tissues, for example, is slower than in young, although the same rate of synthesis is attained eventually[22]. It is not possible, at present, to decide whether such differences in the rates of induction originate in changed nucleic acids, in changed quantities of the various proteins that control gene expression, or in a change in the accuracy with which control proteins are synthesized.

Theoretical Considerations

(a) Organelles. The various theories of clonal ageing are usually supposed to explain modifications of nuclear DNA or cellular protein that occur in cultures as they age. Such theories are equally applicable, however, to changes in the corresponding components of the mitochondria, because mitochondrial DNA and some mitochondrial proteins are synthesized by more or less independent mitochondrial systems. It is possible that mitochondria are more susceptible to the effect of mutation or errors of protein synthesis than are the cells in which they are located. If so, the ageing of the cells might be a consequence of the deterioration of their mitochondria. In the blowfly, ageing is accompanied by the uncoupling of oxidative phosphorylation, but the nature of the molecular changes that occur in the mitochondria has not been determined[23].

(b) Subcultures and cell divisions. The terminal stage in the senescence of a tissue culture is usually identified as the point at which it becomes impossible to split a sample of cells into two equal portions and to regenerate the original number of cells from one of the halves. But each sub-clone does not go through a number of divisions equal to the number of subcultures, for a proportion of non-dividing cells is always present in fibroblast cultures. The relatively small scatter in the number of subcultures obtained before death is, in part, a consequence of an averaging procedure that is implicit in the design of the experiment. As the cultures get older, the proportion of healthy cells decreases, so those cells that can still divide must go through an increasing number of divisions, on average, before the total cell number doubles. This, so to speak, compresses the period of senescence into a few subcultures, although the scatter in the number of cell divisions that can occur before the death of a sub-clone may be much larger. A similar compression may occur in vivo.

The heterogeneity of the cell populations must be borne in mind when interpreting the results of biochemical experiments on ageing cultures. The differences being average properties of young and old cultures may give more information about cells that have ceased to divide than about the cells which are responsible for continued growth. It is important to distinguish progressive changes that accompany the ageing of mitotically competent cells from catastrophic changes that follow cell death (release of lysozomal enzymes, and so on).

(c) Dilution theories[23]. The simplest of all the mechanisms that have been proposed to explain clonal ageing, the dilution of an essential and irreplaceable component of the primary cells, is ruled out on numerical grounds—the molecules would be diluted out long before the fiftieth division. A closely related mechanism, the dilution of an essential component that is synthesized at a rate that just fails to keep pace with cell division (the mitochondria, for example), is compatible with what we know about ageing, but does not seem very plausible. The simple accumulation of damage also fails to account for the ageing of dividing cells, for damaged molecules would be diluted out by newly synthesized material. This last argument does not hold for non-dividing cells, and it is possible that the ageing of neurones, for example, is due to postsynthetic damage to DNA, proteins, or other cellular components.

(*d*) Molecular clocks and programmed ageing. It has been suggested that the ageing of the individual, far from representing an inevitable consequence of the limited accuracy of biological processes, has a positive adaptive value for the species. Thus, well-defined mechanisms may have evolved that guarantee senescence at an appropriate time. These ideas lead naturally to the notion of a molecular clock, a device that counts cell divisions (or elapsed time) and triggers an irreversible ageing process when the optimal life span is approached[25].

Very little is known for certain about the factors determining reproductive life span and longevity in different species. It is even uncertain that ageing is a significant phenomenon in wild populations ; it is said that few animals survive long enough to be able to age. It is not clear that senescence has a positive adaptive value, so it seems unwise, in the absence of experimental evidence, to take for granted the idea of "programmed obsolescence" in man and other mammals[26].

The one great merit of the molecular-clock theory is that it accounts readily for the phenomenon of transformation. If cells are programmed to age, a mutation that interfered with the programme could make them "immortal". Similarly, a transforming virus might interfere with the natural ageing programme. We shall see that the phenomenon of transformation can be explained only with some difficulty on the basis of error theories.

(*e*) Elementary protein error theory. There are two steps in protein synthesis involving discrimination between related molecules. Each activating enzyme must select a unique amino acid and load it onto an appropriate tRNA. A codon of messenger RNA must pair with the anticodon of an appropriate tRNA, but not with any inappropriate anticodon. Each of these processes is associated with a small but non-zero error frequency. Errors in RNA synthesis could also decrease the fidelity of protein synthesis.

Because the proteins that are part of the protein-synthetic apparatus are synthesized in the same way as other cellular proteins, errors in protein synthesis could be self-propagating[27]. Suppose we started with an error-free first-generation protein synthetic apparatus and used it to synthesize activating enzymes, ribosomal proteins, and so on. The second generation protein synthetic apparatus would certainly contain some errors. If we used this error-containing second generation protein synthetic apparatus to synthesize more activating enzymes, and so on, they would contain even more errors.

If the cycle discussed above was repeated many times, each successive generation of protein synthetic apparatus being used to synthesize the components of the next generation, one of two things would happen. Either the error-frequency would converge to a stable non-zero value, or it would diverge. In the former case, no ageing phenomenon would occur, but in the latter the error frequency would, sooner or later, become so large that the cell could no longer function.

When I first discussed this problem I thought that in the absence of cellular selection, the error frequency of protein synthesis would certainly diverge. I no longer believe this to be the case. It is possible that a protein synthetic apparatus containing a relatively small number of errors might be able to synthesize a new protein-synthetic apparatus containing fewer errors. This must have been true at some stage in the evolution of the genetic apparatus, unless cellular selection became important very early. The exist-

ence of enzymes that scavenge error-containing proteins makes it more plausible that a small, stable error frequency can be maintained[28].

In only one case is there direct experimental evidence that the frequency of errors of protein synthesis increases rapidly and leads to cell death. The *leu*-5 mutant of *Neurospora* synthesizes a temperature-sensitive leucine-activating enzyme which, at higher temperatures, allows other amino acids to substitute for leucine[29]. At low temperatures protein synthesis proceeds more or less normally. The *leu*-5 strain, unlike the wild-type, senesces when maintained at 35° C.

Lewis and Holliday have shown that the accuracy of protein synthesis in *leu*-5 falls slightly when this mutant is shifted from 25° C to 35° C, and then remains almost constant for a considerable time. After about 70 h the cells age rapidly, and at the same time the thermolability of glutamic dehydrogenase increases, while its specific activity falls dramatically. The "error-catastrophe" hypothesis fits all of Lewis and Holliday's extensive observations on this system[30].

There is, as yet, no experimental evidence concerning the mechanism of senescence in fungi that age spontaneously in nature, for example, *Podospora*. We have already seen that the origin of thermolabile protein in ageing fibroblast cultures is unclear. If an error catastrophe is the cause of ageing in non-mutant fungi or in mammalian cells in culture, this can only be established by the measurement of the fidelity of protein synthesis.

(*f*) Error theory—more general models[31]. Holliday and Tarrant have remarked[16] that it may not be possible to separate the contributions to cellular ageing caused by errors of protein synthesis from those due to the accumulation of somatic mutations. Errors of protein synthesis must occasionally lead to the formation of "mutator" DNA polymerase molecules which replicate DNA inaccurately. Conversely, some mutations, for example, the *ram* mutations in *E. coli*[32], must reduce the fidelity of protein synthesis. Thus inaccurate protein synthesis and inaccurate DNA synthesis are coupled phenomena.

If the coupling between different causes of ageing is sufficiently strong, we cannot use a conventional description in which the deterioration of one process, say, protein synthesis, is thought of as primary and the cause of the subsequent breakdown of other processes such as DNA replication. Instead we must use a more general description in which the fidelity of each of a set of significant processes is dependent on the fidelity of some or all of the others.

A theory of this kind forms a natural generalization of the simple protein-error theory. It retains the central idea of positive feedback—the greater the number of errors that have accumulated in the macromolecular constituents of the cell, the faster the accumulation of further errors. At the same time it permits one to consider the effect of mutations, relaxed control processes, faulty membranes, and so on, in parallel with errors of protein synthesis, without assigning priority to any one process.

The coupling between different contributions to ageing must also be important in the senescence of whole animals. The deterioration of extracellular components (collagen, say) is likely to change the intracellular environment in such a way as to reduce the fidelity of macromolecular synthesis. The inaccurate synthesis of proteins would surely have important effects on the condition of extracellular components. It follows that extracellular and intracellular mechanisms of ageing are coupled. I believe that the theoretical and experimental analysis of strongly coupled deteriorative processes will do much to clarify our ideas on the nature of mammalian ageing.

(*g*) Advanced mathematical methods. If errors of macromolecular synthesis are established by experiment to contribute to ageing, a more sophisticated mathematical treatment of the subject will be justified. The overall error frequencies and mutation rates which we have discussed will need to be replaced by error matrices, which specify the probability that any given amino acid A_1 (or nucleotide, N_1) is replaced erroneously by a different amino acid, A_2 (nucleotide N_2). It is also likely that simple treatments which utilize difference or differential equation will be replaced by a more general stochastic theory.

(*h*) Transformation. Mammalian cells in tissue culture are transformed by certain viruses, by chemical carcinogens, or spontaneously to permanent or transformed lines which can be subcultured indefinitely. It seems unlikely that DNA replication or protein synthesis is made more accurate by each of these modes of transformation. Thus, at first sight, the phenomenon of transformation seems incompatible with error theories of clonal ageing.

In fact, the situation is more complex. The switch from a "spanned" culture to one that is "immortal" does not necessarily signify a qualitative change in other cellular properties. An elementary mathematical result from the theory of branching-chain processes shows that if cells continue to divide and to produce, on average, more than one viable descendant at division, then cell cultures are potentially immortal. On the other hand, if each cell sooner or later comes to produce on average less than one mitotically competent descendant, the cultures inevitably die out. Thus a "spanned" culture would be made "immortal" by any change that increased the mean number of viable descendants sufficiently. It is important to note that this result is completely independent of the mechanism which leads to the increase in the number of viable descendants.

Consider first a very artificial situation, a culture of *E. coli* growing while exposed to a source of high-energy radiation. Suppose that each cell accumulates a lethal mutation in the time $t_{1/2}$ with probability 0.5. If cells are maintained under conditions which permit doubling in a time less than $t_{1/2}$, cultures are "immortal". Otherwise, they are "spanned". A small change in the medium could, therefore, "transform" *E. coli* in these rather special conditions.

This result cannot be applied directly to error theories of ageing, but it does suggest the sort of explanation of transformation that is needed. It must be postulated that an increased mitotic rate or the escape from contact-inhibition provides a protection against the accumulation of errors because, as in the simple example, it permits more cellular selection. This would be true if, for example, the rate of accumulation of errors (in proteins or nucleic acids) includes a term that is linear in the time (for example, a term describing errors occurring in non-dividing cells) while selection is entirely of a cellular type and occurs only through cell division.

A theory of this kind would be complicated and its predictions would depend crucially on the details of the postulated model of error accumulation, for example, on the ratio of the rates of error accumulation in stationary and non-stationary phases. It is almost certain that a stochastic treatment would be needed to obtain significant results. Some predictions, however, seem general. Transformed cultures, in spite of their so-called immortality, should generate slowly dividing or non-dividing cells. If errors of protein synthesis are relevant, transformed lines should "revert" in the presence of amino acid analogues.

If errors accumulate in somatic cells and finally lead to the cessation of cell division, why does the same process fail to occur in the germ line? The simplest escape from this dilemma is to suppose that protein synthesis is sufficiently accurate in germ-line cells to maintain a constant, low error level. In somatic cells the rate may be higher, or may become higher with age, so that an error catastrophe ultimately develops.

It is also possible that "quality control" processes operate during oogenesis and early development, and lead to the rejection of ova or embryos in which the error level is too high. It is known that, in humans, many of the oocytes that begin to mature later degenerate. This could be the result of cellular selection for "metabolic efficiency".

Future Developments

Techniques are now available that permit the measurement of mutation rates and error frequencies in fibroblast and other cell cultures. We can expect to know fairly soon whether or not these quantities vary with the age of the cultures in the manner predicted by error theories. It is important that information of this kind be accumulated for as wide a range as possible of cell types and animal species.

Such information, while valuable, will not be sufficient to enable us to decide whether or not errors of molecular synthesis are among the primary causes of clonal ageing, or whether they are triggered by some different molecular process. Studies of the division times within individual clones of cells may help to resolve this issue. Protein-error theory, for example, would lead one to expect a positive correlation between the division times of mother and daughter cells, even in young or transformed cultures— molecular clock theories do not require such a correlation.

It is still difficult to measure the fidelity of macromolecular synthesis in the tissues of whole animals, but information of this kind should become available soon. Even when we have this information, the difficulty of determining the relevance of such experiments to ageing will be formidable. A low level of errors in any particular type of cell would not rule out the possibility that ageing is due to error accumulation in some other type of cell. Conversely, if the error level is high in some or all tissues of old animals, this could be the consequence of some earlier triggering event, and might have little or no importance for the pathology of ageing.

A connexion between the ageing of cells and the ageing of animals would be suggested if it could be shown that a process that increases the life span of cells in culture produces a corresponding increase in the longevity of the donor species. The demonstration that genetic diseases characterized by premature ageing are associated with inaccurate DNA, RNA or protein synthesis would also count as evidence in favour of a contribution of errors of synthesis to whole-animal ageing.

I believe that it will be a considerable time before the importance of intracellular errors of macromolecular synthesis for *in vivo* ageing can be assessed. It seems clear that immunological mechanisms and the deterioration of extracellular material play an important part in the ageing of mammals. It may be difficult to determine whether these processes are independent of errors of macromolecular synthesis, coupled to them, or a consequence of them.

I thank Dr A. Comfort, Dr F. H. C. Crick, Dr J. Danielli, Dr R. Dulbecco, Dr R. Holley and Dr R. Holliday for discussion and comment.

[1] Hayflick, L., and Moorehead, P. S., *Exp. Cell. Res.*, 25, 585 (1961).
[2] Hayflick, L., *Exp. Cell. Res.*, 37, 614 (1965).
[3] Martin, G., Sprague, C., and Epstein, C., *Lab. Invest.*, 23, 86 (1970).
[4] Maciero-Coelho, A., *Experientia*, 22, 390 (1966).
[5] Cristofalo, V. J., in *Ageing in Cell and Tissue Culture* (edit. by Holeckova, E., and Cristofalo, V. J.) (Plenum, New York, 1970).
[6] Litwin, J., *Exp. Cell. Res.*, 72, 566 (1972).
[7] Harris, M., *Growth*, 21, 149 (1957).
[8] Peturrson, C., Coughlin, J. I., and Meylon, C., *Exp. Cell. Res.*, 33, 60 (1964).
[9] Krooth, R. S., Shaw, M. W., and Campbell, B. K., *J. Nat. Cancer Inst.*, 32, 1031 (1964).
[10] Hayflick, L., *Amer. J. Med. Sci.* (in the press).
[11] Maciero-Coelho, A., in *Ageing in Cell and Tissue Culture* (edit. by Holeckova, E., and Cristofalo, V. J.) (Plenum, New York, 1970).
[12] Daniel, C. W., *Adv. Gerontol. Res.*, 4, 167 (1972).
[13] Williamson, A. R., and Askonas, B. A., *Nature*, 238, 337 (1972).
[14] Cristofalo, V. J., *Adv. Gerontol. Res.*, 4, 45 (1972).

[15] Kanungo, M. S., *Biochem. Rev.*, **61**, 13 (1970).
[16] Holliday, R., and Tarrant, G. M., *Nature*, **238**, 26 (1972).
[17] Lewis, C. M., and Tarrant, G. M., *Nature*, 316 (1972).
[18] Gershon, H., and Gershon, D., *Proc. US Nat. Acad. Sci.*, **70**, 909 (1973).
[19] Gershon, H., and Gershon, D., *Nature*, **227**, 1214 (1970).
[20] Fulder, S., quoted in Holliday, R., *Humangenetik*, **16**, 83 (1972).
[21] Curtiss, H. J., *Symp. Soc. Exp. Biol.*, **21**, 51 (1967).
[22] Finch, C. E., Foster, J. R., and Mirsky, A. E., *J. Gen. Physiol.*, **54**, 690 (1969).
[23] Tribe, M. A., Ashurst, D. E., *J. Cell. Sci.*, **10**, 443 (1972).

[24] Butschli, O., *Zool. Anz.*, **5**, 64 (1882).
[25] Medvedev, Z. A., *Exp. Gerontol.*, **1**, 227 (1972).
[26] See Comfort, A., *Ageing—The Biology of Senescence*, revised ed. (Routledge and Kegan Paul, London, 1963).
[27] Orgel, L. E., *Proc. US Nat. Acad. Sci.*, **49**, 517 (1963); **67**, 1476 (1970).
[28] Goldberg, A. L., *Proc. US Nat. Acad. Sci.*, **69**, 427 (1972).
[29] Printz, D. B., and Gross, S. R., *Genetics*, **55**, 451 (1967).
[30] Lewis, C. M., and Holliday, R., *Nature*, **228**, 877 (1970).
[31] Medvedev, Z. A., *Symp. Soc. Exp. Biol.*, **21**, 1 (1967).
[32] Gorini, L., *Nature*, **234**, 261 (1971).

Reprinted from *Science* **198**:366–372 (1977)

Testing the Commitment Theory of Cellular Aging

R. Holliday, L. I. Huschtscha,

G. M. Tarrant, T. B. L. Kirkwood

Summary. The commitment theory may explain both the finite lifespan of diploid fibroblasts and the apparent immortality of transformed lines. Potentially immortal cells are assumed on division to generate with some fixed probability cells committed to senesce after a specific number of divisions. During the period between commitment and senescence, cells are assumed to maintain normal growth so that the uncommitted cells are diluted by committed ones and may ultimately be lost in subculturing. A number of predictions of this model are described and experiments strongly supporting the theory are reported. We conclude that the limited growth of diploid fibroblasts is, in effect, an artifact of normal culturing procedures.

It is well known that human diploid fibroblasts cannot be subcultured indefinitely (*1–3*). A long period of normal growth is invariably followed by a senescent phase and, subsequently, death of the whole population; fibroblasts have therefore often been used as a model system for the study of the intrinsic process of cellular aging. Unlike rodent cells, cultured diploid human cells never undergo spontaneous transformation to a permanent cell line. It follows that the whole population in some way becomes committed to cell death.

It has sometimes been suggested that the finite lifespan of human fibroblasts is due to a built-in genetic program and that their cessation of division is due to "terminal differentiation" (*4, 5*). If senescence is the result of a program that specifies the total number of cell divisions, it might be expected that, at any one time, all the cells in a population would be of roughly the same age, that is, at similar points along the programmed pathway from primary culture (phase I) through the period of steady growth (phase II) to senescence (phase III).

R. Holliday is head of the Genetics Division, National Institute for Medical Research, Mill Hill, London NW7 1AA, England; L. I. Huschtscha and G. M. Tarrant are members of the same division. T. B. L. Kirkwood is in the Statistics Section of the National Institute for Biological Standards and Control, Holly Hill, Hampstead, London NW3, England.

This is clearly not the case, since it has been shown that individual cells taken from a population of fibroblasts are extremely variable in their doubling potential (*4, 6*). Nevertheless this heterogeneity is not sufficient to allow the emergence of potentially immortal cells. Another important feature of human fibroblast aging is that there is considerable variation from experiment to experiment in the actual lifespan of parallel cultures of any one strain, as measured in passages or population doublings (*3, 7*).

To explain these observations we previously proposed a commitment theory of fibroblast aging (*8, 9*). Starting with a population of uncommitted cells, which are potentially immortal, we assumed that there was a given probability that cell division will give rise to fibroblasts that are irreversibly committed to senescence and death. These cells initially multiply normally, but after a given number of cell divisions, which we call the incubation period, all the descendants of the original committed cell die out. We showed that if the probability of commitment is reasonably high and the incubation period sufficiently long, then the number of uncommitted cells in the population will progressively decline to < 1 in 10^6, at which time, with normal subculturing regimes, they will inevitably be lost from the population. At this

point, all the remaining cells are committed to senescence, and the population is therefore a mortal one although different cells within it have very different growth potentials. We also showed that a reduction in the probability of commitment or the length of the incubation period could produce an immortal steady state culture, consisting of a small subpopulation of uncommitted cells, a majority of committed cells, and a constant fraction of nonviable cells. We suggested that transformed or permanent lines of mammalian cells may be in such a steady state.

If the theory is correct, then the number of cells in a culture becomes an important factor in determining the longevity of fibroblasts, since the population size is directly related to the probability of losing the last uncommitted cells from the culture. Varying the population size thus provides a means of testing the validity of the theory. Our previous model was a deterministic one and did not take into account the stochastic process of progressive dilution out of a very small number (for example, 1 in 10^5) of uncommitted cells. We have now developed a stochastic model and used it to predict by computer simulation the consequence of drastically reducing population size at different passage levels. We have carried out experiments of this type to test the theory, and these have given results that are in agreement with the computed predictions. In our view, it is hard to explain these observations in any way other than the one we have suggested.

Structure of Fibroblast Populations

Cells can be divided into three categories: uncommitted cells, committed cells, and dead cells. Any difference between dead cells and cells that have ceased to be able to divide is immaterial to discussion of population growth. (We assume that all such cells are passively transmitted, rather than lost, during routine subculture since it is known that many chemical or physical agents that kill cells do not prevent their attachment to solid surfaces. However, this assumption is not essential to the general conclusions we reach.) During the division of an uncommitted cell, we assume a probability P for each daughter cell that it becomes committed and a probability $1 - P$ that it remains uncommitted. We thus have three possible outcomes to the division of an uncommitted cell:

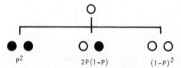

Commitment is assumed to be irreversible so that division of a committed cell always produces two committed cells. We assume that the path from commitment to death is the same for every cell and define the incubation period M as the number of cell divisions that elapse between commitment and death. (In defining M in this way we assume the cell cycle time to be constant, which is certainly an oversimplification. Variation in cell cycle time about a mean value has no radical effect on the general structure of the model.)

The committed cells can then be divided into M distinct subclasses, the first of which contains newly committed cells, the second of which contains cells that have divided exactly once since commitment, and so on. The last subclass contains cells that have divided exactly $M - 1$ times since commitment and that will die after their next division.

The total population thus consists of $M + 2$ classes, one of uncommitted cells, M of committed cells, and one of dead cells. The population structure is determined by the number of cells in each class, with transitions from one class to another occurring at each cell division.

Since the number of cells under consideration is usually large (10^5 to 10^7), it is reasonable to neglect random fluctuations in the numbers of cells becoming committed and to use the resulting deterministic form of the model for a general study of the expected population growth (9).

The process of commitment cannot result in an immediate reduction in cell growth rate since cellular selection would then maintain the population indefinitely. We therefore assume for the present that all live cells divide at the same rate. (This assumption is probably unjustified, since there is evidence that cells divide more slowly as they approach death, but we make it to simplify the mathematical treatment. Slowing down the growth rate toward the end of the incubation period does not significantly alter the conclusions that we reach.)

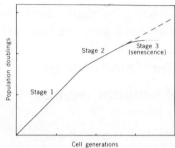

Fig. 1. The commitment theory predicts that an initially uncommitted cell population will double in size with each cell generation (stage 1) until, at the end of the incubation period, the first cells die. The growth rate then falls by an amount determined by P, the probability of commitment. The population subsequently grows at a steady but reduced rate (stage 2). Depending on whether or not the uncommitted cells were lost by dilution before the end of stage 1, the population will either become senescent (stage 3) and die after a finite period in stage 2, or will continue stage 2 growth indefinitely. The scale of cell generations on the abscissa represents the number of cumulative cell divisions by the viable cells in the population.

If we assume that a cell population is initially fully uncommitted and if we allow it to grow without restraint, we expect it to double in size with each successive cell division until, at the $(M + 1)$th division, cell deaths occur as the first committed cells reach the end of the incubation period. Inevitably the growth rate falls. What happens to the population then depends on the value of P. If $P > .5$, each uncommitted cell produces an average of less than one uncommitted daughter so that the uncommitted cells decline to extinction and the population dies out rapidly. If $P < .5$, the cell population growth rate falls to a level determined solely by the value of P, and the population thereafter maintains this reduced but steady growth rate with a stable distribution over the classes. We term the first period of growth, prior to the first cell deaths, stage 1, and the subsequent growth at reduced rate, stage 2 (Fig. 1).

However, in the normal laboratory context populations do not grow without restraint. Populations of size N cells ($\approx 10^6$) are repeatedly split (for example, 1:2) and the subcultures are allowed to grow until they again contain approximately N cells each. (Each 1:2 split is referred to as a passage, and this is approximately the same as one population doubling.) Clearly the total number of subcultures soon becomes prohibitively large if all are kept; and it is usual to

discard all but a few, these being assumed to be representative of the potential total population. The essential feature of this procedure is that at any time no single unit of the population contains more than N cells. During stage 1 growth the accumulation of committed cells dilutes the uncommitted cells even though the potential total number of uncommitted cells may be increasing (for $P < .5$). There may thus come a point where the number of uncommitted cells is so diluted that there is a high probability that individual subcultures contain no uncommitted cells, and are therefore mortal. Alternatively, if a reasonable number of uncommitted cells remains in the individual subcultures at the transition from stage 1 to stage 2, the adoption of a stable distribution of cells over the classes means that no further dilution of the uncommitted cells occurs. Thus, under these circumstances, cultures may continue to grow indefinitely.

For this reason, both the population size N and the incubation period M play important parts in determining the mortality or immortality of the population. For a given value of P, if N is too small or M is too large, the uncommitted cells are likely to be lost during stage 1 and the populations are then certain to be mortal. When this happens, the growth rate will still show the transition from stage 1 to stage 2 growth as before and will remain steady for a number of cell divisions in stage 2. However, as the last cells to become committed approach death, the growth rate will again begin to fall, this time steadily to extinction. We term this final stage of growth stage 3. The transition from stage 2 to stage 3 is not a sharply defined event but corresponds roughly with the appearance of visible senescence in the culture.

The general predictions of the model are thus as follows. For $P > .5$, cultures will grow for a time at a steady rate and will then show a rapid decline followed by death. For $P < .5$, cultures will grow at a steady rate until the first cell deaths occur when they change to a reduced, but again steady, growth rate. Depending on the combination of values of P, M, and N the cultures will then either senesce and die or will grow on indefinitely (Fig. 1). The predicted sudden change in growth rate at the transition from stage 1 to stage 2 is specific to the model and it is interesting that such an effect has been observed by us and by others (10). Indeed a growth experiment performed with MRC-5 fibroblasts (11) some years ago gives a close fit to the model's prediction (9).

Stochastic Predictions

Statistical variation in lifespan is an inherent feature of fibroblast growth experiments. By incorporating the elements of the random processes embodied in our model into a stochastic structure, we have been able to simulate the observed distribution of lifespans and to predict the outcomes of experiments designed to test the validity of the model.

The model contains three main sources of random variation. The first governs the outcome of the division of an uncommitted cell. The second represents the selection of the cells that are to form the subculture. The third determines the "choice" of those cells which are to divide more than once during stages 2 and 3 (when dead cells are present) in order to restore the population size to N. The mathematical nature of these three stochastic elements and the details of the model are to be described (12).

Prediction according to the stochastic version of the model was achieved by computer simulation since its structure was too complex for full mathematical analysis. The computer programs were based on two subroutines. The first, called SPLIT, performed the subcultures and the second, called GROWUP, allowed the subcultures to grow back to the specified size. Both subroutines utilized a common function that generated pseudorandom outcomes to the various stochastic elements of the model. Thus any culture experiment could be simulated with comparative ease by an appropriate se-

quence of calls to the subroutines. All simulations were run on the Hewlett-Packard 3000 computer at the National Institute for Medical Research, London.

Estimation of Parameters

The model is defined in terms of three parameters, N (population size), P (probability of commitment), and M (incubation period). In order to test its validity, it was necessary to estimate values for these parameters and show that they would give satisfactory predictions of experimental results.

Population size, N, is determined by the size of containers used for growing cells. We routinely use 150-milliliter glass Bow bottles: a confluent monolayer contains 1.5×10^6 to 3.0×10^6 cells. We assume N to be equal to 2×10^6.

The probability of commitment, P, may be estimated in two fairly direct ways. The first is to use the observed reduction in growth rate at the transition from stage 1 to stage 2 since this depends only on P. The growth experiment mentioned above gave an estimate of P equal to .275. The second, discussed below and in more detail elsewhere (12) is that the model predicts that the variation in lifespan of mass populations should decrease as P increases. As we shall see, comparison of the observed and predicted distributions of lifespan supported the estimate $P \simeq .275$.

The incubation period, M, is more difficult to determine since, within fairly broad limits, the predictions of the model

are relatively insensitive to changes in M. The major problem is that we do not have any means at present of estimating the true "age" of our cultures in terms of the numbers of population doublings from the time when the population was fully uncommitted, since some of these will have occurred in vivo.

In fact, for this reason, we shall find it necessary in later comparisons and discussion to express different points in the culture lifespan in terms of population doublings or passages "before normal death" (BND). In the growth experiment referred to above, the transition from stage 1 to stage 2 apparently occurred after 42 population doublings in culture. Thus we may place a lower limit of 42 on M. It may further be shown (12) that for $P = .275$, $N = 2 \times 10^6$ we must have $M \geq 55$ if the chance of uncommitted cells surviving into stage 2 is to be negligible. Since increasing M increases the number of cell classes and, thus, the computer time necessary to calculate predictions from the model, we chose to work with the value $M = 55$. We discuss this point below and show that, most probably, M lies in the range 55 to 60.

Verification of Predictions

Lifespans of mass cultures. It has been apparent for some years that a given fibroblast strain such as MRC-5, WI-38, or WI-26 (all derived from fetal lung tissues), does not achieve the same number of population doublings (PD's) in replicate longevity experiments. Our ex-

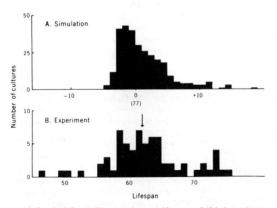

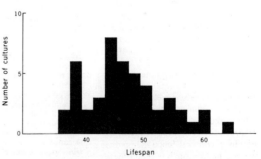

Fig. 2 (left). Histograms of simulated (A) and experimental (B) culture lifespans (population doublings or passages). Three hundred simulations were performed on the assumption of a fully uncommitted initial population ($P = .275$, $M = 55$, and $N = 2 \times 10^6$). The lifespans of the simulated cultures are expressed as deviations from the mean (77 population doublings). The experimental lifespans of 67 independent populations of MRC-5 fibroblasts are expressed in passage numbers, which are very similar to population doublings. The cultures were started from early passage cells and split either 1:2 (one passage) or 1:4 (two passages), according to the procedures described by Thompson and Holliday (3). The histogram includes previously published lifespans, including ten cultures grown at 34°C, and unpublished results, including six cultures grown throughout at 33°C. Cultures grown at these temperatures had the same average lifespan as those at 37°C. One culture, which died at passage 33, probably from an experimental mishap, is excluded. The histogram does not include any of the longevity results shown in Fig. 5. The mean of the experimental lifespans is indicated by an arrow.   Fig. 3 (right). Distribution of lifespan in passage numbers of 47 cultures of the human diploid fibroblast strain WI-38. These results were obtained in Hayflick's laboratory between 1968 and 1972. Cultures were derived from a total of 28 passage 8 ampuls and grown until senescence and death according to procedures described by Hayflick (1).

periments were carried out mainly with MRC-5. This strain was originally characterized by Jacobs *et al.* (*13*) and many passage 8 ampuls are stored in liquid nitrogen. We normally receive MRC-5 cells at passage 10 or 11, and over a 7-year period during which various longevity experiments have been carried out, it is evident that parallel cultures set up from a population derived from one ampul and grown under identical conditions exhibit considerable variation in lifespan. Moreover, the doubling potential of cells derived from different ampuls is also significantly different and cannot be accounted for by possible variation in viable cell number between ampuls. The combined longevity data for 67 populations derived from several different ampuls is shown in Fig. 2B. (It is important to mention that in these experiments the individual populations were set up at an early passage when, according to our model, a significant number of uncommitted cells are present. Thus, the loss of these cells by dilution in any one culture is independent of that in any other culture.)

The mean lifespan is 62 passages and, as can be seen, some experiments give longevities well to either side of the main body of the distribution. A short lifespan may, of course, be the result of unknown culture mishaps, such as inferior batches of serum or medium. Although we routinely test for mycoplasma infection by the method of Russell *et al.* (*14*), very small numbers of particles are not seen and these may sometimes spread rapidly through the population as it becomes senescent and thereby accelerate cell death. It should be noted that the observed variation in the longevity of MRC-5 represents a very real heterogeneity in growth potential. For instance, a difference in lifespan of 10 PD's represents a difference of 2^{10}, that is, a factor of 1000 in growth potential.

In Fig. 2A we represent the lifespan distribution of 300 simulated cultures. The scatter is similar to the experimental results except that the distribution of lifespan is skewed and there are no very short-lived cultures. Since the predicted variance in lifespan is strongly dependent on *P* (*12*) we can limit our estimate of *P* to a range of approximately 0.2 to 0.3, and thus the value *P* = .275 may be fairly accurate. We have also received from L. Hayflick information about the longevity of WI-38 cells reconstituted at different times from passage 8 ampuls. The distribution in lifespan of 47 cultures is shown in Fig. 3. The results are very similar to those shown in Fig. 2, A and B, except that the mean lifespan is much

shorter than MRC-5, and the scatter is somewhat greater.

Effect of bottlenecks. Culture size, *N*, has a critical effect on lifespan since it determines the time at which the uncommitted cells are lost by dilution. Increasing *N* increases the expected lifespan until a size is reached where virtual immortality is possible, although *N* may need to be so large that this may not be seen in laboratory cultures. Conversely, decreasing *N* decreases lifespan. We can thus see that, if we can increase or decrease *N* by a sufficient amount, we may

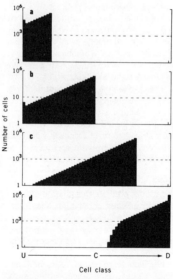

Fig. 4. The expected population profiles, that is, distributions of cells in the classes of the model, at various points during a culture lifespan. Uncommitted cells (*U*) give rise to committed daughter cells, which move steadily to the right through the committed classes (*C*) until at the end of their incubation period, they die (*D*). The broken lines indicate the critical cell number (10^3) below which individual classes of cells may be lost during a 1000-fold reduction in culture size, or bottleneck. (a) After only ten cell generations, the cells in each occupied class are sufficiently numerous to avoid elimination. (b) After 27 cell generations, the uncommitted and recently committed cells are vulnerable to loss, and a reduction in culture lifespan may be expected. Occasional bottlenecks may, however, contain some uncommitted cells and consequently be rejuvenated. (c) After 43 cell generations, the uncommitted cells have been lost by dilution, and bottlenecking will generally eliminate the younger committed cells. Since these have the greatest division potential, a reduction in culture lifespan is expected. This reduction remains constant until the first cell deaths occur since the shape of the profile does not change, the population moving collectively one class to the right with each division. (d) At a late stage in the culture lifespan, cells are accumulating in the dead and final committed classes, and the effect of bottlenecking decreases.

be able to demonstrate a corresponding increase or decrease in lifespan. This, however, presents considerable practical difficulties and so we adopted an alternative approach. Routine cultures were grown normally, but at a certain point in the lifespan were drastically reduced in size, or "bottlenecked," to about 10^3 to 10^4 cells. They were then allowed to grow back up to the normal size *N* and cultured routinely for the remainder of their lifespan.

According to the time at which the bottlenecks were taken, we would predict a variety of effects. Early bottlenecks, when the cells are concentrated in large numbers in the uncommitted and early committed classes (Fig. 4a) should show little effect on lifespan other than increasing the variance. Similarly, late bottlenecks, when the cells are concentrated in the dead and late committed classes (Fig. 4d), will also have little effect. However, intermediate bottlenecks will correspond to times when the cells are distributed over many or all of the classes (Fig. 4, b and c), and there will then be relatively few cells in the uncommitted or earlier committed classes. These cells represent a considerable proportion of the cell division potential of the population, and their premature loss, as is likely to happen when the bottleneck is taken, will thus result in a reduction in lifespan. Particularly important are the uncommitted cells, since these are capable of producing further uncommitted cells and thus have a high division potential. Bottlenecks taken when the uncommitted cells are relatively few in number may contain no uncommitted cells, in which case the lifespan is reduced, or may contain some uncommitted cells in which case by chance the proportion of uncommitted cells in the culture may be temporarily increased. In the latter case, the population is slightly "rejuvenated" and may show an increase in lifespan. Computer simulations of the distribution of lifespans of bottleneck cultures are shown in Fig. 5A where these various predictions may be seen.

Preliminary experiments were carried out with a culture of human fibroblasts obtained from abdominal tissue of a normal 10-week-old fetus. The primary culture was established, with cells passaged in the normal way. Cell counts were made at every subculture so that cumulative cell numbers and population doublings could be calculated exactly. In bottleneck experiments, 10^4 cells at various passage levels were inoculated into Falcon plastic wells (15 mm diameter) and then allowed to grow up to the normal population size of $\sim 2 \times 10^6$. In two experiments shown in Table 1, it can be seen that an early bottleneck had a striking effect on the final lifespan, whereas bottlenecks taken at later times had less or no effect.

These results are in accord with our theoretical expectation. Another prelimi-

nary experiment (results not shown) was carried out with an MRC-5 culture at passage 10 and also passage 20. Bottlenecks of 10^3, 2×10^3, or 10^4 cells were taken and allowed to grow up to the normal culture size, and the subsequent lifespan was determined. We found that cultures derived from the smaller bot-

tlenecks have the shorter lifespan, as we would expect.

The above experiments can be criticized since the stochastic nature of the process of dilution and the loss of uncommitted cells will tend to result in considerable variation in longevity from experiment to experiment. The definitive experiment therefore requires the study of many populations, which in practice is hard to achieve. We have carried out one major experiment with MRC-5 in which 8 to 12 bottlenecks of 2×10^3 cells were taken at each of four PD levels. Results are shown in Fig. 5, together with the longevity of control cultures derived from the same original ampul. (This particular ampul consistently produced mass cultures with an average lifespan of about 54 PD's.) It is clear that the average lifespan of the 8, 13, and 21 PD bottlenecks was reduced by approximately eight PD's, which is close to the reduction seen in the simulated experiments. The likelihood of obtaining these results by chance is negligible. The 31 PD bottlenecks had a significantly greater subsequent lifespan. These overall results strongly support our theory, and more particularly, they indicate that our choice of $P = .275$ may be close to reality.

We did not, however, see a bimodal distribution of longevity with passage 8 bottlenecks, which suggests that at this stage the proportion of uncommitted cells is already as little as about 0.005 percent. If this is so, we can estimate when the population might have consisted entirely of uncommitted cells since approximately 30 cell generations would be required to reach this level, if $P = .275$. We have estimated from the growth of primary cultures that small tissue explants contain 10^3 to 10^4 fibroblasts capable of proliferation. The passage 1 population of 2×10^6 cells would therefore have doubled 8 to 11 times in vitro. The proportion of uncommitted cells in the tissue explant would therefore be approximately 1 percent. MRC-5 cells are derived from a 14-week-old fetus weighing approximately 75 grams. If we assume that 0.1 to 1 percent of the cells are fibroblasts, there would have been approximately 10^8 to 10^9 fibroblasts altogether. Our calculations above suggest that these were derived from a population of primordial uncommitted fibroblasts by approximately 15 PD's. Our estimate of the size of this population is therefore 10^3 to 10^4, a reasonable figure, and we calculate that passage 1 fibroblasts have already achieved approximately 20 to 25 in vivo and in vitro population doublings.

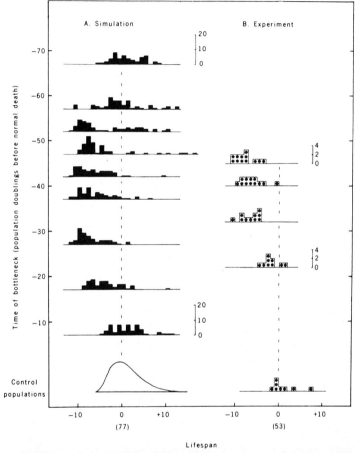

Fig. 5. Histograms of simulated (A) and experimental (B) bottleneck culture lifespans. Each simulated distribution is derived from 50 initially uncommitted cultures whose size was reduced from 2×10^6 to 2×10^3 at the indicated time in the lifespan. These times are denoted as population doublings prior to the mean population doubling level at which control populations died out. The experimental distributions are derived from cultures of MRC-5 cells which experienced a similar 1000-fold reduction in size at various points during their lifespan. To facilitate comparison between simulation and experiment, the same convention has been adopted to describe the times of the bottlenecks. The cells were all derived from one passage 8 ampul and subcultured with a 1:2, 1:4, or 1:8 split ratio (depending on the age of the culture); Eagle basal medium (BME) was used, supplemented with 10 percent fetal bovine serum previously heated at 56°C for 30 minutes, 1 percent nonessential amino acids and antibiotics (penicillin, 100 unit/ml; streptomycin, 100 μg/ml; aureomycin, 10 μg/ml). Confluent cells were harvested with trypsin-versene (3) and counted with a Coulter counter prior to subculture to calculate cumulative population doublings. Bottleneck cultures were established at passages 8, 13, 21, and 31 by placing 2×10^3 cells in 15-mm Falcon plastic wells. These cultures were transferred to containers of increasing size until the normal population of $\sim 2 \times 10^6$ cells was reached, and this was then subcultured normally until death. Cell counts during the buildup of the bottleneck populations show an exponential growth curve, and extrapolation back to zero time demonstrates that most if not all of the cells in the bottleneck were viable. In calculating population doublings we assumed that viability was 100 percent.

We are now in a better position to estimate the value of M, the incubation period. We have reason to believe that the transition from stage 1 to stage 2, that is, the first appearance of dead cells, occurs at about passage 40 for MRC-5. Thus we estimate M to be approximately 60 cell generations.

Variation among clones. Since the proportion of uncommitted cells is extremely small, even in early passage cultures, it would be very hard to obtain clones from uncommitted cells. If they were isolated, they would not give rise to immortal populations although the number of population doublings achieved would often be significantly greater than the average lifespan of mass cultures. Figure 4 illustrates the predicted heterogeneity in doubling potential of individual cells isolated at various points during the lifespan of fibroblast cultures. In each case, there would be considerable scatter in the longevity of clones, with the older populations (Fig. 4d) giving a high proportion of short-lived ones. These effects are well documented by the extensive cloning experiments of Smith and Hayflick (6), with WI-38 cells at the 22 to 41 PD levels. With cells at the ninth PD level (roughly equivalent to Fig. 4b), they also observed great heterogeneity in the doubling potential of individual clones. However, there was a subclass of dividing cells (~ 30 percent) that achieved only one to eight cell divisions. We do not predict the existence of such short-lived cells in young populations; and, since it is known that fibroblasts in isolation are inhibited in cell division, we suggest that this subpopulation is not made up of senescent cells.

Mixed cultures. The growth of mixtures of distinguishable but otherwise similar cell populations allows an important test of the model. If the growth rate of the two cell types is the same, the proportion of each should remain constant. However, when we consider the dilution out of uncommitted cells, it is clear that small numbers of these will often result in random drift to an excess of one cell type over the other. The consequence will be that the composition of the final population will often deviate from the proportion maintained throughout most of the lifespan. In 20 simulations of this experiment, starting with a 1:1 mixture, 11 populations diverged to give 80 to 100 percent cells of one type at the time of senescence. The behavior of six typical simulations is shown in Fig. 6.

We were pleased to discover that experiments of this type had already been carried out by Zavala, Fialkow, and Her-

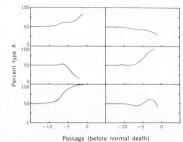

Fig. 6. The predicted behavior of six typical cultures consisting initially of a mixture of equal numbers of uncommitted cells of two distinguishable but otherwise similar types, A and B. For the greater part of the lifespan, the equal proportions of the two cell types are maintained. However, toward the end of the lifespan, the cultures deviate in a variety of ways from this equal distribution. This behavior results from stochastic differences in the losses by dilution of the uncommitted cells of the two types during an earlier part of the lifespan.

ner (15). They established many skin fibroblast cultures from female donors who were heterozygous for glucose-6-phosphate dehydrogenase (G6PD) A and B, which are electrophoretically distinguishable allelic variants. Since G6PD is X linked and only one X chromosome is active, individual cells have either G6PD A or B, and it is well known that this phenotype is completely stable. In most cultures approximately equal numbers of each cell type were present. In a few cases the ratio of A to B diverged steadily through passaging, which indicates that one cell type grew rather more quickly in culture than the other. However, in most cases the ratios of A to B stayed constant

for many population doublings, which showed that there was no selection for one cell type. Our specific prediction was borne out since it was frequently observed that at the end of the lifespan either G6PD A or G6PD B cells became predominant. In the populations where there was no evidence of cellular selection, about half showed divergence to 80 to 100 percent of one cell type during the senescent period (15). We believe that these final populations are derived from a very small number of uncommitted cells which, by chance, consisted only of G6PD A or B types. This "genetic drift" to uncommitted cells of one phenotype occurred, of course, many cell generations prior to senescence and possibly in the skin tissue itself.

Comparable studies have been carried out by Ogden and Micklem (16) with mice in which hematopoietic cells were serially transplanted from an initial donor to successive recipients, which had previously been irradiated to destroy their own hematopoietic cells. The T6 chromosome marker, either heterozygous or homozygous, was used to track the fate of donor cells. When mixtures of the two distinguishable cell types were used, the ratio remained constant for three to four serial transplants. But in the last one to two transplant generations, before the cells finally died out, one or the other type came to predominate. This result is essentially the same as that obtained by Zavala *et al.* (15) and suggests that the dying out of the cells may be due to a stem line of uncommitted, perhaps immortal, cells being diluted out—a possibility that Ogden and Micklem themselves discuss.

Table 1. The longevity of fetal human fibroblasts as a mass culture (~ 2 × 10⁶ cells) and after reduction in population size ("bottlenecks") at different passages. Cells were grown from abdominal wall tissue from a normal 10-week-old fetus. From the rate of growth of the primary culture, it was estimated that the population arose from ~ 10⁴ cells. This can be regarded as a passage-0 bottleneck. Later bottlenecks of 10⁴ cells were taken at intervals during the lifespan of the mass culture, allowed to grow to 2 × 10⁶ cells, and then subcultured until senescence and death. The cells in experiments 1 and 2 were obtained from two fragments of tissue. Media and methods were the same as those described in the legend to Fig. 2, except that cumulative population doublings were calculated from cell counts at each 1:4 split subculture.

Time of bottleneck (passage)	Population doublings		Total lifespan (PD)	Difference from control
	Before	After		
Experiment 1				
0 (control)			45	
10	12.5	18	30.5	−14.5
14	17	22.5	39.5	−5.5
20	21.5	22.5	44	−1
26	28.5	14.5	43	−2
32	35.5	7.5	43	−2
38	40.5	3.5	44	−1
Experiment 2				
0 (control)			40	
10	12.5	17	29.5	−10.5
27	28.5	11.5	40	

124

Conclusions

The commitment theory of fibroblast aging provides an explanation for the following observations:

1) Individual fibroblasts produce clones with variable growth potential.

2) Parallel populations of one fibroblast strain grown under the same conditions also vary considerably in their longevity.

3) Longevity is related to population size, in particular, a 1000-fold reduction in population size at early passage levels, followed by a 1000-fold increase, results in a reduction of final lifespan of about eight population doublings.

4) When cultures contain equal numbers of two distinguishable types of cell, which have stable phenotypes and the same growth rate, one or other cell type frequently becomes predominant at the end of the lifespan.

We do not believe that any other single explanation for all these observations is available.

Our theory may also help to explain the fundamental difference between mortal diploid cells and immortal heteroploid transformed ones since reduction in the incubation period or the probability of commitment can convert a population with finite growth to one with infinite growth potential. However, it also predicts that diploid fibroblast populations are not intrinsically mortal since, if extremely large numbers of cells could be grown, the uncommitted cells would never be lost and would therefore continually replenish the population. The Hayflick limit may, in a sense, be an artifact that is the inevitable consequence of normal culturing procedures.

Implicit in our model are three simplifying assumptions that could be questioned. These are that P, M, and cell division time are the same for all cells. It is important to know what happens if we relax these and assume instead that the values we have used are simply the means of distributions within the population. Varying M or cell division time does not greatly affect our predictions, the only change being a more gradual transition from stage 1 to stage 2. However, varying P can have significant effects on population behavior. If we suppose either that the range of the distribution is sufficient to allow a small number of cells with very small P values or that the population contains—in addition to the majority of cells with P values close to the mean—a small subpopulation with a heritable small P value, we may expect the following: (i) during stage 1 there will remain a slowly declining number of these particular uncommitted cells (subject, of course, to stochastic variations, and, perhaps, loss); (ii) during stage 2 they will slowly increase in numbers while the overall population growth rate is reduced; (iii) the population will apparently die out in stage 3, but there may remain a few uncommitted cells (with small P value), which will continue to divide, and the population will eventually be regenerated, growing out of the plateau at a rate higher than that of stage 2. This prediction agrees with the observations by Todaro and Green (17), who studied the effect of cell density on senescence and transformation in mouse fibroblast cultures. In particular their observations agree with our predictions that culture size should influence the time taken to reach senescence (phase III) and the proportion of cultures showing spontaneous transformation. In interpreting the results of Todaro and Green, we assume that culture vessels of constant size were used, so that varying cell density inevitably changes population size.

So far we have not considered the molecular events which may accompany the process of commitment and the subsequent long incubation period prior to senescence and death. At the outset we explained that the heterogeneity in growth potential of individual cells is incompatible with the setting up of a clock or program in the population as a whole. It is, however, possible that commitment is the setting of such a program, which then runs through the incubation period to cell death. It has not escaped our notice that the value of P we favor, .275, is close to .25, and that mechanisms may occur which result in the segregation of 25 percent programmed cells. For instance, if, in the replication of homologous chromosomes, one daughter chromatid acquires a heritable but recessive epigenetic switch, such as a specific DNA methylation (18), then on average one in four cells will inherit a maternal and paternal chromosome containing that switch.

A completely different explanation of commitment is based on the error theory of aging. Orgel suggested that errors in protein synthesis may lead to an irreversible increase of errors and a final lethal error catastrophe (19). Later he pointed out that errors may simply reach a steady state (20). It is quite possible that such a steady state is unstable, with a given probability (P) that a feedback of errors will begin and eventually lead, many generations later, to senescence and death. Considerable evidence is available which suggests that aging of fibroblasts is accompanied by alterations or defects in enzymes (21, 22), genes (23), chromosomes (24), DNA replication (22, 25), and repair (26), all of which might be expected if errors in macromolecules are accumulating.

Whatever the basis of commitment, if our theory has validity, it opens up the possibility of selectively isolating uncommitted cells, or cell populations enriched with uncommitted cells. In this way, the population might be propagated indefinitely. This would not only be of intrinsic interest to gerontologists, but it would also facilitate a variety of experimental procedures, not least somatic cell genetics, which are at present impeded by the Hayflick limit.

References and Notes

1. L. Hayflick and P. S. Moorhead, *Exp. Cell Res.* 25, 585 (1961); L. Hayflick, *ibid.* 37, 614 (1965).
2. G. M. Martin, C. A. Sprague, C. Epstein, *Lab. Invest.* 23, 86 (1970).
3. K. V. A. Thompson and R. Holliday, *Exp. Cell Res.* 80, 354 (1973).
4. G. M. Martin, C. A. Sprague, T. H. Norwood, W. R. Pendergrass, *Am. J. Pathol.* 74, 137 (1974).
5. L. Hayflick, *Am. J. Med. Sci.* 265, 433 (1973).
6. J. R. Smith and L. Hayflick, *J. Cell Biol.* 62, 48 (1974).
7. L. Hayflick, personal communication.
8. R. Holliday, *Fed. Proc. Fed. Am. Soc. Exp. Biol.* 34, 51 (1974).
9. T. B. L. Kirkwood and R. Holliday, *J. Theoret. Biol.* 53, 481 (1975).
10. J. P. Jacobs, unpublished observations.
11. R. Holliday, results cited in (9).
12. T. B. L. Kirkwood, in preparation.
13. J. P. Jacobs, C. M. Jones, J. P. Baillie, *Nature (London)* 227, 168 (1970).
14. W. C. Russell, C. Newman, D. H. Williamson, *ibid.* 253, 461 (1975).
15. C. Zavala, P. J. Fialkow, G. Herner, in preparation.
16. D. A. Ogden and H. S. Micklem, *Transplantation* 22, 287 (1976).
17. G. J. Todaro and H. Green, *J. Cell Biol.* 17, 299 (1963).
18. R. Holliday and J. E. Pugh, *Science* 187, 226 (1975).
19. L. E. Orgel, *Proc. Natl. Acad. Sci. U.S.A.* 49, 517 (1963).
20. ———, *ibid.* 67, 1476 (1970).
21. R. Holliday and G. M. Tarrant, *Nature (London)* 238, 26 (1972); C. M. Lewis and G. M. Tarrant, *ibid.* 239, 316 (1972); R. Holliday, J. S. Porterfield, D. D. Gibbs, *ibid.* 248, 762 (1974); S. Goldstein and E. Moerman, *N. Engl. J. Med.* 292, 1305 (1975); *Interdis. Top. Gerontol.* 10, 24 (1976).
22. S. Linn, M. Kairis, R. Holliday, *Proc. Natl. Acad. Sci. U.S.A.* 73, 2818 (1976).
23. S. Fulder and R. Holliday, *Cell* 6, 67 (1975).
24. E. Saksela and P. S. Moorhead, *Proc. Natl. Acad. Sci. U.S.A.* 50, 390 (1963); K. V. A. Thompson and R. Holliday, *Exp. Cell Res.* 96, 1 (1975).
25. T. Petes, R. A. Farber, G. M. Tarrant, R. Holliday, *Nature (London)* 251, 434 (1974).
26. R. Hart and R. B. Setlow, *Mech. Age. Devel.* 5, 67 (1976); M. R. Mattern and P. A. Cerutti, *Nature (London)* 254, 450 (1975).
27. We thank L. Hayflick, P. J. Fialkow, C. Zavala, and G. Herner for allowing us to cite their unpublished observations; R. W. Young and J. M. Sowden for assistance in running the computer simulations; J. P. Jacobs for supplying us with early passage cells of MRC-5; V. M. McGuire for providing human fetal tissue; and H. S. Micklem and L. E. Orgel for helpful comments.

12

Reprinted from *Science* **207**:82–84 (1980)

INTRACLONAL VARIATION IN PROLIFERATIVE POTENTIAL OF HUMAN DIPLOID FIBROBLASTS: STOCHASTIC MECHANISM FOR CELLULAR AGING

James R. Smith and Ronald G. Whitney
W. Alton Jones Cell Science Center,
Lake Placid, New York, 12946

Normal human diploid fibroblasts have a limited life-span in vitro (that is, a finite proliferative potential in culture) (*1*). This potential has been shown to be inversely proportional to the age of the tissue donor (*2*) and has been widely studied as a model for cellular aging. Numerous biochemical and morphological changes occur as these cells reach the end of their proliferative potential (*1*). A number of hypotheses have been proposed to explain this phenomenon; however, direct biochemical analysis of cultures has not yet identified the mechanisms responsible for limiting the proliferative potential of normal human cells.

Many investigators (*3-6*) have reported heterogeneity in morphology, interdivision time, and ability for clonal growth among the individual cells of mass cultures of human diploid fibroblasts. At all stages in the life-span in vitro of human diploid fibroblast cultures (WI-38 cells), wide variation in the proliferative potential of isolated clones was observed (*7*). From these experiments, it was not possible to determine whether variability continues to develop during repeated doubling of the culture in vitro or whether the variation is due to heterogeneity among the individual cells in the tissue of origin. Alternatively, the variation might develop during the process of initiating the cells in culture. A more detailed investigation of this variation would provide some insight into the mechanisms determining proliferative potential. In addition, this information would provide a powerful data base against which to test hypotheses. In the experiments reported here, we found that heterogeneity in proliferative potential appears rapidly within a single clone of human diploid fibroblasts. We also found that the proliferative potentials of the two cells from a single mitotic event can differ by as many as eight population doublings (PD's).

To determine the intraclonal variation in proliferative potential, the distributions of doubling potentials were determined for three sets of subclones isolated from a single clone that was initiated from a culture of human embryonic lung cells (Flow 2000) (*8*) in vitro at PD 23. Two hundred cells were isolated at random from the parent clone at 16, 26, and 36 PD's after its initiation as a single cell (see legend to Fig. 1). Each subclone was observed until the limit of its replicative ability was reached, and frequency distributions of proliferative potential (the number of PD's achieved after isolation) were constructed for each set of subclones (Fig. 1). Cultures were considered to be at the end of their proliferative potential when they failed to double in cell number for 2 weeks. At PD 16, one of the subclones was used to initiate a second set of subclones, and the frequency distribution of their proliferative potential was determined (Fig. 1, inset).

Several characteristics of the distributions are noteworthy: (i) the parent clone was more homogeneous with respect to the proliferative potential of its individual cells than are mass cultures of the same cell type (7); (ii) within only 16 doublings after starting from a single cell, the resulting clone contained cells ranging in doubling potential from 0 to 33 PD's (Fig. 1a); and (iii) the distributions are distinctly bimodal except for the one obtained nearest the end of proliferation of the parent clone (Fig. 1c), where the two modes seem to have merged into a single mode.

In view of the rapid development of intraclonal variability in doubling potential, we considered it important to determine the degree of difference in doubling potential between the two cells arising from a single mitotic event. In Fig. 2, the results of these experiments are presented in terms of the normalized frequency distribution of the proliferative potentials of one of the cells of a mitotic pair whose other cell was able to undergo the indicated number of doublings (each histogram combines the results from several mitotic pairs). It is apparent that the proliferative potential of the two cells from a single mitotic event can differ by as many as eight PD's.

Our results agree with those of studies (3–6) in which heterogeneity in interdivision time and in growth rate within individual clones is shown, support the observation (9) that a large fraction of the cells in a normal culture have a very low proliferation potential and therefore cannot form large clones, and support the finding (10) that a bimodality of colony size results from a 1- to 2-week incubation of repeatedly subcloned human diploid cells and HeLa cells. In addition, we have demonstrated that there is marked variation in the remaining proliferative potential among the cells within single clones of human embryonic lung fibroblasts.

The rapid development of intraclonal variation and the large differences between the products of single mitotic events strongly suggest that a stochastic process determines the doubling potential of human diploid cells. As a clone un-

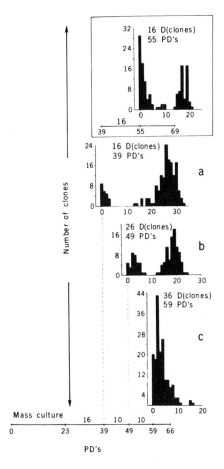

Fig. 1. Intraclonal variation in proliferative potential. Frequency distribution of proliferative potential of subclones isolated from a single clone is shown at (a) 16, (b) 26, and (c) 36 PD's after initiation of the parent clone. The frequency distribution of a secondary subclone is shown in the inset; this set of subclones was isolated from one of the subclones of group (a) 16 PD's after initiation of the primary subclone. The distribution of doubling potential of subclones was determined by the following procedure. Two days after subculture in Eagle's MEM supplemented with 28 mM Hepes buffer and 10 percent fetal bovine serum, a single cell suspension was prepared by trypsinization of the log phase culture. The cells were cultured at low density (1000 to 5000 cells per 60-mm dish) in medium MCDB 102 (16) supplemented with 10 percent fetal bovine serum in tissue culture dishes containing small cover-glass fragments (17). Clones were then isolated and grown as described in (7).

127

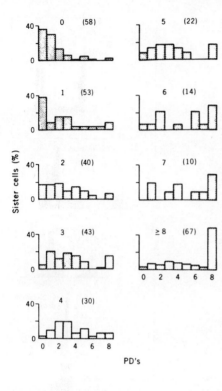

Fig. 2. Variation in proliferative potential between cells arising from single mitotic events. The results are presented as the normalized frequency distributions of doubling potentials of cells whose sister cells were able to undergo the indicated number of doublings. The number of mitotic pairs analyzed are given in parentheses. Cells were cultured in 60-mm dishes containing 5 ml of MCDB 102 or Eagle's MEM in 10 percent fetal bovine serum. There were 50 to 100 cells per dish. The day after seeding, single cells were isolated by cutting narrow grooves (in the form of a square) around them. (Control experiments showed that in 98 of 100 cases, cells were unable to cross the groove cut into the plastic dishes.) During the next 24 hours, the squares were intermittently examined to see if any of the cells within the squares had divided. If a cell had divided, a groove was cut between the two sister cells. After 2 weeks, the dishes were stained in crystal violet (0.5 percent) and the number of doublings achieved by each sister cell was determined by counting with a dissecting microscope. A maximum of 256 cells from either daughter cell of a mitotic pair could be counted with precision, therefore all cells giving rise to 256 or more cells are considered as a single category of ≥ 256 cells.

dergoes doublings in vitro there is a gradual loss in the remaining proliferative potential of its individual cells (the large doubling-potential mode shown in Fig. 1, a to c, decreases), indicating multievent processes. The degree of variability in the doubling potential of the subclones within the large doubling-potential mode is greater than that expected from the variable interdivision time previously observed (11). Furthermore, this variability is probably not due to contact inhibition of cells in the interior portion of the original parent clone, since the proportion of cells incorporating ^{3}H-labeled thymidine during a 24-hour exposure was approximately the same for cells in the center and on the periphery of the clone (70 and 80 percent, respectively). The intrinsic nature of the distribution of cells about the large doubling-potential mode is further argued by the high degree of variability between pairs of cells resulting from one mitosis (Fig. 2). In a series of experiments (12), we found that under the proper clonal conditions, human embryonic lung fibroblasts had the same PD time, proportion of nondividing cells, and proliferative potential when continuously cultured at clonal densities (one to four cells per square centimeter) and at mass-culture densities (1 × 10[4] cells per square centimeter). With respect to the data presented in Fig. 1, this evidence indicates that a subpopulation of cells with very low doubling potential exists in the parental population and must be considered when searching for the mechanisms that limit proliferative potential.

We conclude that two kinds of quantitative events occur in this context: one that results in a gradual decrease in proliferative potential and one that results in an abrupt transition from the larger to the smaller (six or fewer) doubling frequencies; the rate of this transition appears to increase as the proliferative potential of the clone decreases (Fig. 1a and inset).

The information given in this report should be useful for testing hypotheses that seek to explain the limited doubling potential of human diploid cells. It is clear that neither a precise counting mechanism (*13*) nor the commitment theory of aging (*14*) are compatible in their current form with our findings. However, if one considers the involvement of stochastic mechanisms in stem cell differentiation (*15*), our data may be seen as compatible with a process of genetically controlled terminal differentiation, such as has been proposed by Martin *et al.* (*4*) and Bell *et al.* (*5*). The gradual decrease in proliferative potential would also be compatible with a continuous buildup of damage or errors, a process that has been theorized. However, the wide variability in doubling potentials, especially in mitotic pairs, suggests an unequal partitioning of damage or errors at division. The transition of the proliferative potential subset from large to small suggests the occurrence of a more serious species of damage or the crossing of a threshold level of accumulated damage. In terms of somatic mutation theory, the wide variation at each cell division suggests mutation rates (nonlethal but capable of affecting proliferative potential) approaching one per cell division. Clearly, all current hypotheses should be reexamined in the light of our data.

References and Notes

1. L. Hayflick, in *Handbook of the Biology of Aging*, C. E. Finch and L. Hayflick, Eds. (Van Nostrand Reinhold, New York, 1977), p. 159.
2. G. M. Martin, C. A. Sprague, C. J. Epstein, *Lab. Invest.* **23**, 86 (1970); Y. LeGuilly, M. Simon, P. Lenoir, M. Bouree, *Gerontologia* **19**, 303 (1973); E. L. Schneider and Y. Mitsui, *Proc. Natl. Acad. Sci. U.S.A.* **73**, 3584 (1976).
3. P. M. Absher, R. G. Absher, W. K. Barnes, *Exp. Cell Res.* **88**, 94 (1974).
4. G. M. Martin *et al.*, *Am. J. Pathol.* **74**, 137 (1974).
5. E. Bell, L. F. Marek, D. S. Levinstone, C. Merrill, S. Sher, I. T. Young, M. Eden, *Science* **202**, 1158 (1978).
6. G. S. Merz and J. D. Ross, *J. Cell. Physiol.* **74**, 219 (1969); V. J. Cristofalo and B. B. Sharf, *Exp. Cell Res.* **76**, 419 (1973).
7. J. R. Smith and L. Hayflick, *J. Cell Biol.* **62**, 48 (1974).
8. Flow 2000 cells, human embryonic lung fibroblast-like cells, were obtained from Flow Laboratories at population-doubling level 18.
9. J. R. Smith, O. M. Pereira-Smith, E. L. Schneider, *Proc. Natl. Acad. Sci. U.S.A.* **75**, 1353 (1978); J. R. Smith, O. Pereira-Smith, P. I. Good, *Mech. Ageing Dev.* **6**, 283 (1977).
10. G. M. Martin, C. A. Sprague, T. H. Norwood, W. R. Pendergrass, *Am. J. Pathol.* **74**, 137 (1974); A. O. Martinez, T. H. Norwood, J. W. Prothero, G. M. Martin, *In Vitro* **14**, 996 (1978).
11. Based on the data reported by Absher *et al.* (*3*) for a clone with the same average interdivision time as the one we studied (doubling every 20 hours), we calculated that variable interdivision time alone would result in a doubling-potential distribution with a standard deviation of 0.80 PD. This is considerably smaller than the standard deviation of 3.7 PD we observed.
12. J. R. Smith and K. I. Braunschweiger, *J. Cell. Physiol.* **98**, 597 (1979).
13. P. I. Good and J. R. Smith, *Biophys. J.* **14**, 811 (1974).
14. T. B. L. Kirkwood and R. Holliday, *J. Theor. Biol.* **53**, 481 (1975).
15. J. E. Till, E. A. McCulloch, L. Siminovitch, *Proc. Natl. Acad. Sci. U.S.A.* **51**, 29 (1964).
16. R. G. Ham, S. L. Hammond, L. L. Miller, *In Vitro* **13**, 1 (1977).
17. G. M. Martin and A. Taun, *Proc. Soc. Exp. Biol. Med.* **123**, 138 (1966).
18. Supported by NIH grant R01-AG00338 and the W. Alton Jones Foundation.

17 July 1979; revised 10 October 1979

13

Reprinted from Science **221**:964–966 (1983)

EVIDENCE FOR THE RECESSIVE NATURE OF
CELLULAR IMMORTALITY

O. M. Pereira-Smith and J. R. Smith

Abstract. *Fusion of immortal cell lines with normal human fibroblasts or certain other immortal cell lines yields hybrids having limited division potential. Cellular immortality was found to be a recessive phenotype in hybrids. It was also found that at least two separate events in the normal cell genome can result in immortality. In fusions involving certain immortal parent cells, these events can be complemented to result in hybrids with finite division capacity.*

The phenotype of limited division of the normal human cell has been reported to be dominant over the immortal phenotype of HeLa (*1*) and SV40-transformed cells (*2, 3*) in hybrids. We now report that hybrids resulting from the fusion of diploid human fibroblasts with several other immortal cell lines also have limited division potential. Our results suggest that cellular immortality is a result of recessive dysfunctions or alterations in the genetic program that limits the division of normal cells. If immortality can occur via more than one set of dysfunctions or alterations in the normal cell program, it should be possible to identify "complementation groups" of immortal cells, which when fused with each other would yield hybrids having finite division potential. We have found that fusion of certain types of immortal cells result in immortal hybrids, whereas fusion of other combinations of immortal cell lines yield hybrids with finite division capacity. These results suggest that there are at least two complementation groups for cellular immortality.

Somatic cell hybridization has been used extensively for analysis of the dominant versus recessive nature of the phenotypes of tumorigenicity and transformation. The results are controversial since the phenotypes have been suppressed in some hybrids (*4*) and expressed in others (*5*). The transformation phenotypes that have been studied include morphology, density-dependent inhibition of growth, the requirement for serum, and anchorage-independent growth in soft agar and methyl cellulose. The phenotype of immortality (that is, the capacity for indefinite division in culture) has rarely been considered or studied. Nevertheless, the notion that cellular immortality is a dominant trait has been accepted. This idea has been based primarily on the observation that in fusions of various species of normal cells with immortal cells, it is possible to obtain hybrids with indefinite proliferative potential (*4*). Since the objectives (for example, chromosomal analysis, tumorigenicity, or anchorage-independent growth) of previous experiments required extensively proliferating hybrids, the methodology used for hybrid isolation restricted analysis to such hybrid clones. Scant attention was paid to the overwhelming majority of hybrid cells that ceased to proliferate. In the few studies (*1–3*) in which careful attention was given to the proliferative behavior of hybrids, the fusion of normal human cells with the immortal cell lines HeLa or SV40-transformed human fibroblasts resulted in hybrids having finite division potential. These results indicated that the phenotype of immortality was recessive in hybrids. Variant immortal cells occurred in some of the nonproliferating hybrid populations at a frequency of about one in 10^5 cells (*1, 3*). These variant cells probably account for the widespread belief that such cell fusions do not yield hybrid cells with finite division capacity.

We have analyzed the proliferative potential of hybrids from various fusions involving normal, virally transformed, and tumor-derived human cells. The technique we use for hybrid isolation (*3, 6*) allows us to analyze hybrids having a small proliferative potential (fewer than eight population doublings) and a more extensive proliferative potential (more than eight population doublings). We found that the phenotype of limited division is dominant in hybrids obtained after fusion of normal human cells with other human cells, whether the other cells are normal (*6*) or are derived from immortal tumor or immortal SV40-transformed cells (*3*). In the fusions of normal cells with SV40-transformed cells we had found earlier that the hybrids having limited division capacity expressed T antigen (*3*). Using ^{32}P-labeled SV40 DNA to probe Southern blots of Eco RI–cleaved DNA isolated from these hybrids, we found that the viral DNA band patterns in the hybrids and the SV40-transformed parent used for fusion are the same. Thus, the phenotype of limited cell division of the normal cell is dominant over immortality even in the presence of a stably integrated viral genome that is being expressed.

Table 1 summarizes the results obtained from fusion of normal human cells with various immortal human cell lines. The normal cells had a proliferative potential of 12 to 17 population doublings remaining at the time of fusion. The most significant result is that all hybrids analyzed ceased dividing, even after proliferating through approximately 80 population doublings. Since we used immortal cell lines that varied greatly in tissue of origin as well as the process of immortalization (tumor-derived versus virally transformed), it appears that the recessive nature of the phenotype of immortality may be a general phenomenon. Rare variant immortal cells did occur during the period of division cessation in some hybrids, at very low frequencies of about one in 10^5 cells (*7*).

To test whether the fusion of various immortal cell lines might result in complementation to yield hybrids with limited division potential, we fused an SV40-transformed human fibroblast line with various other immortal human cell lines (Table 2). When a oubain-resistant, thioguanine-resistant (O^RTG^R) mutant SV40-transformed cell line was fused with the cell line from which it was derived or with other independently derived SV40-transformed lines, there was no complementation to yield hybrids with limited life-span. All hybrids analyzed could divide indefinitely. This indicates that in the three SV40-transformed cell lines we have studied, immortality is the result of either the same or noncomplementary genetic defects. In all the other fusions, most of the hybrid clones were capable of very limited division (fewer than eight population doublings). We feel confident that these small clones are hybrid clones and not contact feeding colonies because the number of cells inoculated is low (ten cells per square centimeter), and no clones were obtained on the control dishes. Two controls were used (*3*, *6*): parent cells that were mixed and not treated with polyethylene glycol and parent cells individually treated with polyethylene glycol and mixed before inoculation into selective medium. However, since highly aneuploid parents were used in the fusions, the possibility exists that some events

other than true complementation resulted in hybrids with low proliferative potentials. We therefore base our major conclusions on the results obtained with extensively proliferating hybrid clones (that is, those achieving at least 15 population doublings). When HT 1080 or its subclone were fused with SV40-transformed human fibroblasts, the hybrids that achieved more than eight population doublings could divide indefinitely (more than 100 population doublings) and exhibited no complementation for mortality. In all other fusions with SV40-transformed human fibroblasts, there was complementation, and division ceased in all hybrid clones analyzed. In the fusions of HeLa and T98G cells with SV40-transformed cells there was a resumption of division in some instances after foci of dividing cells appeared in the hybrids (*7*).

Fusion with 143B TK$^-$ (thymidine kinase–deficient) cells yielded no clones capable of indefinite proliferation. All clones that had limited division potential showed positive staining for SV40 T antigen by indirect immunofluorescence.

Since the immortal phenotype of 143B TK$^-$ was recessive in fusion with normal cells, and since 143B TK$^-$ yielded hybrids of limited life-span when fused with SV40-transformed cells, we fused 143B TK$^-$ to a thioguanine-resistant mutant of HT 1080, which gave immortal hybrids when fused with SV40-transformed cells. In this fusion experiment (Table 2), three of the five hybrid clones that could be subcultured ceased dividing, and division was not resumed. The other two hybrid clones were able to attain more than 100 population doublings and are therefore considered immortal.

Table 1. Fusion of human diploid fibroblasts with various immortal cell lines. The immortal cell lines used were SV40-transformed fibroblasts GM 639, VA 13, and GM 847 O^R; cervical carcinoma HeLa; fibrosarcomas HT 1080 and 1080 21A (a clone of HT 1080); glioblastoma T98G; and the Kirsten mouse sarcoma virus–transformed osteosarcoma 143B TK$^-$. The percentage of the immortal cell lines able to produce clones with more than eight population doublings (PD) during a 2-week incubation is shown in parentheses. Extensively proliferating hybrids (more than 15 PD) were examined for immortal foci.

Immortal cell line	Human diploid fibroblast	Number of experiments	Hybrids with PD ≥ 8 (%)	Extensively proliferating hybrids	
				PD range before division ceased	Ratio of clones with immortal foci to total clones
GM 639 (90)	GM 1662 O^R	6	30	16 to 64	9/18
VA 13 (90)	GM 1662 O^R	4	30	16 to 69	5/10
GM 847 O^R (60)	CSC 303 G	2	40	16 to 78	10/20
HeLa (75)	GM 1662 O^R	2	20	16 to 21*	18/89*
HT 1080 (70)	GM 1662 O^R	4	7	16, 22	1/2
1080 21A (60)	GM 1662 O^R	2	10	21 to 32	3/4
T98G (85)	GM 1662 O^R	3	4	17 to 64	2/6
143B TK$^-$ (95)	GM 1662 O^R	2	7	16 to 26	0/6

*Data from Bunn and Tarrant (*1*).

Table 2. Fusions of various immortal cell lines. GM 639 O$^RTG^R$ and GM 847 O^R are SV40-transformed human skin fibroblasts. The percentage of cells in immortal cell lines that are able to produce clones with more than eight population doublings (PD) during a 2-week incubation is shown in parentheses.

Immortal cell line 1	Immortal cell line 2	Hybrids with PD ≥ 8 (%)	Extensively proliferating hybrids				Ratio of clones with immortal foci to total clones
			Number of clones achieving given PD range before division ceased				
			15–19	20–29	30–39	40–70	
GM 639 (90)	GM 639 O$^RTG^R$ (80)	100					12/12
VA 13 (90)	GM 639 O$^RTG^R$	100					12/12
VA 13 (90)	GM 847 O^R (50)	100					12/12
GM 639 (90)	GM 847 O^R	100					12/12
HeLa (75)	GM 639 O$^RTG^R$	18	5	3	2	1	2/11
HT 1080 (70)	GM 639 O$^RTG^R$	10					4/4
1080 21A (60)	GM 639 O$^RTG^R$	20					8/8
T98G (85)	GM 639 O$^RTG^R$	20	1	8	3		6/12
143B TK$^-$ (95)	GM 639 O$^RTG^R$	6	4	5		2	0/11
HT 1080 TG^R (60)	143B TK$^-$ (95)	6	2			1	2/5

Since the fusions are intraspecies, and in many cases involve highly aneuploid lines, it is not possible to perform detailed cytogenetic analyses. However, comparisons of total chromosome counts of hybrid clones with those of the parental cell lines indicate varying degrees of chromosome loss in the hybrids. The median chromosome number in the hybrids from the GM 639 × T98G and GM 639 × GM 639 fusions is very nearly the same as the sum of the median number for each of the parental cell lines. In contrast, the median chromosome number of hybrids from fusions of GM 639 with HeLa, HT 1080 21A, and HT 1080 is 15 to 30 less than the sum of the median chromosome numbers of the parents, but in all cases larger than the number for either parent. Thus, it appears that there is no correlation between the extent of chromosome loss and the proliferative potential of the hybrids. In addition, the hybrids are positive for T antigen and therefore retain chromosome 5 of the SV40-transformed parent to which the viral genome has been mapped (8).

We conclude that the phenotype of immortality is recessive in hybrids with normal human cells and that complementation between immortal parent cells can result in hybrids with limited division potential. The data indicate that changes in two or more different events (or sets of events) occurring in the genetic program that limits the division of normal cells can result in immortality. The changes in the normal cell genome that lead to SV40 virus–mediated immortality are different from those occurring in the tumor-derived cell lines studied, with the possible exception of HT 1080 and its subclone; changes leading to immortality in HT 1080 may be the same as those mediated by SV40 virus. The fact that hybrids formed from fusion of 143B TK⁻ cells with SV40-transformed or HT 1080 cells showed limited proliferation is compatible with this possibility.

Our results support the hypothesis that limited proliferation is a result of a rigorously programmed series of events and that immortality is caused by certain changes—recessive in hybrids—in these events. The results argue strongly against the hypotheses that errors in the protein-synthesizing machinery of cells or recessive mutations are responsible for limited division in normal cells (9). The hypotheses suggest that the probability of these events occurring is decreased or eliminated in immortal cells. If this were true, fusion of immortal cell lines should yield only immortal hybrids, which we have experimentally found not

to be the case. Smith and Lumpkin (10) proposed a hypothesis and model involving changes in gene expression to explain the mechanisms that limit the division potential of normal human cells in vitro. On the basis of this model, one would expect that cells could become immortal by at least two different mechanisms. The results we have described are compatible with this model.

References and Notes

1. C. L. Bunn and G. M. Tarrant, *Exp. Cell Res.* **127**, 385 (1980).
2. A. L. Muggleton-Harris and D. W. DeSimone, *Somat. Cell Genet.* **6**, 689 (1980).
3. O. M. Pereira-Smith and J. R. Smith, *ibid.* **7**, 411 (1981).
4. H. Harris, *ibid.* **5**, 923 (1979); E. J. Stanbridge *et al.*, *Science* **215**, 252 (1982); C. J. Marshall and H. Dave, *J. Cell Sci.* **33**, 171 (1978); C. Szpirer and J. Szpirer, *Exp. Cell Res.* **125**, 305 (1980).
5. K. K. Jha, J. Cacciapuoti, H. L. Ozer, *J. Cell. Physiol.* **97**, 147 (1978); C. M. Croce, *Biochim. Biophys. Acta* **605**, 411 (1980).
6. O. M. Pereira-Smith and J. R. Smith, *Somat. Cell Genet.* **8**, 731 (1982).
7. The hybrid cultures were grown in 25-cm² tissue culture flasks. At each subculture the inoculum was 1.25 × 10⁵ cells. Thus, when the hybrids entered the period of slow growth or division cessation, each flask contained about 1 × 10⁵ cells. When foci of dividing cells appeared, there were generally one or two per flask. We therefore estimated the frequency of this occurrence as one in 10⁵ cells. The cultures in which such foci appeared were subcultured through an additional 100 population divisions, at which time we assumed the culture would divide indefinitely.
8. H. Sheam-Pey and R. Kucherlapati, *Virology* **105**, 196 (1980).
9. S. Ohno and Y. Nagai, *Birth Defects* **14**, 501 (1978); L. E. Orgel, *Proc. Natl. Acad. Sci. U.S.A.* **49**, 517 (1963).
10. J. R. Smith and C. K. Lumpkin, *Mech. Age. Dev.* **13**, 387 (1980).
11. We thank R. Kucherlapati and R. Davies for help with Southern blot analysis. This work was supported by NIH grant AG 03262 and the W. Alton Jones Foundation.

24 January 1983; revised 4 April 1983

14

Copyright © 1984 by S. Karger AG, Basel

Reprinted from *Monogr. Devel. Biol.* **17**:60–77 (1984)

The Unsolved Problem of Cellular Ageing

Robin Holliday

National Institute for Medical Research, London, England

Introduction

Thirty years ago, *Medawar* [43] published a review on the ageing of animals under the title 'An unsolved problem in biology'. At that time the study of the ageing of cells or populations of cells in the laboratory had not been initiated, but in the ensuing 30 years, it has become possible to carry out experiments on cells in culture, particularly those derived from human tissue or chick embryos, as well as with several protozoa and fungi.

Following the early work on cultured chick cells by *Carrel* and his associates [for a review, see 75] and the establishment of permanent cell lines from human tumours, it was widely believed that all vertebrate cells capable of division could be propagated indefinitely in culture. This view was challenged by *Hayflick and Moorhead* [14], following the earlier demonstration by *Swim and Parker* [68] that there is a limit to the serial subculture of human connective tissue cells, generally referred to as fibroblasts. *Hayflick* [12] fully documented this phenomenon by carrying out a series of control experiments, which established for the first time that the finite growth of human diploid fibroblasts was intrinsic to the cells themselves and could not be readily explained by environmental factors, such as faulty medium, or the presence of infectious agents. He proposed that these intrinsic changes in fibroblasts are a manifestation of ageing at the cellular level, an interpretation which has been widely, although not universally, accepted. He also clarified some of the basic differences between heteroploid transformed cells, which grow indefinitely, and diploid cells which do not.

The results of *Carrel* with chick cells have never been substantiated [75], and many subsequent studies have demonstrated that chick embryo fibroblasts have an even shorter growth potential in culture than human cells. A remarkable feature of populations of human and chick fibroblasts is the apparent absence of any subpopulation of transformed cells, since if any of these arose during serial subculture, they would survive the senescence of the diploid cells and form permanent lines. Such permanent lines can occasionally be obtained from human diploid cell populations treated with SV40 virus [26], but they consistently emerge from untreated rodent or rabbit cell populations. Three major problems arise from these studies. First, what are the molecular mechanisms which limit the proliferation of cultures of fibroblasts? Second, what is the basis of the difference in growth potential of diploid and transformed cells? Third, what genetic or other factors determine the longevity of cells from different species and the likelihood of spontaneous transformation?

None of these problems has yet been solved. In this critical review I assess some of the theoretical and experimental advances which have so far been made, with particular reference to the studies with human fetal lung fibroblast strain MRC-5 in this laboratory in the last 10 years.

Validity of Hayflick's Experimental Model

Although the finite growth of human fibroblasts has been demonstrated on innumerable occasions in many laboratories, there is continuing controversy surrounding the interpretation that it is due to cellular ageing. *Martin* et al. [41] and *Bell* et al. [1] have proposed that the cessation of growth is due to a process of terminal differentiation in which the cells enter a non-cycling state. There are several reasons for believing this interpretation to be incorrect. Firstly, the fibroblast is already a differentiated connective tissue cell capable of collagen synthesis. In normal development, differentiation is indeed terminal and it is not observed that one type of differentiated cell changes into another. Second, examination of proteins by two-dimensional gels have not revealed other than slight changes in the pattern, although newly differentiated cells might be expected to synthesize significant amounts of one or more new 'luxury' proteins [58]. Third, there is no evidence that late passage cells adopt a uniform, distinct phenotype. Instead, the cells vary considerably in size, shape, number of nuclei and degree of granularity. Their structure deteriorates, and they frequently round up and detach from the surface. Fourth, premature ageing syndromes, or those with karyotype instability or defects in DNA repair, strikingly reduce the longevity of cultured fibroblasts [40, 72, 73]. However, these individuals show, for the most part, normal development and it is hard to see why their cells in culture should be strikingly altered in temporal differentiation. Finally, one might ask why there should be such a long period of normal proliferation (50–70 population doublings) before differentiation takes place. Cultured cells which are known to be predetermined to differentiate, such as epidermal keratinocytes, usually have only limited growth in culture [67].

Some cell biologists have maintained (often in open discussion, rather than in print) that the finite growth of diploid fibroblasts is merely an artifact of laboratory culture conditions, and that if an ideal medium could be found, the cells would grow indefinitely. However, this does not mean that the cells do not age under the conditions which have so far been used; rather it implies that the cells are in a state of 'unbalanced growth'. Unbalanced growth, or the gradual breakdown of the normal cellular homeostatic mechanisms, is itself a valid theory of ageing. As *Hayflick and Moorhead* [14] pointed out long ago, the limited growth could not be due to the simple dilution of an essential cellular constituent, but it might be due to a progressive imbalance in cellular constituents. It is possible, for instance, that the formation of certain organelles does not keep pace with cell division, so that they slowly become depleted [51]. Alternatively, they may be overproduced, until finally there is a lethal imbalance. In *Podospora*, there is evidence that cellular senescence is associated with the proliferation of mitochondria with abnormal DNA sequences [74], but so far, there is no evidence that the same changes occur in cultured vertebrate cells. However, there is some evidence that changes in certain nuclear DNA sequences occur during in vitro ageing [60]. This raises the possibility that ageing could be due to a failure of normal DNA maintenance mechanisms. These suppositions are now testable, using the new procedures of DNA technology. However, even if positive results were obtained, the problem of the origin of these changes and why they occur in some cells and not others would remain unresolved. In the same way, the discovery of the changes in mitochondrial DNA of *Podospora* do not reveal whether this is a cause of cellular ageing or a consequence of some other macromolecular change.

Ageing of Cells and Populations

It is very important to distinguish between the death of individual cells and of whole populations. It was first pointed out by *Orgel* [49] that all cell lineages are mortal. He considered an experiment in which there is no cellular selection: a cell is allowed to divide, a daughter is picked at random and allowed to divide again, and so on, indefinitely. Eventually a dead cell will be obtained. This cell is not necessarily senescent or aged, it may have died, for instance, as a result of a dominant lethal mutation, the probability of which may be constant with time. Cellular ageing must be defined as a decrease in the probability of survival with time. This is manifested in cell lineages of fibroblasts and also populations where all cell lineages die out. However, it is quite possible to have immortal populations which contain certain ageing cell lineages, but which grow indefinitely as a result of the selection of potentially immortal cells. I will return to this question in a later section.

In a remarkable series of experiments on the longevity of clones of human fibroblasts, *Smith* and his associates [64–66] have demonstrated the heterogeneity of growth potential of individual cells. Moreover, daughters from one individual cell also vary greatly in their proliferative potential, thus demonstrating the stochastic, or non-deterministic nature of the ageing process. In addition, it has been demonstrated with fetal lung strains, MRC-5 and WI-38, that parallel populations (set up from cells stored in liquid nitrogen at early passage) vary considerably in the number of passages or population doublings achieved before growth ceases. *Thompson and Holliday* [69] found that 24 parallel cultures of MRC-5 had an average life span of 57 passages, with a standard deviation of 7.5 passages. (It should be noted that a difference of 10 passages represents 2^{10}, or a 1,000-fold range in growth potential.) It is also clear that the ageing of animals depends at least in part on stochastic processes. Populations of inbred laboratory mice in a uniform environment vary considerably in their longevity. For example, in experimental studies on the ageing of the long-lived CBA strain, *Holliday and Stevens* [23] found that the life span of 35 female control mice was 901 days with a standard deviation of 157 days.

Programmed Ageing

The view that fibroblasts are predetermined to differentiate is, of course, closely related to the hypothesis that the ageing of these cells is programmed. The molecular or cellular basis of such a programme has never been precisely formulated, but presumably a cell division counting mechanism is envisaged: when the clock runs out, cell proliferation ceases. It is not obvious how such a programme could explain the stochastic features of ageing which were previously mentioned, or the strong effect of environmental factors, such as temperature [69]. Nor is it clear why the programme would be altered by mutations which effect, for instance, DNA repair capacity. Most important, it is very hard to explain why a programme for ageing should evolve in the first place [33]. Programmed ageing of cells would not be advantageous and may reduce an organism's fitness, in relation to individuals which do not have such a programme.

Nevertheless, the view that ageing is in some way programmed is widely held, and it is capable, in principle, of explaining the difference in growth potential between diploid and transformed cells, if the 'immortalization' of the latter is due to the destruction or bypass of the programme. Testing the hypothesis is difficult because it has never been formulated in precise terms. Nevertheless, if genes for immortalization exist, then an obvious approach would be to try to isolate them by recently developed methods of genetic manipulation and identify the physiological role of their products.

Genetic Control of Ageing and Its Genetic Consequences

Since animal species have a wide range of maximum longevities, there can be no doubt that ageing is genetically determined. The correlation between the life span of an organism and the growth potential of cultured fibroblasts holds for man, chicken and mouse [13] and probably several other species [56]. However, it is often mistakenly supposed that the genetic control of the rate of ageing in vitro or in vivo, implies that ageing must be programmed. This is incorrect, since genetic factors are responsible for a variety of cellular processes which may influence ageing, including the accuracy of protein synthesis, protein turnover, the efficiency of DNA repair, and the activity of enzymes, such as superoxide dismutase, which remove free radicals.

Mutations which influence the rate of ageing provide a powerful means of investigating primary causes. In man, the rare autosomal recessive Werner's syndrome has a pleiotropic effect in accelerating the onset of a variety of the normal symptoms of ageing, and also greatly reduces the growth potential of cultured skin fibroblasts [40, 72]. Although it is a matter of debate whether the mutation really induces premature ageing in toto [39], it would be very surprising if the identification of the nature of the biochemical or metabolic defect did not provide very important information about the origins of some ageing processes, both in vitro and in vivo.

Although the role of somatic mutations in ageing has frequently been discussed, both theoretical arguments and the experimental evidence strongly suggest that these are more likely to be a consequence rather than a cause of ageing [19]. *Thompson and Holliday* [71] and *Hoen* et al. [15] have shown that tetraploid human fibroblasts have the same longevity as diploid ones, which would not be predicted if deleterious recessive mutants played an important role in ageing. The same conclusions can be drawn from longevity studies of animals of different ploidy [42].

There have been few successful attempts to actually measure the frequency of mutations during ageing. *Fulder and Holliday* [8] screened MRC-5 fibroblast populations for rare variants with a significantly enhanced level of glucose-6-phosphate dehydrogenase, which could have arisen from mutation, and showed that their frequency increased exponentially with culture age [7]. *Morley* et al. [44] measured the frequency of human T-lymphocytes which were resistant to 6-thioguanine from donors of different age, and found a strong correlation between this frequency and age. The data were more consistent with an exponential, than a linear increase. Subsequently, *Morley* et al. [45] have cultured human lymphocytes and have demonstrated that the 6-TG resistant cells are indeed mutants. It had previously been shown that chromosome abnormalities increase sharply in the lymphocytes of old individuals [28] and also in late passage human fibroblasts [59, 70]. Although much more information is needed and better methods are required to screen genetic damage during ageing, the evidence to date rather strongly suggests that the ageing process itself may be mutagenic.

Error Propagation

Orgel [49] first pointed out that the transcription and translation apparatus is potentially unstable. Some errors in the synthesis of proteins may feed back into the highly specific pathway of information transfer from DNA to protein, thereby producing additional errors over and above the intrinsic error level. He later pointed out that, depending on the degree of feedback, errors may simply stabilize at a steady-state level, or they may increase with time, with or without cell division, until a lethal error catastrophe is reached [50]. Although *Hoffman* [16] attempted to demonstrate that error catastrophes were unlikely to occur in biological systems, *Kirkwood and Holliday*

[31] showed that *Hoffman's* [16] model was based on false assumptions and that *Orgel's* [49] original formulation was essentially correct. That is, cells could be in a steady state and therefore potentially immortal, or in a meta-stable condition, with a given probability of moving into an unstable one. This provides a possible theory of cellular ageing with the following basic features: (1) The initial changes would be cytoplasmic rather than nuclear, since the errors are in proteins. (2) The increase in errors with time would be essentially exponential; thus the observable or measurable phenotypic effects would accumulate rapidly towards the end of the life span. (3) A variety of secondary consequences would be expected from a build up of errors in proteins [51], for example, errors in DNA replication, defects in membranes and imperfect cation transport, progressive abnormalities in organelles, such as mitochondria and lysosomes. (4) The senescent cells should have a dominant effect in the heterokaryon or hybrid, since the 'young' machinery for protein synthesis would become rapidly contaminated with faulty components from senescent cytoplasm. Clear evidence that hybrids selected from young and senescent fibroblasts have very limited growth has recently been obtained by *Pereira-Smith and Smith* [52].

Although the theory of error propagation rests on a sound basis [for a review, see 30], it should be emphasized that it makes no predictions about the level of errors which might be lethal to cells or organisms, and it is possible that the error catastrophe may only affect a small minority of molecules. Nor, in the absence of exact information, is it possible to make meaningful predictions about the rate of increase of errors. It is obviously essential to discover whether error catastrophes can actually occur in biological systems. The first fairly direct evidence came from studies with the mutant leu-5 with *Neurospora crassa*. The phenotype of this strain strongly indicates that it has an altered leucyl tRNA synthetase, with reduced specificity, with the result that it incorporates incorrect amino acids at leucine codons, especially at elevated temperature [54]. The mutant grows continuously at 25 °C, but when shifted to 35 or 37 °C it continues to grow at a constant rate for approximately 3 days and then dies. *Lewis and Holliday* [35] followed the effects on the enzyme glutamic dehydrogenase during this 3-day period and showed that the ratio of inactive cross-reacting material to active enzyme increased dramatically during the last 12 h. The results strongly suggested that the shift in temperature caused an initial increase in protein errors, which was followed 2–3 days later by a secondary effect due to protein error feedback.

In experiments with *Escherichia coli*, conflicting results were initially obtained. *Edelmann and Gallant* [4] and *Gallant and Palmer* [9] claimed to have increased errors in protein synthesis 50-fold by adding streptomycin to the medium, but they observed very little effect on growth and no 'error catastrophe'. On the other hand, *Branscomb and Galas* [2] had previously shown that low concentrations of streptomycin could result in the progressive synthesis of altered β-galactosidase, as judged by its heat lability, together with the cessation of cell division. These experiments have been extended by *Rosenberger* [57], who explored the conditions under which streptomycin affected growth and viability, and has developed, in effect, a model system for studying the finite growth of populations of *E. coli*. At appropriate concentrations of streptomycin, cells continued to grow normally for up to 20 generations, but during this time protein errors increased exponentially, as judged by the suppression of a nonsense codon of β-galactosidase. When growth ceases, there is massive cell death. The fact that populations of *E. coli* can in one environment grow indefinitely, whilst in another they show behaviour very comparable to the Hayflick limit to fibroblast growth, raises the possibility that immortal and mortal populations of animal cells depend on intrinsic differences in error propagation.

Attempts to Measure Error Frequencies in Mammalian Cells

The protein error theory of ageing makes many predictions, which in principle are testable. In practice, it has proved to be very difficult to devise adequate tests. Several papers have been published which purport to disprove the theory, but close examination of the methods used shows that they are inadequate. The basic requirement is to first measure the intrinsic, or spontaneous, level of errors of protein synthesis in normal young cells. Once this is done, the same methods can then be used to measure changes in error levels, if any, during the process of ageing.

Harley et al. [10, 11] and *Wojtyk and Goldstein* [76, 77] have recently claimed that protein errors do not increase during the in vitro ageing of human fibroblasts. In one series of experiments the misincorporation of leucine into polyphenylalanine was measured using cell-free extracts from cultures of different age with poly-U as synthetic message [76, 77]. The percentage of leucine misincorporation (0.03–0.4) was much lower than has been observed in other laboratories, and this was probably due to the addition of suboptimal concentrations of the two amino acids. Indeed, the poly-U system yields widely different error levels with eukaryotic ribosomes, depending on the experimental conditions used [for a review, see 34]. *Wojtyk and Goldstein*

[77] reported that error frequencies actually *decline* during ageing, but as these results are based on an 800-fold range in the rate of incorporation of phenylalanine by cell-free extracts, they cannot be taken seriously. *Buchanan* et al. [3] optimized the experimental conditions and found the amount of leucine misincorporation is about 1% in cell-free extracts from young or senescent fibroblasts. This is, of course, much higher than the in vivo level of protein errors [5], and it is therefore not altogether surprising that any differences between young and old cells would be obscured by what is, in effect, an in vitro artifact. In a second series of experiments, two-dimensional gels were used to look for small subfractions of protein molecules with altered charge [10]. None were seen in protein extracts from senescent cells. Since the method does not detect any errors in proteins from young cells, it is not possible to know whether or not the error level changes in old ones. The authors also measured error levels in cells treated with histidinol. This analogue blocks the normal charging of hystidyl tRNA and from earlier studies with bacterial cells it is known that it results in the misincorporation of wrong amino acids (particularly glutamine) at histidine codons. Under these conditions of histidine starvation, old and young cells produced roughly the same proportion of altered protein, from which the authors conclude that the *spontaneous* level of errors is the same! The theory [11] on which this latter conclusion is based is very simplistic, in that it assumes that the levels of histidyl tRNA in young and old cells are the same under all conditions and also that ribosomes from old cells, if ambiguous, would have increased affinity for non-cognate amino acids but unaltered affinity for cognate amino acids. Neither of these assumptions are justified.

Much stronger evidence against the error theory comes from experiments with viruses. *Holland* et al. [17] infected young and old populations of fibroblasts with vesicular stomatitis, herpes simplex or polioviruses and found no specific differences in yield, heat stability, or, in the case of poliovirus, mutation frequency. This suggests that the pathways of macromolecule information transfer are unimpaired during senescence. However, nothing is known about the effects of cellular errors on these viruses, and it would have been advantageous to find out if treatment of cells with RNA base or amino acid analogues had any measurable effects. *Fulder* [6] examined the reversion frequency of three ts mutants of herpes simplex virus which were grown in young and senescent MRC-5 and found that in one case, reversion frequency was elevated in senescent cells; in another, it was reduced and in a third, it was unchanged. Experiments with virus probes clearly need to be interpreted with great caution.

Considerable indirect evidence in favour of the error theory has accumulated over the last 10 years, particularly from studies with MRC-5 in this laboratory. DNA polymerase α was chosen for detailed investigations of fidelity, because it is a key enzyme in information transfer. It is known that amino acid substitutions reduce the accuracy of DNA polymerase in bacteriophage T4 [48], so a prediction of the theory would be that the enzyme from senescent cells should have reduced accuracy in an appropriate in vitro assay system. This was first tested by *Linn* et al. [37], who found that polymerase α from senescent cells was several times less accurate than the same enzyme from young cells. Much more detailed studies were subsequently carried out by *Murray and Holliday* [47] with enzyme from populations of cells of increasing age. Several template primer systems were used and misincorporation of an incorrect deoxynucleotide triphosphate was measured in the presence of Mg^{++} or Mn^{++}. A variety of control experiments established that the errors seen were indeed due to mistakes made by DNA polymerase α. In addition, evidence was obtained that DNA polymerase γ also lost fidelity during cellular ageing.

In earlier studies, the heat stability of glucose-6-phosphate and 6-phosphogluconate dehydrogenases (G-6-PD and 6-PGD) had been measured in cell-free extracts from MRC-5 cultures, and it was found that late passage cultures always contained a significantly increased fraction of heat-labile enzyme [24]. Although this result has been confirmed by several other laboratories, the interpretation has remained controversial [for a review, see 25]. The results are, however, fully consistent with the possibility that the misincorporation of amino acids increases in senescent cultures, especially as it is known that naturally occurring variants of G-6-PD (new alleles arising from amino acid substitutions) are often heat-labile, and that agents which are known to reduce the fidelity of protein synthesis, such as 5-fluorouracil (5-FU) or paromomycin (Pm) [3], increase the proportion of heat-labile enzyme [22, 24, 25].

Pm has also recently been shown to accelerate many of the normal features of ageing [22]. MRC-5 cells grown in the presence of the antibiotic are unaffected for many generations of growth, but then adopt the morphological characteristics of senescence much sooner than control cultures, including the accumulation of autofluorescent 'age pigments'. Moreover, the long-term effects of Pm are not removed by returning cells to normal medium. It was also shown that as the cells age, they become progressively more sensitive to the effects of Pm. These results, together with earlier experiments with 5-FU, certainly confirm one of the major predictions of the error theory.

Phenotype of Senescent MRC-5 Fibroblasts

The human fetal lung strain MRC-5 was originally characterized by *Jacobs* et al. [27], who showed that it maintains a constant karyotype and rate of growth for 40–50 population doublings. Subsequently, the growth rate slows down and cultures usually die out between 55 and 70 population doublings. It is a striking fact that when cultures are monitored throughout their in vitro life span, significant changes in the phenotype are normally detected only in the last 10–20 population doublings. These experiments are listed in table I, where $Y \rightarrow O(E)$ indicates that the result is more compatible with an exponential change in the particular phenotypic characteristic under study, than with a linear increase throughout the life span. This is particularly well-documented for the increase in autofluorescence [55] and variants with high G-6-PD activity [7]. In some cases, only young and senescent cultures were compared (Y and O) and a significant phenotypic difference was observed between them. Experiments with phosphoglucose isomerase produced negative results. *Shakespeare and Buchanan* [62] used immunological methods, which would detect about 15% of inactive cross-reacting material, but none was detected in senescent cultures. However, almost all the other results with MRC-5 either directly support, or are compatible with, the general error theory of cellular ageing.

Unified Theory of Cellular Ageing

In almost all animals, somatic tissues have a finite life span, whereas germ-line lineages are potentially immortal. Although cultured cells are derived from somatic tissue, they differ from those in vivo, because there is ample opportunity for selection of long-lived or potentially immortal populations. The fact that this does not happen in human fibroblast cultures must mean that all the cells are committed to senescence. The commitment theory of cellular ageing was proposed by *Kirkwood and Holliday* [32] and *Holliday* et al. [20] to account for the major biological features of in vitro ageing, and also provide a possible basis for the difference between diploid and transformed populations. Two of these features of ageing are the well-documented variability in life spans of both populations and clones, and the observation that in populations containing two distinguishable cell phenotypes, one frequently becomes predominant during senescence, even in the absence of a selective growth advantage [20, 78]. It was proposed that prior to the

143

Table I. The phenotype of senescent MRC-5 fibroblasts, based on experiments in the Mill Hill laboratories

Phenotypic characteristics	Cells examined	Reference
1 Increased proportion of heat-labile G-6-PD and 6-PGD	$Y \rightarrow O \ (E)$[1]	*Holliday and Tarrant* [24]
2 Increased proportion of lactic dehydrogenase cross-reacting material	$Y \rightarrow O \ (E)$	*Lewis and Tarrant* [36]
3 Reduced activity and fidelity of DNA polymerase α	Y and O[2]	*Linn* et al. [37]
Reduced activity and fidelity of DNA polymerase α Reduced fidelity of DNA polymerase γ	$Y \rightarrow O \ (E)$	*Murray and Holliday* [47] *Murray* [46]
4 Slower rate of replicon elongation	Y and O	*Petes* et al. [53]
5 Increased protein turnover and lysosomal proteolytic activity	Y and O	*Shakespeare and Buchanan* [61, 63]
6 Increased frequency of variants with elevated G-6-PD	$Y \rightarrow O \ (E)$	*Fulder and Holliday* [8] *Fulder* [7]
7 Increased polyploidy and chromosomal abnormalities	$Y \rightarrow O \ (E)$	*Thompson and Holliday* [70]
8 Increased autofluorescence	$Y \rightarrow O \ (E)$	*Rattan* et al. [55]
9 Greater sensitivity to paromomycin	$Y \rightarrow O$	*Holliday and Rattan* [22]

[1] $Y \rightarrow O$ indicates that observations were made at intervals throughout the in vitro life span. (E) indicates that the change seen is more compatible with an exponential than a linear increase. In two experiments (6 and 8), the data very strongly suggest an exponential increase.
[2] Y and O indicates that early and late passage cultures were compared.

establishment of primary fibroblast cultures, diploid cells are potentially immortal, or uncommitted. However, during cell division they give rise with a given probability (P) to committed cells, which continue to divide for many generations, but finally die out. The number of divisions between commitment and death is defined as the incubation period (M) and to facilitate the mathematical treatment this is assumed to be constant. (This model was derived from the experimental evidence for random commitment and a constant incubation period in very detailed studies of the ageing of populations of cells of the fungus, *Podospora* [38]. However, the fibroblast model does not depend on a constant value for M; it could vary considerably in different cell lineages.) It follows that the proportion of uncommitted cells will progressively decline and, if M is sufficiently long, these cells will eventually be lost from the population. It can be shown that if P $\sim$ 0.25 and M $\sim$ 55 generations, then cultures will inevitably die out. The model makes the surprising prediction that *population size* will influence longevity, and this has been confirmed by 'bottleneck' experiments, in which a transient reduction in the culture size is introduced at various levels [20, 21]. Computer simulations show that the stochastic features of fibroblast ageing and the sorting out of mixed populations are explained by 'random drift' in the loss of the final small number of uncommitted cells. The model is also supported by the observed growth rate of fibroblast populations. A rapid period of proliferation, with very few non-cycling cells, is succeeded by a period of constant slower growth, with approximately 20% non-cycling cells, and finally by senescence and death [21].

In molecular terms, we propose that the potentially immortal diploid cells are in a metastable state, with regard to the stability of their translation apparatus, and have a fairly high probability ($\sim$ 0.25) of initiating error propagation. The build up of errors during the incubation period (M) is presumed to be slow and does not kill the cells until an average of $\sim$ 55 generations have elapsed. It is now quite clear that the accurate synthesis of macromolecules does not just depend on enzyme specificity, but also on a range of editing or proof-reading processes, which consume energy [for a review, see 5]. It has been pointed out by *Kirkwood* [29] and *Kirkwood and Holliday* [33] that each species must balance the advantages of the accurate synthesis, maintenance, replacement or repair of macromolecules, against the metabolic cost. The resources necessary for prolonged survival of cells may be better diverted into rapid growth to sexual maturity and reproduction. This will increase overall fitness, but the result is the eventual ageing and death of the soma. Thus the error theory is able to explain evolution of ageing of cells and organisms, whereas other theories of ageing are unable to do so.

Cells will grow indefinitely if the feedback of errors is below a critical level, since then the probability of commitment becomes very low or zero. The theory proposes that germ-line cells are intrinsically more accurate than somatic cells, although it must also be borne in mind that there may be special mechanisms which select out defective cells or lineages [18, 49]. It is possible that transformed cells are also more accurate than diploid somatic cells, perhaps because they are physiologically more similar to proliferating embryonic cells, and can therefore escape from in vitro senescence. However, contrary to the assertion of *Harley* et al. [10], the error theory does not predict that this is necessarily the case. They showed that under conditions of histidine starvation, transformed cells are more error-prone. If transformed cells are also in a metastable state ($p \sim 0.25$), but errors escalate more quickly ($M < 50$ generations), then it is easy to show that populations will contain constant proportions of uncommitted, committed and dead cells, and will grow indefinitely [32].

The unified theory suggests answers to the three major problems raised in the introduction and its predictions can, in principle, be tested experimentally by comparing cells from short- and long-lived species, and somatic and germ-line cells. However, further advances will depend on the development of better methods for measuring accuracy in macromolecule synthesis. At the cellular level, it might be possible to prevent the dilution-out of uncommitted cells and thereby obtain immortal populations of normal diploid cells. It would be even more important to devise experimental conditions which would convert a population of transformed cells with indefinite growth into one which aged and died out. The implications of such a discovery would be far-reaching.

References

1 Bell, E.; Marek, L.F.; Levinstone, D.S.; Merrill, C.; Sher, S.; Young, I.T.; Eden, M.: Loss of division potential in vitro: aging or differentiation? Science *202:* 1158–1163 (1978).

2 Branscomb, E.W.; Galas, D.J.: Progressive decrease in protein synthesis accuracy induced by streptomycin in *E. coli.* Nature, Lond. *254:* 161–163 (1975).

3 Buchanan, J.H.; Bunn, C.L.; Lappin, R.I.; Stevens, A.: Accuracy of in vitro protein synthesis: translation of polyuridylic acid by cell free extracts of human fibroblasts. Mech. Age. Dev. *12:* 339–353 (1980).

4 Edelmann, P.; Gallant, J.: On the translational error theory of ageing. Proc. natn. Acad. Sci. USA *74:* 3396–3398 (1977).

5 Fersht, A.R.: Enzyme editing mechanisms and the genetic code. Proc. R..Soc. *212:* 351–379 (1981).

6 Fulder, S.J.: Spontaneous mutations in ageing human cells: studies using a herpes virus probe. Mech. Age. Dev. *6:* 271–282 (1977).

7 Fulder, S.J.: Somatic mutations and ageing of human cells in culture. Mech. Age. Dev. *10:* 101–115 (1978).

8 Fulder, S.J.; Holliday, R.: A rapid rise in cell variants during the senescence of populations of human fibroblasts. Cell 6: 67–73 (1975).

9 Gallant, J.; Palmer, L.: Error propagation in viable cells. Mech. Age. Dev. 10: 27–38 (1979).

10 Harley, C.B.; Pollard, J.W.; Chamberlain, J.W.; Stanners, C.P.; Goldstein, S.: Protein synthetic errors do not increase during ageing of cultured human fibroblasts. Proc. natn. Acad. Sci. USA 77: 1885–1889 (1980).

11 Harley, C.B.; Pollard, J.W.; Stanners, C.P.; Goldstein, S.: Model for messenger RNA translation during amino acid starvation applied to the calculation of protein synthetic error rates. J. biol. Chem. 256: 10786–10794 (1981).

12 Hayflick, L.: The limited in vitro lifetime of human diploid cell strains. Expl Cell Res. 37: 614–636 (1965).

13 Hayflick, L.: The cellular basis of human ageing; in Finch, Hayflick, Handbook of the biology of aging, pp. 159–186 (Van Nostrand-Reinhold, New York 1977).

14 Hayflick, L.; Moorhead, P.S.: The serial cultivation of human diploid cell strains. Expl Cell Res. 25: 585–621 (1961).

15 Hoen, H.; Gryant, E.M.; Johnston, P.; Norwood, T.H.; Martin, G.M.: Non-selective isolation, stability and longevity of hybrids between normal human somatic cells. Nature, Lond. 258: 608–609 (1975).

16 Hoffman, G.W.: On the origin of the genetic code and the stability of the translation apparatus. J. molec. Biol. 86: 349–362 (1974).

17 Holland, J.J.; Kohne, D.; Doyle, M.F.: Analysis of virus replication in ageing human fibroblast cultures. Nature, Lond. 245: 316–319 (1973).

18 Holliday, R.: Growth and death of diploid and transformed human fibroblasts. Fed. Proc. 34: 51–55 (1975).

19 Holliday, R.; Kirkwood, T.B.L.: Predictions of the somatic mutation and mortalisation theories of cellular ageing are contrary to experimental observations. J. theor. Biol. 93: 627–642 (1981).

20 Holliday, R.; Huschtscha, L.I.; Tarrant, G.M.; Kirkwood, T.B.L.: Testing the commitment theory of cellular ageing. Science 198: 366–372 (1977).

21 Holliday, R.; Huschtscha, L.I.; Kirkwood, T.B.L.: Further evidence for the commitment theory of cellular ageing. Science 213: 1505–1508 (1981).

22 Holliday, R.; Rattan, S.I.S.: Evidence that paromomycin induces premature ageing in human fibroblasts; in Sauer, Cellular ageing; Monogr. devl Biol. 17, pp. 221–233 (Karger, Basel 1984).

23 Holliday, R.; Stevens, A.: The effect of an amino acid analogue p-fluorophenylalanine on longevity of mice. Gerontol. 24: 417–425 (1978).

24 Holliday, R.; Tarrant, G.M.: Altered enzymes in ageing human fibroblasts. Nature, Lond. 238: 26–30 (1972).

25 Holliday, R.; Thompson, K.V.A.: Genetic effects on the longevity of cultured human fibroblasts. III. Correlations with altered glucose-6-phosphate dehydrogenase. Gerontol. 29: 89–96 (1983).

26 Huschtscha, L.I.; Holliday, R.: The limited and unlimited growth of SV40 transformed cells from human diploid MRC-5 fibroblasts. J. Cell Sci. 63: 77–99 (1983).

27 Jacobs, J.P.; Jones, C.M.; Baillie, J.P.: Characteristics of a human diploid cell designated MRC-5. Nature, Lond. 227: 168–170 (1970).

28 Jacobs, P.S.; Brunton, W.; Court Brown, W.M.: Cytogenetic studies of leucocytes on the

general population subjects of ages 65 years and more. Ann. hum. Genet. *27:* 353–365 (1964).

29 Kirkwood, T.B.L.: Evolution of ageing. Nature, Lond. *270:* 301–304 (1977).

30 Kirkwood, T.B.L.: Error propagation in intracellular information transfer. J. theor. Biol. *82:* 363–382 (1980).

31 Kirkwood, T.B.L.; Holliday, R.: The stability of the translational apparatus. J. molec. Biol. *97:* 257–265 (1975).

32 Kirkwood, T.B.L.; Holliday, R.: Commitment to senescence: a model for the finite and infinite growth of diploid and transformed human fibroblast in culture. J. theor. Biol. *53:* 481–496 (1975).

33 Kirkwood, T.B.L.; Holliday, R.: The evolution of ageing and longevity. Proc. R. Soc. *205:* 531–546 (1979).

34 Laughrea, M.: Speed-accuracy relationships during in vitro and in vivo protein biosynthesis. Biochemie *63:* 145–168 (1981).

35 Lewis, C.M.; Holliday, R.: Mistranslation and ageing in *Neurospora.* Nature, Lond. *228:* 877–880 (1970).

36 Lewis, C.M.; Tarrant, G.M.: Error theory and ageing in human diploid fibroblasts. Nature, Lond. *239:* 316–318 (1972).

37 Linn, S.; Kairis, M.; Holliday, R.: Decreased fidelity of DNA polymerase activity isolated from ageing human fibroblasts. Proc. natn. Acad. Sci. USA *73:* 2818–2822 (1976).

38 Marcou, D.: Notion de longévité et nature cytoplasmique due déterminant de la sénescence chez quelques champignons. Annls. Sci. nat. Bot. *12:* 653–764 (1961).

39 Martin, G.M.: Genetic syndromes in men with potential relevance to the pathology of ageing; in Bergsmead, Harrison, Genetic effects on ageing, pp. 5–39 (Liss, New York 1978).

40 Martin, G.M.; Sprague, C.A.; Epstein, C.J.: Replicative lifespan of cultivated human cells: effect of donor's age, tissue and genotype. Lab. Invest. *23:* 86–92 (1970).

41 Martin, G.M.; Sprague, C.A.; Norwood, T.H.; Pendergrass, W.R.: Clonal selection, attenuation and differentiation in an in vitro model of hyperplasia. Am. J. Path. *74:* 137–154 (1974).

42 Maynard-Smith, J.: The causes of ageing. Proc. R. Soc. *157:* 115–127 (1962).

43 Medawar, P.B.: An unsolved problem in biology (Lewis, London 1952); reprinted in 'The Uniqueness of the Individual' (Methuen, London 1957).

44 Morley, A.A.; Cox, S.; Holliday, R.: Human lymphocytes resistant to 6-thioguanine increase with age. Mech. Age. Dev. *19:* 21–26 (1982).

45 Morley, A.A.; Trainor, K.J.; Seshadri, R.; Ryall, R.G.: Measurement of in vivo mutations in human lymphocytes. Nature, Lond. *302:* 155–156 (1983).

46 Murray, V.: Properties of DNA polymerases from young and ageing human fibroblasts. Mech. Age. Dev. *16:* 327–344 (1981).

47 Murray, V.; Holliday, R.: Increased error frequency of DNA polymerases from senescent human fibroblasts. J. molec. Biol. *146:* 55–76 (1981).

48 Muzyczka, N.; Poland, R.L.; Bessman, M.J.: Studies on the biochemical basis of spontaneous mutation. I. A comparison of the deoxyribonucleic acid polymerases of mutator, antimutator, and wild type strains of bacteriophage T4. J. biol. Chem. *247:* 7116–7122 (1970).

49 Orgel, L.E.: The maintenance of the accuracy of protein synthesis and its relevance to ageing. Proc. natn. Acad. Sci. USA *49:* 517–521 (1963).

50 Orgel, L.E.: The maintenance and accuracy of protein synthesis and its relevance to ageing; a correction. Proc. natn. Acad. Sci. USA *67:* 1476 (1970).

148

51 Orgel, L.E.: Ageing of clones of mammalian cells. Nature, Lond. *243:* 441–445 (1973).

52 Pereira-Smith, O.M.; Smith, J.R.: The phenotype of low proliferative potential is dominant in hybrids of normal human fibroblasts. Somatic Cell Genet. *8:* 731–742 (1982).

53 Petes, T.D.; Farber, R.A.; Tarrant, G.M.; Holliday, R.: Altered rate of DNA replication
in ageing human fibroblast cultures. Nature, Lond. *251:* 434–436 (1974).

54 Printz, D.B.; Gross, S.R.: An apparent relationship between mistranslation and an altered
leucyl tRNA synthetase in a conditional lethal mutant of *Neurospora* crassa. Genetics *55:*
451–467 (1967).

55 Rattan, S.I.S.; Keeler, K.D.; Buchanan, J.H.; Holliday, R.: Autofluorescence as an index
of ageing in human fibroblasts in culture. Biosci. Rep. *2:* 561–567 (1982).

56 Röhme, D.: Evidence for a relationship between longevity of mammalian species and lifespans of normal fibroblasts in vitro and erythrocytes in vivo. Proc. natn. Acad. Sci. USA
74: 4876–4880 (1981).

57 Rosenberger, R.F.: Streptomycin-induced protein error propagation appears to lead to
cell death in *Escherichia coli.* IRCS med. Sci. *10:* 874–875 (1982).

58 Sakagami, H.; Mitsui, Y.; Murota, S.; Yamada, M.: Two-dimensional electrophoretic
analysis of nuclear acidic proteins in senescent human diploid cells. Cell Struct. Funct. *4:*
215–225 (1979).

59 Saksela, E.; Moorhead, P.S.: Aneuploidy in the degenerative phase of serial cultivation
of human cell strains. Proc. natn. Acad. Sci. USA *50:* 390–395 (1963).

60 Shmookler Reis, R.J.; Goldstein, S.: Loss of reiterated DNA sequences during serial passage of human diploid fibroblasts. Cell *21:* 739–750 (1980).

61 Shakespeare, V.; Buchanan, J.H.: Increased degradation rates of protein in ageing human fibroblasts and in cells treated with an amino acid analog. Expl Cell Res. *100:* 1–8
(1976).

62 Shakespeare, V.; Buchanan, J.H.: Studies on phosphoglucose isomerase from cultured
human fibroblasts: absence of detectable ageing effects on the enzyme. J. cell. Physiol. *94:*
105–116 (1978).

63 Shakespeare, V.A.; Buchanan, J.H.: Increased proteolytic activity in ageing human fibroblasts. Gerontol. *25:* 305–313 (1979).

64 Smith, J.R.; Hayflick, L.: Variation in the life-span of clones derived from human diploid
cell strains. J. Cell Biol. *62:* 48–53 (1974).

65 Smith, J.R.; Whitney, R.G.: Intraclonal variation in proliferative potential of human
diploid fibroblasts: stochastic mechanism for cellular ageing. Science *207:* 82–84 (1980).

66 Smith, J.R.; Pereira-Smith, O.M.; Schneider, E.L.: Colony size distribution as a measure
of in vivo and in vitro aging. Proc. natn. Acad. Sci. USA *75:* 1353–1356 (1978).

67 Sun, T.T.; Green, H.: Differentiation of the epidermal keratinocyte in cell culture: formation of the cornified envelope. Cell *9:* 511–521 (1976).

68 Swim, H.A.; Parker, R.F.: Culture characteristics of human fibroblasts propagated
serially. Am. J. Hyg. *66:* 235–243 (1957).

69 Thompson, K.V.A.; Holliday, R.: Effect of temperature on the longevity of human fibroblasts in culture. Expl Cell Res. *80:* 354–360 (1973).

70 Thompson, K.V.A.; Holliday, R.: Chromosome changes during the in vitro ageing of
MRC-5 human fibroblasts. Expl Cell Res. *96:* 1–6 (1975).

71 Thompson, K.V.A.; Holliday, R.: The longevity of diploid and polyploid human fibroblasts: evidence against the somatic mutation theory of cellular ageing. Expl Cell Res. *112:*
28–287 (1978).

149

72   Thompson, K.V.A.; Holliday, R.: Genetic effects on the longevity of cultured human fibroblasts. I. Werner's syndrome. Gerontol. *29:* 73–82 (1983).
73   Thompson, K.V.A.; Holliday, R.: Genetic effects on the longevity of cultured human fibroblasts. II. DNA repair deficient syndromes. Gerontol. *29:* 83–88 (1983).
74   Viermy, C.; Keller, A.M.; Begel, O.; Belcour, L.: A sequence of mit DNA is associated with the onset of senescence in a fungus. Nature, Lond. *297:* 157–159 (1982).
75   Witkowski, J.A.: Dr. Carrel's immortal cells. Med. Hist. *24:* 129–142 (1980).
76   Wojtyk, R.I.; Goldstein, S.: Fidelity of protein synthesis does not decline during ageing of cultured human fibroblasts. J. cell. Physiol. *103:* 299–303 (1980).
77   Wojtyk, R.I.; Goldstein, S.: Clonal selection in cultured human fibroblasts: role of protein synthetic errors. J. Cell Biol. *95:* 704–710 (1982).
78   Zavala, C.; Fialkow, P.J.; Herner, G.: Evidence for selection in cultured diploid fibroblast strains. Expl Cell Res. *117:* 137–144 (1978).

Part III

PROTEINS AND AGING

Editor's Comments
on Papers 15 Through 22

The most direct way to measure the accuracy of protein synthesis is to select a protein that lacks a particular amino acid and then determine the misincorporation of that amino acid. In practice, the experiment is very hard to do, because one must be certain that radioactive counts from the misincorporated amino acid are from the purified material, rather than from contaminating proteins, almost all of which, of course, will contain the amino acid. A simple calculation shows that the protein must be purified far more highly ($\sim$99.9%) than is usually necessary for structural or enzyme studies. The method has nevertheless been used to measure the misincorporation of amino acids into proteins or specific peptides from several sources (Kirkwood, Holliday, and Rosenberger, 1984). However, the only successful attempt to measure age-related changes in the misincorporation of an amino acid has been carried out by Medvedev and Medvedeva (1978). They used histone H1, which does not contain methionine, and were able to obtain sufficiently pure material to detect the misincorporation of ^{35}S-methionine. The results show that there is a small but significant increase in the error level in the liver and pancreas of old, inbred CBA mice, in comparison with that in the same tissues of young animals, but only a preliminary report of this work has been published. An attempt to use the same system with cultured human fibroblasts ran into difficulties, because it was impossible to obtain sufficient quantities of purified H1 histone to detect misincorporation in either young or old cell cultures (Buchanan and Stevens, 1978).

Another method depends on the use of cell-free extracts to measure the accuracy of translation of an RNA message. So far, only experiments with polyU have been carried out, and these suffer from the disadvantage that the frequency of misincorporation of leucine in place of phenylalanine is rather high (Laughrea, 1981). This means that any differences between cell-free extracts from young and old cells may be obscured by the artificially high error level in this in vitro system. For example, Buchanan et al. (1980) showed that the misincorporation of leucine is about 1%, perhaps a hundredfold higher than the expected in vivo substitution of leucine for phenyalanine. No differences in error level were seen in cell-free extracts from early- or late-passage human fibroblasts. It was shown that the aminoglycoside antibiotic, paromomycin, strikingly decreased the fidelity of translation, from which it can be concluded that ribosomes from senescent cells are not comparable with ribosomes to which paromomycin is bound. Wojtyk and Goldstein (1980, 1982) confidently assert that they have measured the accuracy of protein synthesis in human fibroblasts, using the polyU in vitro system and that this does not change and may even increase during aging. Scrutiny of their methods shows, however, that they have neither optimized the concentrations of the correct and incorrect amino acids nor standardized their assay (Laughrea, 1981; Buchanan et al., 1980), as shown by the fact that the activity of their extracts varied by eight-hundredfold.

Another technique that can be used to study alterations in

proteins depends on the appearance of so-called stutter spots in two-dimensional gels. A change in charge, either by misincorporation of an amino acid or by a postsynthetic modification, will alter the positions of molecules on the gel relative to the major spot. In Paper 15, Harley and associates used this technique to look for protein errors in normal and transformed human fibroblasts, as well as Werner's syndrome and progeria fibroblasts. No stutter spots were seen in proteins such as actin, from either young or senescent cells, or from Werner's and progeria cells. Since the method does not measure the intrinsic error level in young cells, it is not possible to know whether there is any change during aging, although it is clear that proteins in senescent cells cannot be laden with errors, or post-synthetic changes, for that matter. The authors also measured error levels in cells treated with histidinol. This analogue blocks the normal charging of histidyl tRNA, and from earlier studies with bacterial cells, it is known that this results in the misincorporation of wrong amino acids (particularly glutamine) at histidine codons (Parker et al., 1978). Under these conditions of histidine starvation, old and young cells produced roughly the same proportion of altered protein, from which the authors conclude that the spontaneous level of errors is the same. The theory on which this latter conclusion is based is simplistic in that it assumes that the levels of histidyl tRNA in young and old cells are the same under all conditions and also that ribosomes from old cells, if ambiguous, would have increased affinity for noncognate amino acids, but unaltered affinity for cognate amino acids (Harley et al., 1981). Neither of these assumptions is justified. It is unfortunate that Paper 15, although important, has a misleading title, and it will be widely cited as a "disproof" of the protein error theory. Perhaps the most interesting result is the finding that under the same conditions of histidinol starvation, transformed fibroblasts produce more errors than normal diploid ones. The authors incorrectly conclude, however, that this result is also contrary to the error theory, because they presume that the theory predicts that permanent lines should have increased accuracy in protein synthesis. The commitment theory makes it clear that an increase in the rate of error accumulation (i.e., decrease in the parameter M) could convert a population with limited growth potential into a permanent line (Holliday, 1975; Kirkwood and Holliday, 1975; Paper 11).

Since methods have not been developed for measuring the fidelity of transcription or translation in cultured mammalian cells, indirect methods have often been employed. Paper 16 documents an early study with MRC-5 fetal lung fibroblasts, in which a significant reduction in the heat stability of the enzymes glucose-6-phosphate dehydrogenase (G6PD) and 6-phosphogluconate dehydrogenase (6PGD) was detected during the last 15 passages of their lifespan. Many naturally occurring variants of G6PD are known in humans and about half of them are heat labile (McKusick, 1978). Therefore, random amino acid substitutions should also produce a proportion of heat-labile molecules. Evidence that the unstable fraction is due to errors in synthesis came from studies with the RNA base analogue, 5-fluorouracil, and the antibiotic, paromomycin, both of which would be expected to increase ambiguity in translation. Young cells grown in the presence of these compounds accumulate a level of heat-labile G6PD similar to that in untreated senescent cells, and they also die out prematurely (Paper 16; Holliday, 1975; Holliday and Rattan, 1984). In Paper 16 it is also shown that the altered fraction of enzyme is preferentially able to use the analogue substrate, deoxyglucose-6-phosphate, and studies of naturally occurring variants have shown that these sometimes have altered substrate specificity. Although Pendergrass, Martin, and Bornstein (1976) were unable to detect altered G6PD in senescent skin fibroblasts, several other laboratories have been able to do so. The interpretation of these results is controversial. Although they are consistent with the possibility that the appearance of heat-labile G6PD is due to errors in protein synthesis (for a review, see Holliday and Thompson, 1983), it is possible that postsynthetic changes in the enzyme molecules are responsible for the loss of stability during aging.

Very detailed studies with another enzyme, triosephosphate isomerase (TPI), have established the occurrence of postsynthetic alterations in structure during senescence. These results are documented by Gracy and his associates in Papers 17 and 18. It is shown that acidic isozymes of TPI accumulate in aging erythrocytes and lens (where no protein synthesis occurs). Structural studies on these isozymes demonstrate that they are the result of specific deamidation of two asparagines to aspartic acid. TPI is a dimeric enzyme and the determination of its three-dimensional structure shows that these labile asparagines are situated at subunit contact sites (Banner et al., 1975). The introduction of negative charges, therefore, causes the enzyme to disassociate more readily into the inactive monomer. It is thought that altered molecules are degraded by proteases in normal cells and that they accumulate in aging cells or tissues (for reviews, see Rothstein, 1975; Gracy, 1983). The observations on TPI document for the first time the chemical basis of one of these alterations.

Studies on TPI have been extended to the fibroblast system, and

Paper 18 demonstrates that senescent cells also accumulate labile enzyme. Previously, it had been shown that Werner's syndrome cells, which have a very short in vitro lifespan (Paper 9) accumulate head-labile G6PD in early-passages cultures (Holliday, Porterfield, and Gibbs, 1974). Observations on changes in TPI in Werner's syndrome and also progeria fibroblasts are documented in Paper 18. The results provide further evidence that fibroblasts from individuals that age prematurely also age more rapidly in vitro, as judged by the accumulation of altered enzyme. Paper 18 also reviews previous studies on the appearance of other altered enzymes in Werner's syndrome fibroblasts or senescent normal fibroblasts. It is interesting that experiments with phosphoglucose isomerase, which is a monomeric enzyme, have not revealed any accumulation of altered molecules in senescent fibroblast populations (Shakespeare and Buchanan, 1978), whereas all the enzymes in which changes have been seen are polymeric. This finding supports the hypothesis that the enzyme instability is mainly due to a weakening of a subunit interaction, leading to the formation of inactive monomers.

It is particularly informative to examine enzymes that are involved in information transfer. An obvious choice is DNA polymerase, since methods have been worked out for measuring its fidelity in replicating defined templates in vitro, and it is known that alterations in primary structure can reduce fidelity (Kornberg, 1980). The first study (Paper 19) was carried out with early- and late-passage MRC-5 cells using a variety of template primers, and the misincorporation of incorrect nucleotides was measured in most cases in the presence of Mn^{++}, which amplifies the error level. DNA polymerase α from senescent cells was found to be 2 to 10 times less accurate than the enzyme from young cells. A sharp decline in DNA polymerase was seen in very late-passage cells, and in this connection it is interesting that Petes et al. (1974), using ^{3}H thymidine fiber autoradiography, showed that the rate of replicon elongation significantly decreased in senescent cells. Paper 20 documents more detailed investigations in which the fidelity of DNA polymerase was examined throughout the lifespan of MRC-5, using assays carried out in the presence of the normal cation, Mg^{++}. A loss of fidelity was consistently observed in several assay systems, one of which measures the fidelity of DNA polymerase γ, and this loss was particularly marked during Phase III of fibroblast growth. Much of this paper is concerned with various important controls that effectively rule out a variety of possible experimental artifacts. These experiments do not prove that erroneous DNA polymerase is due to alterations in primary structure. It is known that DNA polymerases are prone to proteolytic cleavage and that the products retain activity (Hubscher et al., 1981). It is therefore possible that the senescent cells contain inaccurate forms of the enzyme that are absent, or less common, in normal cells. Furthermore, Krauss and Linn (1982) obtained evidence that young cells held confluent for many days accumulate less accurate forms of DNA polymerase. It was clear that additional studies are necessary to find out whether the loss out whether the loss of accuracy of DNA polymerase during aging is related to changes in primary structure of the enzyme.

There is evidence from both prokaryotes and eukaryotes that abnormal protein molecules are selectively degraded by proteases (for a review, see Goldberg and St. John, 1976). It is indeed possible that a major function of protein turnover in higher organisms is the removal of molecules that may have harmful physiological effects, and in any theory of error propagation, the important role of proteases has to be taken into account (see Paper 3). In general, it might be expected that the production of an increased proportion of altered protein would result in a shorter average half-life of total protein. Independent studies in Papers 21 and 22, in which three different methods are employed, confirm that Phase III fibroblasts do indeed degrade proteins more rapidly than young cells. Paper 21 shows that in early Phase III cultures only short-lived proteins are degraded more rapidly, but in late Phase III this is true both in short-lived and long-lived proteins. In both studies an important control was carried out, which showed that proteins containing an amino acid analogue (canavanine, fluorephenylalanine or azetidine carboxylic acid) were degraded more rapidly than normal proteins. There was some evidence from the experiments in Paper 21, however, that the degradation of analogue-containing proteins was less rapid in Phase III than in younger cultures. This suggests that the cells may have reduced ability to remove abnormal molecules selectively, which would, of course, be an important factor in accelerating senescence.

At first sight, the observations that senescent cells degrade proteins more rapidly than young cells seem at variance with the interpretation of the experiments with TPI reported in Paper 17. The accumulation of isozymes of TPI is thought to be due to a failure of normal proteolysis in aged cells. If errors and/or postsynthetic changes are increasing during senescence, however, then the extent of proteolysis will initially increase (as documented in Papers 21 and 22), but as a consequence, the pathways for removing altered proteins may become saturated, thus leading to their accumulation. A further complication is that the aging process is quite likely to affect the regulation of the synthesis of proteases, as well as their catalytic activity (see, for example, Shakespeare and Buchanan, 1979, and for a general review,

Makrides, 1983). Whatever the correct interpretations, it is clear from many studies that abnormalities in protein metabolism are closely associated with the aging process.

REFERENCES

Banner, D. W., A. C. Bloomer, G. A. Petsko, D. C. Phillips, C. I. Pogson, I. A. Wilson, P. H. Corran, A. J. Furth, J. D. Milman, R. E. Offord, J. D. Priddle, and S. G. Waley, 1975, Structure of Chicken Muscle Triose Phosphate Isomerase Determined Crystallographically at 2.5Å Resolution, Using Amino Acid Sequence Data, *Nature* **255:**609-614.

Buchanan, J. H., C. L. Bunn, R. I. Lappin, and A. Stevens, 1980, Accuracy of *in vitro* Protein Synthesis: Translation of Polyuridylic Acid by Cell Free Extracts of Human Fibroblasts, *Mech. Ageing Dev.* **12:**339-353.

Buchanan, J. H., and A. Stevens, 1978, Fidelity of Histone Synthesis in Cultured Human Fibroblasts, *Mech. Ageing Dev.* **7:**321-334.

Goldberg, A. L., and A. C. St. John, 1976, Intracellular Protein Degradation in Mammalian and Bacterial Cells: Part 2, *Annu. Rev. Biochem.* **45:**747-803.

Gracy, R. W., 1983, Epigenetic Formation of Isozymes: The Effect of Ageing, in *Isozymes: Current Topics in Biological and Medical Research*, M. C. Rattazzi, ed., vol. 7, Alan R. Liss, New York, pp. 187-201.

Harley, C. B., J. W. Pollard, C. P. Stanners, and S. Goldstein, 1981, Model for Messenger RNA Translation during Amino Acid Starvation Applied to the Calculation of Protein Synthetic Error Rates, *J. Biol. Chem.* **256:**10786-10794.

Holliday, R., 1975, The Growth and Death of Diploid and Transformed Human Fibroblasts, *Fed. Proc.* **34:**51-55.

Holliday, R., J. S. Porterfield, and D. D. Gibbs, 1974, Premature Ageing and Occurrence of Altered Enzyme in Werner's Syndrome Fibroblasts, *Nature* **248:**762-763.

Holliday, R., and K. V. A. Thompson, 1983, Genetic Effects on the Longevity of Cultured Human Fibroblasts III. Correlations with Altered Glucose-6-Phosphate Dehydrogenase, *Gerontology* **29:**89-96.

Holliday, R., and S. I. S. Rattan, 1984, Evidence That Paromomycin Induces Premature Ageing in Human Fibroblasts, *Monogr. Dev. Biol.* **17:**221-233. Karger, Basel.

Hubscher, U., A. Spanos, W. Albert, F. Grumm, and G. R. Banks, 1981, Evidence That High Molecular Weight Replicative DNA Polymerase is Conserved during Evolution, *Proc. Natl. Acad. Sci. (U.S.A.)* **78:**6771-6775.

Kirkwood, T. B. L., and R. Holliday, 1975, Commitment to Senescence: A Model for the Finite and Infinite Growth of Diploid and Transformed Human Fibroblasts in Culture, *J. Theor. Biol.* **82:**363-382.

Kirkwood, T. B. L., R. Holliday, and R. F. Rosenberger, 1984, Stability of the Cellular Translation Process, *Int. Rev. Cytol.* **92:**93-132.

Kornberg, A., 1980, *DNA Replication,* Freeman, San Francisco.

Krauss, S. W., and S. Linn, 1982, Changes in DNA Polymerases α, β and γ during the Replicative Life Span of Cultured Human Fibroblasts, *Biochemistry* **21:**1002-1009.

Laughrea, M., 1981, Speed-accuracy Relationships during *in vitro* and *in vivo* Protein Biosynthesis, *Biochemie* **63:**145-168.

Makrides, S. C., 1983, Protein Synthesis and Degradation during Ageing and Senescence, *Biol. Rev.* **58:**434-422.

McKusick, V. A., 1978, Human Glucose-6-Phosphate Dehydrogenase Variants, in *Mendelian Inheritance in Man,* 5th ed., John Hopkins University Press, Baltimore, pp. 732-745.

Medvedev, Zh. A., and M. N. Medvedeva, 1978, Use of HI-Histone to Test the Fidelity of Protein Synthesis in Mouse Tissues, *Biochem. Soc. Trans.* **6:**610-612.

Parker, J., J. Pollard, J. D. Friesen, and C. P. Stanners, 1978, Stuttering: High Level Mistranslation in Animal and Bacterial Cells, *Proc. Natl. Acad. Sci. (U.S.A.)* **75:**1091-1094.

Pendergrass, W. R., G. M. Martin, and P. Bornstein, 1976, Evidence Contrary to the Protein Error Hypothesis for *in vitro* Senescence, *J. Cell. Physiol.* **87:**3-14.

Petes, T. D., R. A. Farber, G. M. Tarrant, and R. Holliday, 1974, Altered Rate of DNA Replication in Ageing Human Fibroblast Cultures, *Nature* **251:**434-436.

Rothstein, M., 1975, Ageing and Alteration of Enzymes—a review, *Mech. Ageing Dev.* **4:**325-338.

Shakespeare, V. A., and J. H. Buchanan, 1978, Studies on Phosphoglucose Isomerase from Cultured Human Fibroblasts: Absence of Detectable Ageing Effects on the Enzyme, *J. Cell. Physiol.* **94:**105-116.

Shakespeare, V. A., and J. H. Buchanan, 1979, Evidence for Increased Proteolytic Activity in Ageing Human Fibroblasts, *Gerontology* **25:**305-313.

Wojtyk, R. I., and S. Goldstein, 1980, Fidelity of Protein Synthesis Does Not Decline during Ageing of Cultured Human Fibroblasts, *J. Cell. Physiol.* **103:**299-303.

Wojtyk, R. I., and S. Goldstein, 1982, Clonal Selection in Cultured Human Fibroblasts: Role of Protein Synthetic Errors, *J. Cell. Biol.* **95:**704-710.

15

Reprinted from *Proc. Natl. Acad. Sci. (U.S.A.)* **77**:1885–1889 (1980)

Protein synthetic errors do not increase during aging of cultured human fibroblasts

(mistranslation/transformation/progeria/genetic code/error catastrophe)

C. B. HARLEY*†, J. W. POLLARD‡§, J. W. CHAMBERLAIN‡, C. P. STANNERS‡, AND S. GOLDSTEIN*¶

*Departments of Biochemistry and Medicine, McMaster University, Hamilton, Ontario L8N 3Z5, Canada; and ‡Department of Medical Biophysics and the Ontario Cancer Institute, 500 Sherbourne Street, Toronto, Ontario M4X 1K9, Canada

Communicated by John W. Littlefield, January 3, 1980

The error catastrophe theory of cellular aging postulates that errors in protein synthesis lead to a protein synthetic machinery with progressively lower fidelity and to the eventual accumulation of a lethal proportion of aberrant proteins (1). A corollary of this hypothesis is that the abrogation of senescence by transformation of fibroblasts into permanent lines should be associated with a rate of translational errors that is lower than that of senescent cells.

A major difficulty in testing these predictions has been the lack of a rapid and direct measure of error frequencies in intact eukaryotic cells. Parker *et al.* (2) have developed a method in which the synthesis of error-containing proteins can be induced in bacterial and animal cells by amino acid starvation. They proposed that as a result of specific amino acid substitutions after starvation for particular amino acids, proteins were synthesized with altered isoelectric points. These proteins appeared on two-dimensional polyacrylamide gels as a series of satellite spots trailing the native protein spots in the isoelectric focusing dimension (2, 3), a phenomenon known as stuttering (2).

We now describe the development of this system into a quantitative assay and its application to the measurement of error frequencies in various cultured human fibroblasts. Our findings are contrary to the error catastrophe theory of aging: error frequencies of late-passage cells and cells from old donors or subjects with progeria and Werner syndromes are not increased compared to early-passage cells from young donors whereas the error frequency of a simian virus 40 (SV40)-transformed permanent line is significantly increased.

MATERIALS AND METHODS

Cell Culture. Strains used are listed in Table 1. "Early passage" denotes cultures with less than 50% of the life span completed (4) and with greater than 80% of cells capable of proliferation by the thymidine labeling index (13). "Late passage" denotes cultures with greater than 90% of the life span completed (7) or with a thymidine labeling index of 20% or less. Cells were grown on plastic surfaces at 37°C in an atmosphere

of 95% air/5% CO_2 with either Eagle's minimal essential medium supplemented with 15% (vol/vol) fetal calf serum (GIBCO), nonessential amino acids, glucose, and pyruvate (14) or α-minimal essential medium (α-MEM) (15) containing asparagine-H_2O at 50 μg/ml and 15% fetal calf serum (GIBCO). Cells were routinely tested for mycoplasma contamination by the fluorescent stain DAPI (16) with uniformly negative results.

Amino Acid Starvation and Labeling Conditions. Cells were subcultured into 35-mm dishes (Corning) at 3–5 × 10⁴ cells per dish. After 2–4 days, cells were rinsed once with α-MEM (15) lacking histidine and methionine and supplemented with 0.2 mM phenylalanine and 5% undialyzed fetal calf serum and then incubated for 40 min in this medium containing either 0.1 mM histidine (unstarved conditions) or 2–20 mM histidinol (Calbiochem) (starved conditions). These media were then replaced with identical media containing either 2–15 μCi of [³H]phenylalanine per ml (New England Nuclear, 20 Ci/mmol; 1 Ci = 3.7 × 10¹⁰ becquerels) for measurement of protein synthetic rates or 5–150 μCi of [³⁵S]methionine per ml (New England Nuclear, > 500 Ci/mmol) for labeling of proteins for analysis by two-dimensional gel electrophoresis. To measure the rate of protein synthesis, we rinsed ³H-labeled cells rapidly with ice-cold phosphate-buffered saline and precipitated them directly on the dish with 5% (vol/vol) trichloroacetic acid at 10-min intervals for a total labeling period of 40 min. Cells were rinsed thoroughly with 5% trichloroacetic acid and H_2O and solubilized with 0.4% sodium deoxycholate in 0.1 M NaOH. Samples were taken for measurement of radioactivity in a liquid scintillation counter or for measurement of protein by the method of Lowry *et al.* (17). For two-dimensional gel analysis, ³⁵S-labeled cells were rinsed rapidly at the end of the 40-min labeling period with ice-cold phosphate-buffered saline and solubilized by addition of 400 μl of lysis buffer A (18) at room temperature. Ten minutes later, the lysate was removed from the dish and frozen at −70°C until

Abbreviations: α-MEM, α-minimal essential medium; SV40, simian virus 40.
† Present address: Division of Biological Sciences, University of Sussex, Brighton, England BN1 9QG.
§ Present address: Department of Biochemistry, Queen Elizabeth College, University of London, London, England W8 7AH.
¶ To whom reprint requests should be addressed.

ERRATUM

The equation on page 1886 should read:

$$p' = \sum_{i=1}^{9} iS_i/9(\gamma + \beta + \sum_{i=1}^{9} S_i) = \sum_{i=1}^{9} iS_i/9$$

Table 1.   Description of cell types

Tissue of origin	Cell strain	Donor age, years	Ref.
Lung	WI38	Fetus	4
	WI38-SV40 (VA13A)	Fetus	5
	MRC5	Fetus	6
Skin	A2	11	7
	GM37 (GM0037A)	18	8
	JO88	76	9
	JO69	69	9
	P5 (progeria)	9	10
	P18 (progeria)	5	11
	WS2 (Werner syndrome)	37	12
	WS4 (Werner syndrome)	41	Goldstein, unpublished

processed as described (18) using all extracts centrifuged at 8000 × *g* for 3 min.

Comparison of protein synthetic error rates in different cell types critically depends on accurately measuring the rate of protein synthesis, and the validation of these methods has been discussed in detail (19, 20). In brief, a small quantity of radioactively labeled amino acid was added to medium containing a large amount of unlabeled precursor. After a very short lag period, linear incorporation of the isotope into protein was observed, indicating that the intracellular pool of aminoacyl-tRNA rapidly reaches a constant specific activity equal to that of the amino acid in the medium (21, 22).

Measurement of total protein synthesis and analysis of error-containing proteins also depend on the rate of protein turnover during the 40-min labeling period. However, significant protein degradation of native or substituted radioactive actins was not seen when the labeled cultures were subsequently incubated in nonradioactive medium for up to 1.3 hr (21).

Two-Dimensional Polyacrylamide Gel Electrophoresis. Supernatants containing 30–80 μg of proteins were layered on pre-run isoelectric focusing gels (1.6% pH 4–6 ampholine, 0.4% pH 3.5–10 ampholine; LKB). Gels from the second (NaDodSO$_4$ M_r) dimension (18) were dried with or without impregnation with 2,5-diphenyloxazole (23) and exposed to x-ray film (Kodak XR-1).

Quantification of Autoradiograms. Actin was chosen as a reference protein for three reasons: it is the major protein synthesized in human fibroblasts, it has a known number of histidine residues, and its stutter spots can be readily resolved and quantified. The actin regions of autoradiograms were scanned with a Joyce–Loebl microdensitometer with a slit height that encompassed the largest spot scanned (21). Selected points on the scan were given coordinates and areas were integrated numerically with a computer programmed to correct for nonlinearity of the x-ray film at saturation. The fraction of the total area under the scan was determined for native actin (γ and β) and each stutter spot S_i (*i* represents the number of substitutions) (see Fig. 4). These fractions were used to calculate the fraction of substituted sites or the error frequency (*p'*):

$$p' = \sum_{i=1}^{9} iS_i/9(\gamma + \beta + S_i) = \sum_{i=1}^{9} iS_i/9.$$

Because there are nine histidine residues in both γ- and β-actin (24, 25), the possible range of *i* is 1–9. However, the more highly substituted actins were seen only with extreme starvation, and in most cases *i* did not exceed 4 or 5. The valleys between peaks of scans showing clear stutter spots were identified to assign integration boundaries between the native and substituted actins. The observed differences in pI between stutter spots (0.06–0.09 pH units) agreed well with the theoretical value of

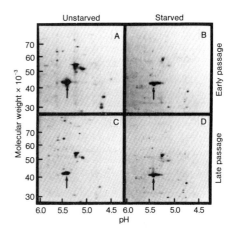

FIG. 1.   Autoradiograms of two-dimensional gels of [^{35}S]methionine-labeled proteins. Early-passage (*A* and *B*) and late-passage (*C* and *D*) human fibroblasts (A2) were grown in medium lacking methionine (*A* and *C*) or in medium lacking histidine and methionine and containing histidinol at 20 mM (*B*) or 10 mM (*D*). The arrow indicates the position of β-actin. The pH gradient was determined from the isoelectric focusing gel; M_rs were estimated from proteins of known M_rs coelectrophoresed in the second dimension.

about 0.08 based on the amino acid composition of actin (unpublished data), and this value was used to assign integration boundaries when the spots were not well resolved.

Optical densities obtained by scanning autoradiographic spots containing a nonuniform distribution of grain densities in the dimension of the slit height are not strictly proportional to the absolute number of grains present. However, as suggested by O'Farrell (18), the relative areas under the scans agreed with the relative intensities determined by the more accurate but time-consuming method of quantitative roster scanning (21).

RESULTS

Fig. 1 shows autoradiograms of newly synthesized proteins resolved by two-dimensional gel electrophoresis from early- and late-passage human skin fibroblasts. The error catastrophe theory predicts that late-passage cells will have increased levels of mistranslation which should be visualized as an altered pattern of proteins, particularly in the isoelectric focusing dimension. Fig. 1 *A* and *C* shows that under normal (unstarved) conditions most proteins form a single discrete spot. This suggests that extensive heterogeneity is not present in either the charge or molecular weight of proteins from late-passage cells. When extracts of labeled proteins from early- and late-passage cells were coelectrophoresed, virtually all the proteins were identical in molecular weight and isoelectric point (unpublished data).

To induce detectable levels of mistranslation, we starved cells for histidine. Due to the relatively high rate of protein degradation in animal cells (19), acute depletion of the intracellular amino acid pool cannot be achieved by removal of histidine from the medium alone. However, histidine starvation can be effected by treating cultures with histidinol in histidine-free medium. Histidinol competes with histidine for the active site on the histidyl-tRNA synthetase but does not become ligated to histidine tRNA (26). Proteins synthesized under these conditions had a trail of stutter spots leading from the native form towards the acidic end of the gel (Fig. 1 *B* and *D*). This has been interpreted to be the consequence of substituting glutamine (a neutral amino acid) for histidine (a basic amino acid), resulting in new proteins with similar molecular weights but reduced isoelectric points (2). In order to induce stutter spots that could be readily quantified, protein synthetic rates were required that were as low as 2–10% of unstarved rates. All cultures of human fibroblasts required similar concentrations of histidinol (5–20 mM) to reduce the synthetic rates to these levels. Histidinol itself does not produce stuttering; in the presence of normal concentrations of histidine, histidinol had no effect on either the protein synthetic rate or the pattern of spots seen on two-dimensional gel autoradiograms (21). Furthermore, in studies with a temperature-sensitive histidyl-tRNA synthetase mutant of CHO cells, cells starved for histidine at the nonpermissive temperature in the absence of histidinol synthesized native and error-containing proteins similar to those of wild-type cells after histidine starvation with histidinol (unpublished data).

The patterns of proteins synthesized under unstarved conditions were essentially identical in all untransformed cell strains listed in Table 1 in both the actin region (Fig. 2 *Left*) and the remainder of the autoradiogram (not shown). With histidine starvation, all of these cell strains, regardless of culture age or donor age, exhibited similar levels of mistranslation at equivalent degrees of inhibition of protein synthesis (Fig. 1 *B* and *D*; Fig. 2 *Right*).

A corollary to the error catastrophe theory is that transformed, immortal cells should have an error rate for protein synthesis less than that of senescent cells. Therefore, we compared the degree of mistranslation upon histidine starvation of SV40-transformed human fibroblasts and the parental diploid strain. Fig. 3 shows the actin region of autoradiograms of gels from WI38 and its SV40-transformed counterpart. Contrary to expectation, transformation was associated with a greatly increased error rate in protein synthesis. Similar results were obtained with SV40 transformants of two other human cell strains (21) and of mouse 3T3 cells (unpublished data).

To quantify mistranslation, we scanned actin regions of the autoradiograms with a densitometer (Fig. 4). The error frequency (fraction of substituted histidine sites) was calculated by integrating the scans and weighting the fractional areas that corresponded to aberrant proteins by the number of substituted histidine sites (see *Materials and Methods* and Table 2). The fraction of substituted sites calculated from Fig. 4 was greatly increased in WI38-SV40 compared to WI38 (Table 2).

A theoretical model of mistranslation during amino acid starvation suggests that a linear relationship exists between error frequency and the inverse of the relative rate of protein synthesis (Table 3; unpublished data). Therefore, the error frequencies calculated from autoradiograms were plotted against the reciprocal protein synthetic rates (Fig. 5). In WI38, WI38-SV40, and all other cell strains tested, the error frequency

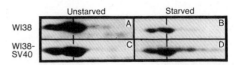

FIG. 3. Autoradiograms of [³⁵S]methionine-labeled proteins in the actin regions of two-dimensional gels of proteins synthesized by WI38 (*A* and *B*) or SV40-transformed WI38 (*C* and *D*). Cells were grown in medium lacking methionine (*A* and *C*) or in medium lacking methionine and histidine and containing histidinol at 10 mM (*B*) or 5 mM (*D*). Lines are drawn through β-actin.

increased as the protein synthetic rate decreased. Late-passage WI38 cells did not have elevated error frequencies compared to early-passage WI38 cells. In contrast, WI38-SV40 cells had dramatically elevated error frequencies compared to their untransformed counterparts (Fig. 5).

Table 3 shows the mean error frequencies of the cells after normalization of mistranslation to the protein synthetic rate. Late-passage cells from fetal, young, or old normal donors did not have elevated error frequencies compared to early-passage cells. In fact, two of the cell strains at late passage had reduced error frequencies ($P < 0.05$). Furthermore, cells from old donors or subjects with progeria or Werner syndrome did not have elevated error frequencies compared to young normal donors. However, the elevated error frequency of WI38-SV40 cells was highly significant when compared to their untransformed counterparts or all diploid fibroblasts combined ($P < 0.0001$). Table 3 also lists the life spans of the cell strains [maximal population doublings *in vitro* (4)]. If the life span *in vitro* were influenced by the inherent error frequency, one might expect to find an inverse relationship between the total number of population doublings of a cell strain and the calculated level of mistranslation. Although a 3-fold range in error frequency was observed among normal cell strains, no such correlation was found by either linear regression or analysis of variance.

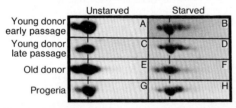

FIG. 2. Autoradiograms of [³⁵S]methionine-labeled proteins in the actin regions of two-dimensional gels. Cells were grown in complete medium lacking methionine (*A*, *C*, *E*, and *G*) or in medium lacking histidine and methionine and containing histidinol at 20 mM (*B*) or 10 mM (*D*, *F*, and *H*). Cell strains include early-passage (*A* and *B*) and late-passage (*C* and *D*) A2 fibroblasts, old donor (JO69) (*E* and *F*), and the premature aging syndrome progeria P5 (*G* and *H*). Lines are drawn through β-actin. Similar results were obtained with the other untransformed cell strains listed in Table 1.

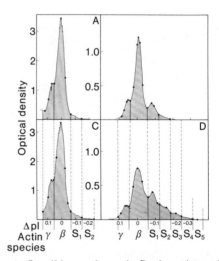

FIG. 4. Quantifying error frequencies. Densitometric traces from the actin region of autoradiograms of proteins synthesized by WI38 (*A* and *B*) and WI38-SV40 (*C* and *D*) are shown for unstarved (*A* and *C*) and starved conditions (10 mM histidinol) (*B* and *D*). The units from the densitometer scan have been scaled to show optical density and ΔpH relative to β-actin. Coordinates were assigned to points on the scans (●); fractions of total area corresponding to native actin (γ and β) and substituted actins S_1, S_2, etc. were calculated (Table 2).

Table 2.   Quantitation of error frequencies in WI38 and WI38-SV40 cells

Cell strain	Rel. rate of protein synthesis	Fraction of the total area in actin spots							Fract. of aberrant protein	Error freq.
		γ	β	S_1	S_2	S_3	S_4	S_5		
WI38:										
Unstarved	1.0	0.106	0.886	0.0085						
Starved	0.059	0.078	0.815	0.095	0.012	0.0005			0.108	0.0084
WI38-SV40:										
Unstarved	1.0	0.153	0.805	0.030	0.010					
Starved	0.051	0.079	0.547	0.192	0.104	0.053	0.023	0.002	0.374	0.0684

Data were obtained from analysis of scans shown in Fig. 4. The fraction of aberrant protein is ΣS_i. The error frequency (fraction of substituted sites) is $p' = \Sigma i S_i / 9$. A background of 0.005 obtained from averaging the calculated fraction of substituted sites in the unstarved state was subtracted from all determinations of p'.

DISCUSSION

The error catastrophe theory of aging (1) has received some support, but the evidence has been indirect (27–30). Our direct measurements of mistranslation in cultured human fibroblasts do not support the error catastrophe theory of aging in four respects. (*i*) Under normal, unstarved conditions old cells (late-passage cells from young donors and cells from old, progeric, or Werner syndrome donors) have proteins that are essentially indistinguishable from young (early-passage) cells of normal donors. (*ii*) Under conditions of histidine starvation, old cells have a similar or reduced level of mistranslation compared to young cells. (*iii*) No correlation exists between error frequency and replicative life span of cultured diploid cells. (*iv*) SV40-transformed (immortal) cells have elevated error frequencies during histidine starvation compared to either early- or late-passage mortal cells. Thus, although our comparisons of error rates apply strictly to actin synthesis, they rule out a general increase in the error frequency of protein synthesis during aging of the cell types studied here.

We have assumed that protein synthetic errors are induced during amino acid starvation by mispairing of pyrimidine bases in the third position of the codon–anticodon interaction (2). Both the occurrence and direction of stuttering in animal and bacterial cells during starvation for eight separate amino acids provide strong evidence for this interpretation (refs. 2, 3, and 31; unpublished data). Direct proof that ribosomal ambiguity is involved in such errors was recently shown by suppression of stuttering in streptomycin-resistant bacteria (32). Streptomycin resistance involves mutation of a ribosomal protein that reduces mistranslation by heightening the specificity of tRNA binding (33). Thus, these data confirm the role of the ribosome–tRNA interaction in the synthesis of aberrant protein during amino acid starvation and rule out alternative explanations such as tRNA misacylation or post-translational modifications.

Are comparisons of error frequencies at drastically reduced rates of protein synthesis relevant to error frequencies under normal conditions? In other, especially cell-free systems, conditions that augment mistranslation such as antibiotics, high magnesium concentrations, low temperature, high pH, and organic solvents, appear simply to amplify the ambiguity inherent in the protein synthetic apparatus (33, 34). Also, our values for mistranslation during various degrees of amino acid starvation give estimates of error frequencies at the histidine codon during normal (unstarved) conditions (35) that are consistent over a 5-fold range of protein synthetic rates (unpublished data). Finally, the derived error frequencies (Table 2) are similar to data reported at other codons measured by independent means (36–38). Thus, we believe that the error frequencies measured here during histidine starvation provide an accurate estimate of mistranslation under normal conditions.

Although a small increase (e.g., less than 30%) in the error frequency of old cells may have gone undetected, it is unlikely that this would be biologically significant. In bacteria, error

frequencies 20-fold greater than normal can be induced without loss of viability (39); *Drosophila*(40) and cultured human cells (41) can tolerate substantial incorporation of amino acid analogs into protein without reduction of their life span; large differences in error frequency among the normal cell strains do not correlate inversely with life span (Table 3). For example, the calculated error frequency of A2 is about twice as large as that of WI38, but its life span *in vitro* is about 20% longer. Indeed, SV40-transformed cells, which are immortal, have the greatest

Table 3.   Error frequencies of cultured human cells

Cell type	Replicative life span*	Error frequency $\times 10^4$ (n)[†]	
		Early passage	Late passage
Young donors			
WI38	55	0.6 ± 0.1 (7)	0.4 ± 0.1 (4)[§]
MRC5	65	1.5 ± 0.2 (2)	
A2	65	1.2 ± 0.2 (10)	0.9 ± 0.2 (8)[§]
GM37	50	1.6 ± 0.1 (3)	
Mean		1.0 ± 0.1 (22)	0.7 ± 0.2 (12)
Old donors			
JO69	50	0.8 ± 0.2 (3)	0.8 ± 0.1 (3)
JO88	48	1.0 ± 0.3 (4)	
Progeria			
P5	42	1.2 ± 0.2 (6)	
P18	53	1.3 ± 0.3 (4)	
Werner syndrome			
WS2	37	1.3 ± 0.3 (2)	
WS4	ND	0.5 ± 0.1 (3)	
Mean of old, progeria, & Werner syndrome donors		1.1 ± 0.1 (22)[‡]	0.8 ± 0.1 (3)
Mean of diploid cells		1.1 ± 0.1 (44)	0.8 ± 0.1 (15)
Immortal cells			
WI38-SV40		2.8 ± 0.2 (10)[¶]	

* Maximal number of population doublings (7). ND, not determined.

[†] The error frequency p at normal protein synthetic rates is derived by normalizing the fraction of substituted sites p' at reduced protein synthetic rates to the calculated step time of ribosomes at the histidine codon (see ref. 3). This step time was directly related to the relative protein synthetic rate r (normal rate/starvation rate) by computer simulations of protein synthesis (ref. 21; unpublished data). The equation relating the error frequency during normal protein synthesis to p' and r is $p = p'/12.53$ $(r - 0.92)$. Data are means $\pm$ SEM of n separately analyzed extracts of proteins labeled during histidine starvation.

[‡] The hypothesis that early-passage cells from old, progeria, or Werner syndrome donors have an error frequency greater by 33% than that of early-passage cells from young donors is rejected ($P < 0.05$).

[§] Late-passage cells from young donors have an error frequency less than that of the corresponding early-passage cells ($P < 0.05$).

[¶] SV40-transformed WI38 cells have an elevated error frequency compared to WI38 ($P < 0.0001$) or all diploid cells combined at early or late passage ($P < 0.0001$).

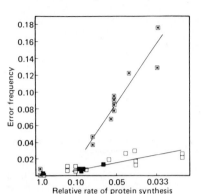

FIG. 5.   Error frequencies are plotted against the reciprocal protein synthetic rate for early-passage WI38 (□), late-passage WI38 (■), and WI38-SV40 ✕. Error frequencies represent the fraction of substituted histidine sites p'. Each point represents the error frequency determined by quantifying the scan of the autoradiogram from a single gel.

error frequencies of any of the human cells analyzed. Therefore, it seems that an elevated level of mistranslation is not involved in generating the aging phenotype.

Others have found similar evidence contrary to the error catastrophe theory. Wilson *et al.* (42) and Engelhardt *et al.* (43) have demonstrated that young and old cells *in vivo* and *in vitro* have nearly identical patterns of labeled proteins resolved by two-dimensional gel electrophoresis, which is now confirmed by our results under normal conditions. Other work in human fibroblasts has shown that the error rate of protein synthesis directed by poly(uridylic acid) does not increase with extracts from late-passage cells or cells from old, progeric, and Werner syndrome donors (44).

Our observations with WI38-SV40 are consistent with the idea that the synthesis of aberrant proteins may be involved in malignant transformation, carcinogenesis, or tumor progression (45, 46). For example, error-containing proteins could lead to an elevated mutation rate or abnormal regulation of gene expression with ultimate loss of growth control. We have found further support for the correlation between malignant transformation in general and increased error frequency, but this correlation has not been entirely consistent because there are exceptions to the rule. It has also been suggested that transformation might affect variables leading to stabilization of error rates, thus preventing the putative error catastrophe and establishing immortality at an elevated error frequency (47). Our observations on WI38-SV40 cells are in accord with this idea, but the relatively low error frequencies of old cells are not.

The data represent direct measurements of mistranslation in human cells. Although our results indicate that SV40-transformed cells have an elevated error frequency, they do not support the error catastrophe theory of aging.

This work was supported by grants to S.G. and C.P.S. from the Medical Research Council of Canada (MT3515 and MT1877) and to C.P.S. from the Natl. Cancer Institute of Canada and the National Institutes of Health (NOI-CP-4331). C.B.H. is a recipient of the National Research Council of Canada Science Scholarship. J.W.P. is a research fellow of the Natl. Cancer Institute of Canada. J.W.C. is a recipient of a Studentship from the Medical Research Council of Canada.

1. Orgel, L. E. (1963) *Proc. Natl. Acad. Sci. USA* **49**, 517–521.
2. Parker, J., Pollard, J. W., Friesen, J. D. & Stanners, C. P. (1978) *Proc. Natl. Acad. Sci. USA* **75**, 1091–1095.
3. O'Farrell, P. H. (1978) *Cell* **14**, 545–557.
4. Hayflick, L. (1965) *Exp. Cell Res.* **37**, 614–636.
5. Girardi, A. J., Jensen, F. C. & Koprowski, H. (1965) *J. Cell. Comp. Physiol.* **65**, 69–83.
6. Jacobs, J. P., Jones, C. M. & Baille, J. P. (1970) *Nature (London)* **227**, 168–170.
7. Harley, C. B. & Goldstein, S. (1978) *J. Cell. Physiol.* **97**, 509–517.
8. *The Human Genetic Mutant Cell Repository List of Genetic Mutants* (1978) United States Dept. of Health, Education and Welfare (NIH Publ. No. 80-2011).
9. Goldstein, S., Moerman, E. J., Soeldner, J. S., Gleason, R. E. & Barnett, D. M. (1979) *J. Clin. Invest.* **63**, 358–370.
10. Goldstein, S. & Moerman, E. J. (1975) *N. Engl. J. Med.* **292**, 1306–1309.
11. Goldstein, S. & Harley, C. B. (1979) *Fed. Proc. Fed. Am. Soc. Exp. Biol.* **38**, 1862–1867.
12. Yatscoff, R. W., Goldstein, S. & Freeman, K. B. (1978) *Somatic Cell Genet.* **4**, 633–645.
13. Cristofalo, V. J. (1976) *Gerontology* **22**, 9–27.
14. Goldstein, S. & Littlefield, J. W. (1969) *Diabetes* **18**, 545–549.
15. Stanners, C. P., Eliceiri, G. L. & Green, H. (1972) *Nature (London) New Biol.* **230**, 52–54.
16. Russell, W. C., Newman, C. & Williamson, D. H. (1975) *Nature (London)* **253**, 461–462.
17. Lowry, O. H., Rosebrough, N. J., Farr, A. L. & Randall, R. J. (1951) *J. Biol. Chem.* **193**, 265–275.
18. O'Farrell, P. H. (1975) *J. Biol. Chem.* **250**, 4007–4021.
19. Baxter, G. & Stanners, C. P. (1978) *J. Cell. Physiol.* **96**, 139–145.
20. Stanners, C. P., Adams, M. E., Harkins, J. L. & Pollard, J. W. (1979) *J. Cell. Physiol.* **100**, 127–138.
21. Harley, C. B. (1979) Dissertation (McMaster University, Hamilton, Ont., Canada).
22. McKee, E. E., Cheung, J. Y., Rannels, D. E. & Morgan, H. E. (1978) *J. Biol. Chem.* **253**, 1030–1040.
23. Bonner, W. M. & Laskey, R. A. (1974) *Eur. J. Biochem.* **46**, 83–86.
24. Collins, J. H. & Elzinga, M. (1975) *J. Biol. Chem.* **250**, 5915–5920.
25. Vandekerckhov, J. & Weber, K. (1978) *Eur. J. Biochem.* **90**, 451–462.
26. Hansen, B. S., Vaughan, M. H. & Wang, L. J. (1972) *J. Biol. Chem.* **247**, 3854–3857.
27. Baird, M. B., Samis, H. V., Massie, H. R. & Zimmerman, J. A. (1975) *Gerontology* **21**, 57–63.
28. Morrow, J. & Garner, O. (1979) *Gerontology* **25**, 136–144.
29. Gershon, D. (1979) *Mech. Ageing Dev.* **9**, 189–196.
30. Rothstein, M. (1979) *Mech. Ageing Dev.* **9**, 197–202.
31. Gallant, J. & Foley, D. (1980) in *Ribosomes: Structure, Function and Genetics*, eds. Chambliss, G., Craven, G. R., Davies, J., Kahan, L. & Nomura, M. (University Park Press, Baltimore, MD), pp. 615–638.
32. Parker, J. & Friesen, J. D. (1980) *Mol. Gen. Genet.*, in press.
33. Gorini, L. (1974) in *Ribosomes*, eds. Nomura, M., Tissieres, A. & Lengyle, P. (Cold Spring Harbor Laboratory, Cold Spring Harbor, NY), pp. 791–803.
34. Woese, C. R. (1967) in *Progress in Nucleic Acid Research and Molecular Biology*, eds. Davidson, J. N. & Cohen, W. E. (Academic, New York), Vol. 7, pp. 107–172.
35. Harley, C. B., Goldstein, S., Pollard, J. W. & Stanners, C. P. (1979) *Fed. Proc. Fed. Am. Soc. Exp. Biol.* **38**, 328 (abstr.).
36. Loftfield, R. B. (1963) *Biochem. J.* **89**, 82–92.
37. Loftfield, R. B. & Vanderjadt, D. (1972) *Biochem. J.* **128**, 1353–1356.
38. Edelmann, P. & Gallant, J. (1977) *Cell* **10**, 131–137.
39. Gallant, J. & Palmer, L. (1979) *Mech. Ageing Dev.* **10**, 27–38.
40. Shmookler Reis, R. J. (1976) *Interdiscip. Top. Gerontol.* **10**, 11–23.
41. Ryan, J. M., Duda, G. & Cristofalo, V. J. (1974) *J. Gerontol.* **29**, 616–662.
42. Wilson, D. L., Hall, M. E. & Stone, G. C. (1978) *Gerontology* **24**, 426–433.
43. Engelhardt, D. L., Lee, G. T. Y. & Moley, J. (1979) *J. Cell Physiol.* **98**, 193–198.
44. Wojtyk, R. I. & Goldstein, S. (1980) *J. Cell. Physiol.*, in press.
45. Loeb, L. A., Springgate, C. F. & Battula, N. (1974) *Cancer Res.* **34**, 2311–2321.
46. Hirsch, G. P., Holland, J. M. & Popp, R. A. (1978) in *Birth Defects. Orig. Art. Ser.*, ed. Bergsma, D. (Liss, New York), **14**, 431–448.
47. Holliday, R. (1975) *Fed. Proc. Fed. Am. Soc. Exp. Biol.* **34**, 51–55.

Altered Enzymes in Ageing Human Fibroblasts

R. HOLLIDAY & G. M. TARRANT

National Institute for Medical Research, Mill Hill, London NW7 1AA

Diploid human fibroblasts accumulate heat labile enzymes during the final stages of their life-span in culture. The RNA base analogue 5-FU induces premature senescence, which is preceded by the appearance of altered enzyme. These observations support Orgel's "error catastrophe" theory of ageing.

Hayflick and Moorehead[1,2] first demonstrated that diploid human fibroblasts have a limited life-span in culture, an observation which has subsequently been confirmed in many other laboratories. It has also been clearly shown that cultured fibroblasts from young donors have a longer life-span than those from old individuals[2,3]. A rare inherited condition in man, Werner's syndrome, leads to premature ageing with clinical symptoms very similar to those in the normal age range[4]. Martin *et al.*[3] have demonstrated that fibroblasts from four patients with Werner's syndrome could be subcultured only a few times before the cells died. These observations strongly suggest that the mortality of cultured fibroblasts is in some way related to the mortality of the individual. They do not show that the cause of fibroblast ageing is necessarily the main cause of normal ageing of the whole organism.

Clonal senescence has also been described and examined in detail in certain microorganisms, particularly fungi.

Evidence has been obtained with certain mutant strains of *Neurospora* that senescence is accompanied by the synthesis of defective protein molecules, with a rapid increase in their production just before growth ceases[5,6].

These results provide evidence for Orgel's theory of ageing[7], which suggests that one cause of ageing could be the accumulation of errors in protein synthesis, leading to a gradual but irreversible breakdown in the accuracy of the protein synthesizing machinery and finally to a lethal "error catastrophe". We demonstrate here that the clonal senescence of human fibroblasts is accompanied by the appearance of a proportion of heat labile enzyme. Moreover, by this biochemical test we can predict when a culture will become visibly senescent.

The Growth of MRC-5

We used the male foetal lung fibroblast strain MRC-5, which has been fully characterized[8]. It grows somewhat faster than the better known WI-38 and its life-span in culture appears to be longer. Most commonly, cells became visibly senescent after about fifty subcultures and ceased growth after 60–65 subcultures. Cells were grown at 37° C in Eagle's basal medium containing 10% foetal calf serum and 200 units of penicillin and 200 μg/ml. of streptomycin. Cultures were trypsinized and split 1:2 twice weekly. If a culture had not become confluent after 3 or 4 days, the medium was changed. The age of a culture was measured by its passage number rather than by chronological time.

Senescent cultures have slow growing cells of very variable size which often have granular cytoplasm. They tend not to line themselves up in the normal parallel arrays and do not reach the usual cell density when confluent. Part of the culture may achieve confluency, whereas other parts remain patchy. Very little is known about the viability of cells in fibroblast cultures of increasing age. We presume that there is considerable variation in the degree of senescence of cells in any ageing culture and that the more healthy cells will continually be selected from a heterogeneous population. It is often observed that a culture appears to have stopped growing, but that after one or more changes of medium it gradually becomes confluent and can then be subcultured several more times before growth finally ceases. In other words, a culture does not necessarily proceed inexorably and steadily to senescence and death ; it may well pass through phases of increasingly retarded growth with intervening phases of more normal growth.

Heat Labile Enzyme

Orgel's theory predicts that in ageing cells there should be a mixture of normal and abnormal protein and that the abnormal fraction should itself be heterogeneous, as the errors introduced would be a random collection of amino-acid substitutions. In such a situation, standard methods for detecting enzyme heterogeneity, such as starch or poly-acrylamide gel electrophoresis, are very unlikely to detect the altered fraction, which would be dispersed in many positions in the gel. It has been shown that random amino-acid substitutions frequently make an enzyme easily denaturable by heat. For example, Langridge[9,10] studied missense mutations of β-galactosidase where serine or glutamine was substituted for ochre or amber mutations. Of fifty-two serine substituted mutations in different positions in the protein, sixteen had a half-life at the inactivating temperature at least ten times less than the wild type enzyme, and a further fifteen had a half-life at least two times less. Of fifty-six naturally occurring human variants of glucose-6-phosphate dehydrogenase, twenty-seven are listed as being heat labile[11].

In ageing *Neurospora* cultures all the glutamic dehydrogenase became heat labile[6], but it would be surprising if mammalian cells were as tolerant of abnormal protein molecules. In order to detect a few per cent of heat labile enzyme, it is necessary to use an enzyme which is normally

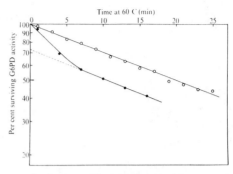

Fig. 2 Inactivation at 60° of G6PD obtained from young and senescent cultures. ○, Passage 23; ●, passage 61.

homogeneous and which therefore loses activity exponentially during heat treatment. (Mixtures of isoenzymes would not be expected to decay exponentially.) It is also essential to have an accurate assay. Of eight enzymes we have so far examined in crude extracts obtained from fibroblasts, only two have fulfilled these criteria: glucose-6-phosphate dehydrogenase (EC 1.1.1.49) and 6-phosphogluconate dehydrogenase (EC 1.1.1.44).

Glucose-6-phosphate Dehydrogenase

G6PD consists of a single polypeptide chain coded for by a gene on the X chromosome[12,13]. The enzyme is in a stable and active multimeric form only in the presence of NADP[14,15]. Activity is measured by the rate of formation of NADPH2 in the presence of the substrate (details of cell free extracts and the enzyme assay are given in the legend to Fig. 1). We carefully checked that with the assay conditions used, none of the increase in absorption at 340 nm is due to the activity of 6-phosphogluconate dehydrogenase[17]. To determine the fraction of heat labile enzyme, it is essential to measure as accurately as possible the activity before heat treatment. This initial measurement was carried out in triplicate ; the extract was then heated to 59° C or 60° C and samples removed and assayed at 2 min intervals. Almost all of our determinations are based on 12–15 assays and we believe that the estimate of the heat labile fraction of enzyme is correct to within 3–5% of the total initial activity. When both inactivating temperatures were used, the difference in the estimate of the heat labile fraction was never greater than this. The inactivation at 59° C of G6PD in three extracts from cultures of different age is shown in Fig. 1. The enzyme from the passage 22 cells had no heat labile fraction. The passage 48 cells showed no morphological symptoms of senescence, but by passage 51 they were visibly senescent. The passage 61 cells showed all the symptoms of senescence and the line died out at passage 63. More often the inactivation of enzyme at 60° C was followed: examples are shown in Fig. 2. The passage 61 cells were from the very last subculture and insufficient extract was available for the normal number of assays. Nevertheless, it is quite clear that these senescent cells contain about 27% heat labile altered enzyme, the largest proportion we have detected. Fig. 3 summarizes a large number of results obtained with extracts from cells of different ages. In general, young cells have about 5% heat labile enzyme, whereas old cells have 15–25%. The variation in the amount of heat labile enzyme in old cells is not due primarily to inaccuracy in measurement, but rather to the considerable variation in the degree of senescence which has already been referred to. The closed circles in Fig. 3 are the successive determinations of the proportion of heat labile enzyme in a single line cultured from the period before senescence until

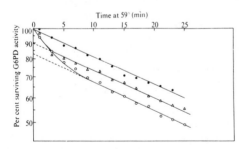

Fig. 1 Inactivation at 59° of glucose-6-phosphate dehydrogenase obtained from young, middle aged and senescent cells. Confluent cultures were rinsed with phosphate buffer saline (Dulbecco A) and harvested with trypsin. The cells were washed twice with 0.005 M Tris/HCl in 0.15 M NaCl (pH 7.4). The pellet was resuspended in 0.05 M Tris/HCl (pH 8.0) containing 10^{-3} M mercaptoethanol, 10^{-3} M EDTA, 4×10^{-3} M ε-amino-n-caproic acid and 10^{-3} M NADP, and the cells disrupted by sonication. Cell debris were removed by centrifuging at 18,000 r.p.m. at 4° C for 30 min. The supernatant (approximately 0.6 ml. per 20-oz. culture bottle) was assayed for G6PD activity. The assay consisted of 0.02 ml. enzyme and 0.78 ml. 0.25 M Tris/HCl buffer (pH 8.6) containing 9.5×10^{-3} M glucose-6-phosphate (Na salt), 3.2×10^{-4} M NADP and 1.9×10^{-2} M MgCl₂. The rate of increase in absorbance at 340 nm was measured with a Pye Unicam SP1800 spectrophotometer, each assay being run for 1 min. The extraction and assay procedures are based on those of Steele[16].
●, Passage 22; △, passage 48; ○, passage 61.

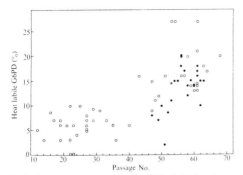

Fig. 3 Estimated proportion of heat labile G6PD in cultures of different ages. Each point is based on an experiment like those in Figs. 1 and 2. ○, Measurements with assorted cell lines of different age; ●, successive measurements on one cell line which finally died at passage 63. More than one point at the same passage number indicates either that both 59° C and 60° C were used to determine the heat labile fraction, and/or that cultures of the same passage number were examined on different days.

death. (More than one point at each passage number indicates that the enzyme was examined at both 59° C and 69° C, or that cells of the same passage number were examined before and three or more days after a change of medium.) In this experiment and on other occasions it was possible to predict when a culture will become visibly senescent by the appearance of significantly more than 10% of heat sensitive enzyme. The fraction of heat labile enzyme then rapidly increased to passage 56 and thereafter fluctuated. Orgel's theory predicts that the amount of altered protein should rise rapidly immediately before death; but there are two reasons why this might not be observed. One which has already been discussed is the heterogeneity of cultures due to the continual selection of the healthiest cells. The other is that the heat labile enzyme we are looking at is likely to have a shorter half-life *in vivo* than normal enzyme[18]. As cells become senescent, they grow much more slowly and make less protein, so altered enzyme will tend to disappear from such cultures more rapidly than normal enzyme.

Altered Substrate Specificity

G6PD can act on the analogue substrate 2-deoxyglucose-6-phosphate with low efficiency, the rate of the reaction being about 5% of that with the normal substrate. A number of naturally occurring mutant forms of the enzyme can use dG6P with very much higher efficiency[11], and errors in protein synthesis should also produce molecules with this property. But an amino-acid substitution which alters the substrate specificity would also in many cases make the enzyme heat labile. One would therefore expect a greater

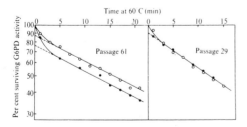

Fig. 4 Inactivation at 60° of G6PD with the normal substrate (○) or the analogue substrate, 2-deoxyglucose-6-phosphate (●), using extracts from young and senescent cultures. In the assay mix the concentration of dG6P (Na salt) was 10^{-2} M.

initial heat inactivation of the enzyme reacting with dG6P than with the normal substrate. After the altered molecules had decayed, the ratio of activities should remain constant. With cell free extracts from old cells this is precisely what we observed (Fig. 4). With young cells, the activity with the two substrates decays throughout the heat treatment at the same rate, even though a small proportion of heat labile enzyme is present. This strongly suggests that the heat labile enzyme in old cells is qualitatively different from that observed in extracts from young cells.

6-Phosphogluconate Dehydrogenase

This enzyme, 6PGD, is in the same pathway as G6PD. It, too, is measured by the rate of formation of NADPH₂ in the presence of the substrate, but the amount of activity is only about one tenth that of G6PD, and at the inactivating temperature used, 52° C, there is initially a slight increase of enzyme activity. For both these reasons the determination of the fraction of heat labile enzyme is less accurate than with G6PD, perhaps to within 5–10% of the total initial activity. We do not know whether young cells have a small fraction of heat labile enzyme, but senescent cells may contain at least 25%. Fig. 5 shows experiments which demonstrate this. If all the proteins in senescent cells are being synthesized inaccurately there would, of course, be a correlation between the proportion of heat labile G6PD and 6PGD. Fig. 6 shows that such a correlation can be demonstrated.

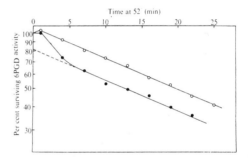

Fig. 5 Inactivation at 52° of 6-phosphogluconate dehydrogenase obtained from young and senescent cultures. The substrate was 7×10^{-4} M 6-phosphogluconic acid (Ba salt), otherwise the cell free extract and enzyme assay was the same as with G6PD. ○, Passage 25; ●, passage 64.

Experiments with 5-Fluorouracil

If ageing is due to the accumulation of errors in protein synthesis, then agents which directly or indirectly introduce mistakes in transcription or translation should induce premature senescence. It has been previously shown that amino-acid analogues can have this effect in *Drosophila*[19], *Podospora*[5] and *Neurospora*[6]. The base analogue 5-fluorouracil (FU) is incorporated into RNA and introduces errors in transcription or translation[20], and it has been shown in *E. coli*[21] and *U. maydis*[22] that heat labile enzyme is synthesized under these circumstances. We have examined the effects of growing fibroblasts in very low concentrations of FU. Concentrations lower than 1 μg ml.⁻¹ FU (in the form of the Tris aminomethane salt) allow cell division, and a growth experiment with 0.25 μg ml.⁻¹ is shown in Fig. 7. At each subculture the trypsinized cells were counted and split 1:4. Initially the cells were at passage 22: they grew vigorously in the medium containing FU for about ten cell doublings, finally reaching the equivalent of passage 32 before growth ceased. The control eventually reached passage 67. In this and other experiments the cells initially appeared to be normal, but after subculturing a few times the characteristic morphological symptoms of senescence appeared. It is, in fact, hard to distinguish cells grown for a few generations in FU from those which have become senescent in its absence.

Table 1 Estimated Proportion of Heat Labile G6PD in Fibroblasts grown in Medium containing 5-Fluorouracil and in Control Cultures

	Passage No.	Days of growth	Heat labile enzyme (%) With FU	Control
Experiment 1	19	3	11	10
0.25 µg ml.⁻¹ FU	21	10	10	—
	23	17	7	—
	25	24	8	8
	27	32	15	8
	29*	39	17	6
Experiment 2	17	5	5	7
0.125 µg ml.⁻¹ FU	19	9	9	7
	21	13	6	6
	23	16	10	6
	25	20	20	10
	27	27	21	8

* FU treated cells ceased growth at passage 31.

It might be expected that the addition of FU to cultures would immediately result in the synthesis of altered proteins, but no greater proportion of heat labile G6PD was initially observed in treated cells than in untreated controls. After a few generations of growth, however, heat labile enzyme appeared and the growth rate then fell. We have not followed the changes in the fraction of heat labile enzyme in cultures treated with FU in full detail, but the results of two preliminary experiments are summarized in Table 1. In both cases, a proportion of heat labile G6PD significantly higher than in the controls appeared after ten generations of growth in FU. The results strongly suggest that the analogue is having an indirect effect on the formation of altered enzyme, and this could be due to the development of an error catastrophe in protein synthesis.

Nature of the Altered Enzyme

The altered proteins we have observed in extracts from senescent cultures need not necessarily have been derived from faulty transcription or translation. It is possible that the heat labile enzymes are formed as a result of somatic mutation. In this case a minority of cells would be synthesizing only altered enzyme and the rest normal enzyme. We have no method at the moment for detecting single cells or colonies which produce entirely heat labile enzyme, but it is hard to believe that at least 25% of the population is

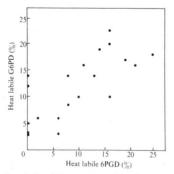

Fig. 6 Correlation between the estimated amount of heat labile G6PD and 6PGD in cultures of various ages. Each point is based on two experiments like those in Figs. 1, 2 and 5, carried out with the same cell free extract.

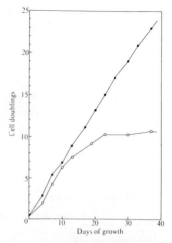

Fig. 7 Growth of MRC-5 cells in normal medium and in medium supplemented with 0.25 µg/ml. 5-fluorouracil (in the form of the Tris aminomethane salt). The experiment was started with passage 22 cells. At each subculture the cells were trypsinized, suspended in 4 ml. medium and counted with a haemocytometer. One ml. of suspension was added to each of two 8-oz. medical flats containing 20 ml. medium. When the FU treated cells became senescent, several bottles were inoculated and counted at intervals. None became confluent. ●, Control; ○, +5FU.

mutant for the genes coding for G6PD and 6PGD, for then every senescent cell would contain extremely large numbers of mutations, and many of these would be lethal and prevent growth long before the culture finally dies. Furthermore, if somatic mutations are simply accumulating in time, one would expect a gradual steady increase in the proportion of heat sensitive enzyme throughout the life-span, rather than the rapid rise during the last fifteen passages which is observed. There is, however, preliminary evidence that mutations do become frequent in senescent cultures (L. K. Hohmann and S. J. Fulder, personal communication) and, as we shall see, they may well play a role in ageing—but the rate of their occurrence is insufficient to account for the heat labile enzymes we observe.

Another possibility is that post-synthetic alteration of the enzyme might occur, for example, by the loss of amide groups; or they may become partially denatured. If protein synthesis or turnover is slower in old cells, there may be more time for the enzymes to be altered and thus a higher proportion of heat sensitive molecules would be seen. We have attempted to test this by depriving young cultures of serum, or by adding cycloheximide, and then preparing cell free extracts 2–3 days later. We detected no increase in heat labile enzyme; nor does prolonged storage at 5° C have this effect. The fact that two enzymes behave very similarly during ageing strongly argues against changes in heat lability as a result of post-synthetic modification. It is also unlikely that post-synthetic changes of this type alter the substrate specificity of the heat labile fraction (Fig. 4). Finally, it is hard to see how only prolonged growth in medium containing FU should have this effect. The most likely explanation of our results is that the heat sensitive enzymes are formed as a result of the breakdown of the fidelity of protein synthesis and the misincorporation of amino-acids into polypeptide chains.

Our results are in agreement with some other studies. Lewis and Tarrant (in preparation) compared the amount of lactic dehydrogenase enzyme activity with enzyme cross-reacting material (CRM) in MRC-5 cells throughout the life-span. A few passages before death, the enzyme/CRM

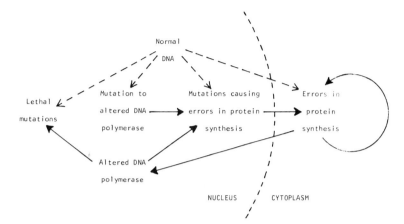

Fig. 8 Schema for senescence.

ratio decreased rapidly, suggesting that a considerable proportion of inactive enzyme was being synthesized. Wang *et al.*[23] found that phosphatases from late passage WI-38 fibroblasts responded in a different way to inhibitors from enzymes from young cells, and concluded that altered enzymes may be present.

If 25% heat labile G6PD is due to misincorporation of amino-acids, it is interesting to estimate what the error frequency might be. The molecular weight of the polypeptide chain is about 43,000 (ref. 14), perhaps 450 amino-acids. If we assume that most random amino-acid substitutions do not inactivate the enzyme and that about half of them make the enzyme heat labile (see above), then approximately one polypeptide in two contains an incorrect amino-acid—a misincorporation frequency of 10^{-3}. (This calculation leaves aside the question of the heat stability of enzyme molecules containing a mixture of normal and altered subunits.) Young fibroblasts usually contain about 5% heat labile G6PD, which could be due to an error frequency of about 2×10^{-4}, but the experiments on substrate specificity make it more likely that this small proportion of heat labile enzyme is formed by postsynthetic modification in all populations. Loftfield[24] estimated that the error frequency in the synthesis of chicken ovalbumin could not be higher than 3×10^{-4} per amino-acid residue.

Causes of Cellular Death

If altered proteins are synthesized in senescent fibroblasts, this does not necessarily mean that the cells are finally killed by the presence of an error catastrophe in protein synthesis. In previous studies of ageing with *Neurospora*[6], large numbers of mutations were detected in a strain, *leu-5*, which has an altered activating enzyme and synthesizes defective proteins, and the question arose whether these mutations were one of the causes of the ageing which is observed in this strain. It has also been shown that amino-acid analogues and FU are mutagenic[22], although it is not easy to see that they could act directly on DNA. It was therefore suggested that in all these cases mutations arose because faulty DNA polymerase molecules cause errors in genetic replication, as is known to be the case in bacteriophage T4[25]. Fig. 8 illustrates some general principles which may underlie the overall mechanism of ageing.

This schema will account for the genetic control of ageing, as both the initial error frequency in protein synthesis and the mutation rate would be strongly influenced by genetic factors. The less common the events marked by dashed arrows are, the greater the stability of the cell and its lifespan will be. If either an Orgelian increase in protein errors begins to develop or a chance mutation reduces the accuracy of protein synthesis, the cell will move into a metastable

condition in which any of the events marked by solid arrows can occur with increasing frequency. In the inevitable and irreversible change to a completely unstable state, the altered DNA polymerase may play a crucial role and the cell could eventually be killed either by an error catastrophe or by lethal mutation.

With regard to the initiating event, experiments on clonal ageing in fungi[26,27] and amoeba[28] show that the determinant of senescence is transmitted by the cytoplasm, not the nucleus, suggesting that slow buildup in errors in protein synthesis may be the primary cause of ageing[5]. It would, however, be very surprising if different organisms, or even different types of cell in the same organism, did not vary considerably in the relative importance of the nuclear and cytoplasmic changes which lead to the initiation and development of senescence.

We thank Dr Leslie Orgel for advice and encouragement, Dr George Sensabaugh for many helpful suggestions, and Mr J. P. Jacobs for supplying early passage MRC-5 cells.

Received April 18, 1972.

[1] Hayflick, L., and Moorehead, P. S., *Exp. Cell Res.*, **25**, 585 (1961).
[2] Hayflick, L., *Exp. Cell Res.*, **37**, 614 (1965).
[3] Martin, G., Sprague, C., and Epstein, C., *Lab. Invest.*, **23**, 86 (1970).
[4] Epstein, C. J., Martin, G. M., Schultz, A. L., and Motulsky, A. G., *Medicine*, **45**, 177 (1966).
[5] Holliday, R., *Nature*, **221**, 1224 (1969).
[6] Lewis, C. M., and Holliday, R., *Nature*, **228**, 877 (1970).
[7] Orgel, L. E., *Proc. US Nat. Acad. Sci.*, **49**, 517 (1963).
[8] Jacobs, J. P., Jones, C. M., and Baillie, J. P., *Nature*, **227**, 168 (1970).
[9] Langridge, J., *J. Bacteriol.*, **96**, 1711 (1968).
[10] Langridge, J., *Mol. Gen. Genet.*, **103**, 116 (1968).
[11] Yoshida, A., Buetler, E., and Motulsky, A. G., in *Mendelian Inheritance in Man* (edit. by McKusick, V. A.), 565a (Johns Hopkins University Press, Baltimore, 1970).
[12] Kirkman, H. N., *Advances in Human Genetics* (edit. by Harris, H., and Hirschhorn, K.), **2**, 1 (1971).
[13] Yoshida, A., *Biochem. Genet.*, **2**, 237 (1968).
[14] Yoshida, A., *J. Biol. Chem.*, **241**, 4966 (1966).
[15] Bonsignore, A., Cancedda, R., Nicolini, A., Damiana, G., and De Flora, A., *Arch. Biochem. Biophys.*, **147**, 493 (1971).
[16] Steele, M. W., *Biochem. Genet.*, **4**, 25 (1970).
[17] Glock, G. E., and McLean, P., *Biochem. J.*, **55**, 400 (1953).
[18] Goldberg, A. L., *Proc. US Nat. Acad. Sci.*, **69**, 427 (1972).
[19] Harrison, B. J., and Holliday, R., *Nature*, **213**, 990 (1967).
[20] Mandel, H. G., *Progr. Mol. Subcell. Biol.*, **1**, 82 (1969).
[21] Naono, S., and Gros, F., *CR Acad. Sci.*, **250**, 3527 (1960).
[22] Lewis, C. M., and Tarrant, G. M., *Mutation Res.*, **12**, 349 (1971).
[23] Wang, K.-M., Rose, N. R., Bartholomew, E. A., Balzer, M., Berde, K., and Foldvary, M., *Exp. Cell Res.*, **61**, 357 (1970).
[24] Loftfield, R. B., *Biochem. J.*, **89**, 82 (1963).
[25] Drake, J. W., Preparata, R. M., Allen, E. F., Forsberg, S. A., and Greening, E. O., *Nature*, **221**, 1128 (1969).
[26] Marcou, D., *Ann. Sci. Nat. Bot.*, **12**, 653 (1961).
[27] Jinks, J. L., *Extrachromosomal Inheritance* (Prentice Hall, New Jersey, 1964).
[28] Muggleton, A., and Danielli, J. F., *Exp. Cell Res.*, **49**, 116 (1968).

17

Reprinted from *Mech. Age. Devel.* **17**:151–162 (1981)

MOLECULAR BASIS FOR THE ACCUMULATION OF ACIDIC ISOZYMES OF TRIOSEPHOSPHATE ISOMERASE ON AGING

Pau M. Yuan*, John M. Talent, and Robert W. Gracy**

Departments of Chemistry and Biochemistry, North Texas State University/Texas College of Osteo-pathic Medicine, Denton, Texas 76203 (U.S.A)

(Received March 12, 1981)

SUMMARY

Triosephosphate isomerase exhibits acidic electrophoretic subforms in many tissues and these isozymes appear to increase during aging of erythrocytes and the eye lens. Incubation of the pure enzyme under mild alkaline conditions results in the generation of acidic forms which are identical to those found *in vivo*. Structural analysis of these isozymes from both *in vivo* and *in vitro* studies showed that they are the result of deamidation of two specific asparagines (Asn-15 and Asn-71). These labile asparagines are located in the subunit–subunit contact sites, and the deamidations introduce a total of four new negative charges in the contact site. The positions of the new aspartic acid residues are juxtaposed, thus creating charge–charge interactions which cause the dimeric enzyme to dissociate more readily. These studies (1) explain the molecular basis for the acidic isozymes observed in many tissues, (2) show that the deamination process is spontaneous and requires no intrinsic cell factors, (3) show that the deamination occurs in a sequential fashion with the deamidation of Asn-71 preceding the deamidation of Asn-15, and (4) suggest that proteolytic degradation processes may become altered during aging resulting in the accumulation of the deamidated intermediates of the normal catabolic process.

INTRODUCTION

Electrophoretic multiplicity of triosephosphate isomerase (TPI; EC 5.3.1.1.) has been reported by several laboratories over the past decade [1–10]. The formation of anodal subforms during the aging of human erythrocytes both *in vitro* [2] and *in vivo* [11] suggested that at least some of the subforms resulted from post-synthetic modification of the primary enzyme. On the other hand, the observation of a new electrophoretic form during transformation of lymphocytes and fibroblasts supported the idea that the multiplicity might be due to the expression of different genes. Recently, two groups of iso-

---

*Present address: Department of Immunology, City of Hope, Duarte, California, U.S.A.
**To whom correspondence should be addressed.

zymes, *viz*. TPI-B and TPI-A, were isolated from human placenta in this laboratory [12]. Although tryptic fingerprints of the two isozymes revealed several apparently unique peptides [12], a detailed structural analysis of the protein primary structure failed to reveal further evidence supporting the multiple gene theory [13].

In this communication we show that four acidic subforms from TPI-B can be generated *in vitro* by incubating the primary isozyme (TPI-B) in mild acid or base. These subforms are generated sequentially and are indistinguishable from the isozymes found in extracts from a variety of human tissues. The analyses of tryptic peptides showed that the multiple forms found *in vivo* or generated *in vitro* are due to the specific deamidation of two specific asparagine residues. The deamidations seem to occur *in vivo* as the first step in the scheduled catabolism of the enzyme.

MATERIALS AND METHODS

Human tissues were obtained from autopsy or by standard venipuncture. All tissues were judged to be normal and were used immediately. Bovine trypsin (DCC-treated, spec. act. 7500 U/mg) was obtained from Sigma Chemical Company. All other chemicals and solvents were analytical grade and were further purified by recrystallization or redistillation.

Human triosephosphate isomerase (TPI) was isolated according to the method described previously [12]. Rabbit muscle and chicken muscle TPI were prepared by a similar procedure described for human enzyme [13]. Enzyme assays were performed at 30 °C using glyceraldehyde-3-phosphate as the substrate [14]. Tryptic peptide analysis was carried out first by a primary separation with gel filtration which resolved the total peptides into seven pools, followed by two-dimensional, thin-layer fingerprinting. Details of the structural analysis methods have been published elsewhere [13].

Alkaline disc slab gels (7.5% resolving gel/3% stacking gel) were prepared according to the methods described previously [12]. Specific TPI activity stains were carried out using coupled formazan precipitation stain [9].

The three-dimensional model of chicken triosephosphate isomerase [15] was purchased from Vis-Aid Devices (Winnipeg, Manitoba, Canada) at a scale of 1 Å = 1.5 cm and an average of 0.5° for angle deviation such that every bond is ± 1% scale. The effects of differences in primary structure of the human and chicken enzyme on secondary and tertiary structure were evaluated by use of Chou and Fasman rules [16].

RESULTS

TPI isozymes from human tissues

Figure 1 shows the triosephosphate isomerase zymogram from various human tissues after electrophoresis on polyacrylamide gels. Kidney, heart, placenta, and liver extracts exhibit similar patterns (*i.e.* the presence of several clear anodal subforms).

171

Lymphocytes contained only TPI-B. However, mitogen-induced lymphoblasts exhibited another isozyme co-migrating with the third component found in the other tissues. Two subforms were detected in freshly prepared red blood cell extracts. When erythrocytes were fractionated according to age, it was found that, although young cells contain predominantly TPI-B, the more acidic forms accumulated in the old erythrocytes. Similar observations have been reported by Turner [11].

Generation of acidic subforms in vitro

By incubating pure TPI-B in acidic or alkaline conditions the acidic subforms were generated (Fig. 2). The sequential formation of these subforms was observed using both protein and enzyme activity stains of the gels. The electrophoretic migration and isoelectric focusing profiles of these subforms generated *in vitro* were identical to the migration of the isozymes found in tissue extracts. Prolonged incubation at more alkaline pH resulted in significant denaturation as indicated by streaking instead of tight banding patterns. Rabbit muscle TPI exhibited similar electrophoretic changes induced by incubation at alkaline pH. Chicken TPI, on the other hand, appeared as a single electrophoretic

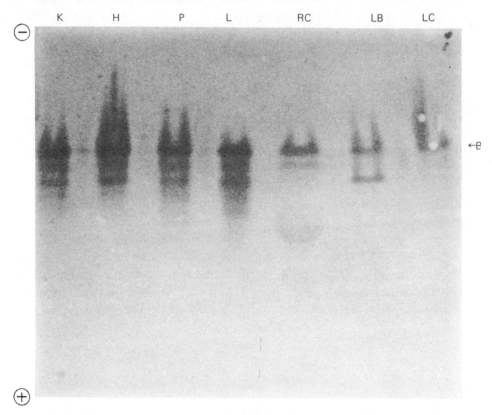

Fig. 1. Electrophoretic patterns of TPI from human tissues. Specific enzyme activity stains were carried out after alkaline slab-gel electrophoresis as described in the text. K = kidney; H = heart; P = placenta; L = liver; RC = red blood cells; LB = lymphoblasts; LC = lymphocytes.

ACTIVITY STAIN

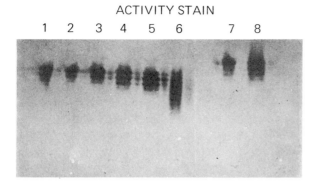

PROTEIN STAIN

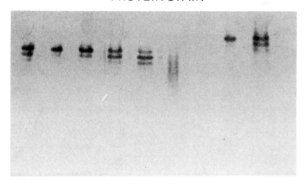

Fig. 2. *In vitro* formation of the acidic forms of TPI-B. Human TPI-B (0.2 mg/ml) was incubated in: (1) pH 6.0, 0.05 M sodium borate, 1 week, 4 °C; (2) pH 7.6, 0.05 M triethanolamine·HCl, 2 weeks, 4 °C; (3) pH 10.0, 0.05 M sodium borate, 1 week, 4 °C; (4) pH 11.0, 0.05 M sodium borate, 1 week, 4 °C; (5) pH 10.0, 0.05 M sodium borate, 2 weeks, 4 °C; (6) pH 11.0, 0.05 M sodium borate, 2 weeks, 4 °C. Rabbit TPI (0.2 mg/ml) was incubated in: (7) pH 7.6, 0.05 M triethanolamine·HCl, 2 weeks, 4 °C; (8) pH 11.0, 0.05 M sodium borate, 1 week, 4 °C. The electrophoresis gels were stained for TPI catalytic activity (top) and total protein (bottom).

form and no evidence of electrophoretic subforms was observed even after prolonged incubation in alkali or acid.

Peptide fingerprint analyses

A detailed analysis of human and rabbit-TPI tryptic peptides has been conducted recently in this laboratory [13]. Utilizing homology peptide mapping and comparison of the human and rabbit peptides, 90% of the primary sequence of the human enzyme has been established [13]. Figure 3 shows the peptide fingerprints of pool IV and pool VI derived from gel filtration. Two pairs of peptides, H-IV-3/H-IV-6 and H-VI-2/H-VI-4, were found, and each respective pair had identical amino acid compositions after acid hydrolysis (Table I). Structural analysis revealed that peptides H-IV-3 and H-IV-6 were identical and corresponded to the tetrapeptide in sequence position 14–17. The only difference was that H-IV-6 contained an asparagine rather than an aspartic acid. Similarly, peptides H-VI-2 and H-VI-4 corresponded to the large peptide at sequence position 69–85, the on-

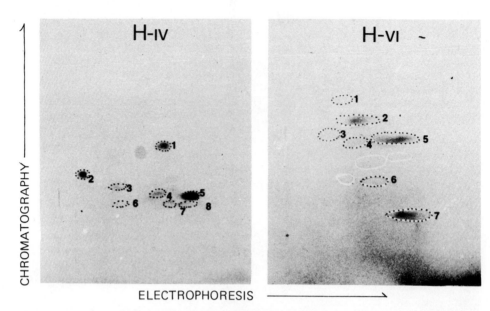

Fig. 3. Tryptic fingerprints of pool IV and pool VI from human TPI G-25 column chromatography. In figure H-IV at the left, the compositions of peptides 3 and 6 proved to be identical (see Table I). Peptide 6 represents the deamidated form of peptide 3. In the fingerprint from pool VI (at the right), the compositions of peptides 2 and 4 were found to be identical (see Table I). Peptide 4 is the deamidated form of peptide 2.

TABLE I

AMINO ACID COMPOSITIONS OF THE DEAMIDATED PEPTIDES

Amino acids	Nanomoles of amino acid/nanomoles of peptide			
	H-IV-3	H-IV-6	H-VI-2	H-VI-4
Asx	1.08 (1)	0.88 (1)	1.23 (1)	1.04 (1)
Thr			1.60 (2)	1.70 (2)
Ser			0.92 (1)	1.01 (1)
Glx			1.45 (1)	1.40 (1)
Gly	0.92 (1)	1.30 (1)	3.01 (3)	3.36 (3)
Ala			1.16 (1)	1.30 (1)
Val			0.80 (1)	1.08 (1)
Met	1.01 (1)	0.84 (1)	0.61 (1)	0.82 (1)
Ile			1.50 (1)	1.34 (1)
Leu			0.86 (1)	0.88 (1)
Tyr				
Phe			0.84 (1)	0.70 (1)
His				
Lys			1.20 (1)	1.01 (1)
Trp				
Arg	0.99 (1)	0.88 (1)		

Amino acid compositions were obtained from 24-hour acid hydrolysates of recovered peptides eluted from thin-layer fingerprints. The peptides were isolated as described in ref. 13 and designated as H = human; the roman numeral represents the primary pool from gel filtration, and the last digits are from the two-dimensional fingerprint (see Fig. 3). Values in parentheses are integral numbers of the residues established from the sequence studies [13].

174

ly difference being that peptide 4 was the result of a deamidation of Asn-71. The two de-amidated peptides were observed in peptide fingerprints of the enzyme isolated from normal human tissues as well as from subforms of TPI which were generated *in vitro*. From detailed structural studies it was found that the only residues which readily under-go deamidation either *in vivo* or *in vitro* are Asn-15 and Asn-71.

Sequential deamidation

　　In vivo most human tissues exhibit two minor subforms of TPI and in some cases three additional more-acidic subforms in lesser concentration. Thus, it seemed possible that a sequential deamidation might occur rather than Asn-15 and Asn-71 deamidating with the same half-life. *In vitro* deamidation studies corroborated this and suggested that Asn-71 deamidates at a faster rate than Asn-15. For example, by quantification of the recovered peptides, it was found that 33% of Asn-71 was deamidated under conditions where only 13% of Asn-15 was deamidated (Table II). Similar results were found in the analyses of rabbit peptides. Chicken TPI has an asparagine at position 15 but does not have an Asn at position 71 (Lys replacement). Since chicken TPI exhibited no deamida-tion under these conditions (Fig. 4), it is possible that the deamidation of Asn-71 is a prerequisite to the deamidation of Asn-15.

Location of Asn-15 and Asn-71 in the subunit–subunit contact sites

　　Figure 5 shows a three-dimensional model of TPI. The model is shown for only the peptide backbone and without the side groups. As seen from the model, Asn-71 from one subunit is juxtaposed to Asn-15 of the other subunit. The deamidation of Asn-15 and Asn-71 thus introduces four new negative charges into the subunit–subunit contact sites.

Subunit dissociation and stability

　　The introduction of two sets of charge–charge repulsions into the subunit contact sites could contribute to alterations in the stability of the dimeric form of the enzyme. In order to examine this quantitatively, the enzyme was treated under conditions that are known [17] to result in the dissociation of the dimeric enzyme into subunits. Figure 6 shows that the deamidated form of the enzyme is much more susceptible to dissociation

TABLE II

QUANTIFICATION OF DEAMIDATION RATES

Peptide	Sequence position	Deamidated residue	Recovery (nmoles)	Percentage deamidation
H-IV-3/H-IV-6	14–17	Asn-15	1.61	13
H-VI-2/H-VI-4	69–85	Asn-71	0.93	33
R-IV-3/R-IV-6	14–17	Asn-15	2.05	26
R-VI-2/R-VI-4	69–84	Asn-71	0.97	45

H = human TPI peptides, R = those from rabbit. Peptide isolation and designation are as in ref. 13 and as shown in Table 1 and Fig. 3.

175

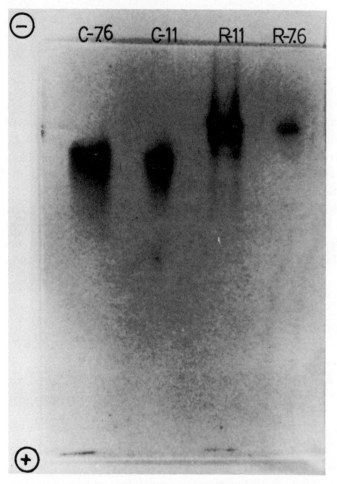

Fig. 4. Effect of the alkaline-induced deamidation of TPI from rabbit and chicken. The pure enzymes were incubated for 1 week at 4 °C at either pH 7.6 or pH 11.0 as described in Fig. 2. Gels were stained for total protein with Coomassie Blue. The samples are as follows: C-7.6 = chicken TPI incubated at pH 7.6; C-11 = chicken TPI incubated at pH 11; R-11 = rabbit TPI incubated at pH 11; R-7.6 = rabbit TPI incubated at pH 7.6.

in the presence of guanidinium chloride than is the nondeamidated form. Moreover, the higher the degree of deamidation, the greater is the tendency of the enzyme to dissociate.

DISCUSSION

Alkaline pH has been observed to cause the generation of acidic subforms of carbonic anhydrase [18] and purine nucleoside phosphorylase [11]. The chemical mechanism of the *in vitro* modification of protein has been shown to involve a progressive

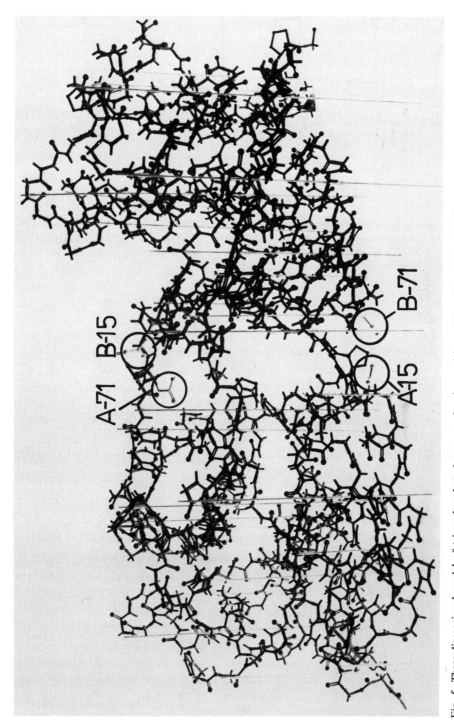

Fig. 5. Three-dimensional model of triosephosphate isomerase showing positions of Asn-71 and Asn-15. The two subunits are shown with only the peptide backbone except for the side-chains of Asn-71 and Asn-15 from each subunit. The subunit designated "A" is on the left and the subunit designated "B" is to the right.

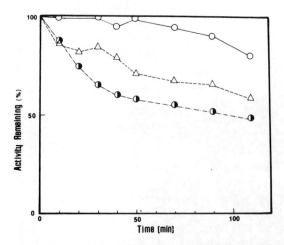

Fig. 6. Dissociation of native and deamidated TPI in guanidinium chloride. Dissociation titrations were carried out as described by Sawyer *et al.* [18]. The figure shows the time course of dissociation and activity loss in 0.75 M guanidinium chloride with 50 mM triethanolamine (pH 8.0) at 30 °C. (o), TPI-B; (△), deamidated band 1; (◐), deamidated band 2 (TPI-A).

deamidation of labile asparagines and glutamines [17]. The *in vitro* generated subforms of human triosephosphate isomerase were indistinguishable from the forms found in aged red blood cells and lens [7, 11, 19]. The electrophoretic and isoelectric focusing patterns, peptide fingerprints and amino acid compositions of the *in vivo* and *in vitro* generated subforms were identical. A sequential formation of progressively acidic subforms at the expense of the primary enzyme indicates that the deamidation is not a random process. Both the human and rabbit enzymes have the labile asparagines at the same positions in the primary sequences. Based on the quantification of deamidated and amidated peptides determined by amino acid analyses, it appears that Asn-71 is more labile than Asn-15. Chicken TPI has a lysine residue at position 71 rather than an asparagine, and no deamidation is observed. Even under harsh conditions the deamidation of Asn-15 did not occur in the chicken enzyme. Thus, it appears that the new aspartic acid at position 71 participates in the deamidation of Asn-15.

Figure 7 shows a proposed schematic representation of the molecular basis of the multiple electrophoretic forms of the enzyme. An electrophoretic pattern consisting of five components is predicted based on the assumption that the half-life of Asn-15 is significantly longer than the half-life of Asn-71. This model also presents a logical explanation for the findings that the second component can be dissociated and reassociated into the first three components [5, 6]. This model is thus consistent with the general observation of five bands of active TPI after electrophoresis.

The two sets of asparagines are in regions of random coil in close proximity, and located at the surface of the subunit–subunit contact sites. To examine the effects of the deamidation on the subunit–subunit interaction sites, low concentrations of guanidinium chloride were used to bring about the dissociation of TPI according to the method described by Sawyer *et al.* [18]. Under these conditions the denaturant brings about a

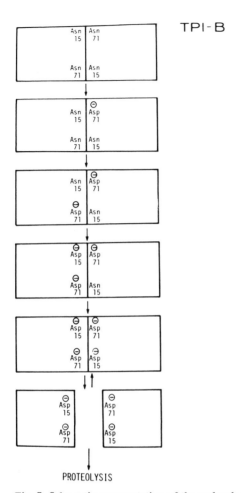

TPI-B

Fig. 7. Schematic representation of the molecular basis of TPI multiple forms and possible mechanism for normal degradation of the enzyme.

dissociation of the dimer into monomers. A significant difference was found in the dissociation titration curves of the partially deamidated sample and the primary enzyme. A similar instability was also demonstrated for the deamidated forms isolated directly from placenta [12]. The deamidation also corroborates the previous observation of lability of the more acidic subform found in lymphoblasts and fibroblasts [2]. Thus, it appears that the specific deamidations of Asn-71 and Asn-15 occur *in vivo* and, although having little effect on the catalytic activity, destabilize the subunit—subunit interactions. In this way a spontaneous dissociation of the enzyme would be favored when all four asparagines were deamidated. The susceptibility of the dissociated enzyme to proteolytic digestion is much greater than that of the native enzyme. While the precise proteases responsible for normal *in vivo* catabolism of the enzyme are unknown, it is known that *in vitro* the native dimer is extremely resistant to protease digestion. For example, the enzyme can be incubated with trypsin or chymotrypsin at 37 °C overnight with no changes or loss in catalytic activity, but when dissociated tryptic hydrolysis proceeds rapidly [13].

Spontaneous deamidation has been proposed as a general mechanism which could trigger the first step in protein catabolism [20]. The rates of deamidation appear to be highly dependent on the sequence of amino acids surrounding specific asparagines or glutamines. In contrast, others [21, 22] have suggested the existence of some active factor which is capable of deamidating specific asparagines. The present study in which identical specific deamidation occurs *in vivo* and *in vitro* with the pure enzyme supports the former concept of a spontaneous deamidation. The study also suggests that the specificity of deamidation is a function of neighboring groups which are the consequence of tertiary structure.

A variety of studies have shown that during the aging process deamidated forms of proteins accumulate (for review see ref. 22). Pushkina [23] reported the increased content of aspartic acid and decreased content of asparagine in aged rat brain, liver and heart. Cytochrome *c* [24], aldolase [25], α-crystallin [26], carbonic anhydrase [18] and nucleoside phosphorylase [11] are examples of proteins where specific deamidations have been found to accompany the aging process. However, as Dreyfus *et al.* [22] have pointed out, a clear mechanism of how deamidation is related to a step in the catabolism of the protein has not emerged in any case. Triosephosphate isomerase thus appears to provide the first explanation of how deamidation can initiate a series of events leading to its catabolism.

Reiss and Rothstein [27, 28] have proposed the following scheme for the catabolism of proteins:

$$\text{Protein synthesis} \longrightarrow \text{Normal enzyme} \longrightarrow \text{Altered enzyme} \longrightarrow \text{Denatured enzyme} \longrightarrow \text{Proteolysis}$$

The observations made in this study support this general scheme and suggest that the deamidated form of TPI is the "altered enzyme" in the scheme. The accumulation of this during aging does not necessarily represent an increased rate of formation of altered enzyme from the normal enzyme but could be due to a defect in the process at a later point in the scheme. For example, a defect in the activity or specificity of proteases to hydrolyze the deamidated TPI could account for the accumulation of the deamidated forms of TPI during aging. Obviously such a defect at the level of the proteases could also account for the accumulation of other "altered proteins" which have been observed during aging in many other systems.

ACKNOWLEDGMENTS

This work was supported in part by grants from the National Institutes of Health (AM14638 and AGO1274) and the Robert A. Welch Foundation (B-502).

REFERENCES

1 P. M. Burton and S. G. Waley, The active centre of triose phosphate isomerase. *Biochem. J., 100* (1966) 702–710.
2 M. V. Kester, E. L. Jacobson and R. W. Gracy, The synthesis of a labile triosephosphate isomerase isozyme in human lymphoblasts and fibroblasts. *Arch. Biochem. Biophys., 180* (1977) 562–579.

180

3 I. L. Norton, P. Pfuderer, D. C. Stringer and F. C. Hartman, Isolation and characterization of triosephosphate isomerase from rabbit muscle. *Biochemistry, 9* (1970) 4952–4958.

4 E. W. Lee, J. A. Barrisco, M. Pepe and R. Snyder, Purification and properties of liver triosephosphate isomerase. *Biochim. Biophys. Acta, 242* (1971) 261–267.

5 H. Rubinson, M. Vodovar, M. C. Meienhofer and J. C. Dreyfus, A unique electrophoretic pattern of triosephosphate isomerase in human cultured fibroblasts. *FEBS Lett., 13* (1971) 290–292.

6 W. K. G. Kreitsch, P. G. Pentche and H. Klingenburg, Isolation and characterization of the isoenzymes of rabbit-muscle triosephosphate isomerase. *Eur. J. Biochem., 23* (1971) 77–85.

7 T. H. Sawyer, B. E. Tilley and R. W. Gracy, Studies on human triosephosphate isomerase II. Nature of the electrophoretic multiplicity in erythrocytes. *J. Biol. Chem., 247* (1972) 6499–6505.

8 J. Peters, D. A. Hopkinson and H. Harris, Genetic and non-genetic variation of triosephosphate isomerase isozymes of human tissues. *Ann. Hum. Genet., 36* (1973) 297–312.

9 R. M. Snapka, T. H. Sawyer, R. A. Barton and R. W. Gracy, Comparison of the electrophoretic properties of triosephosphate isomerases of various species. *Comp. Biochem. Physiol. B, 49* (1973) 733–741.

10 M. V. Kester and R. W. Gracy, Alteration of human lymphocyte triosephosphate isomerase during blastogenesis. *Biochem. Biophys. Res. Commun., 65* (1975) 1270–1277.

11 B. M. Turner, Post translational alterations of human erythrocyte enzymes. *Isozyme vol. I, Molecular Structure,* C. L. Markert (ed.), Academic Press, New York, London, 1975, pp. 781–795.

12 P. M. Yuan, R. N. Dewan, M. Zaun, R. E. Thompson and R. W. Gracy, Isolation and characterization of triosephosphate isomerase isozymes from human placenta. *Arch. Biochem. Biophys., 198* (1979) 42–52.

13 P. M. Yuan, J. M. Talent and R. W. Gracy, Elucidation of the sequence of human triosephosphate isomerase by homology peptide mapping. *Biochim. Biophys. Acta,* in press.

14 R. W. Gracy, Triosephosphate isomerase from human erythrocytes. *Methods Enzymol., 41* (1975) 442–447.

15 D. W. Banner, A. C. Bloomer, G. A. Petsko, D. C. Phillips, C. I. Pogson, I. A. Wilson, P. H. Corran, A. J. Furth, J. D. Milman, R. E. Offord, J. D. Priddle and S. G. Waley, Structure of chicken muscle crystallographically at 2.5 Å resolution. *Nature, 255* (1975) 609–614.

16 P. Y. Chau and G. D. Fasman, Empirical predictions of protein conformation. *Ann. Rev. Biochem., 47* (1978) 251–276.

17 T. H. Sawyer and R. W. Gracy, Ligand binding and denaturation titration of free and matrix-bound triosephosphate isomerases. *Arch. Biochem. Biophys., 169* (1975) 51–57.

18 S. Funakoshi and H. F. Deutsch, Human carbonic anhydrases II. Some physiochemical properties of native isozymes and similar isozymes generated *in vitro. J. Biol. Chem., 244* (1969) 2436–3446.

19 H. Skala-Rubinson, M. Vibert and J. C. Dreyfus, Electrophoretic modifications of three enzymes in extracts of human and bovine lens. *Clin. Chim. Acta, 70* (1976) 385–390.

20 A. B. Robinson, J. H. McKerrow and P. Cary, Controlled deamidation of peptides and proteins – an experimental hazard and a possible biological timer. *Proc. Natl. Acad. Sci. U.S.A., 66* (1970) 753–757.

21 F. Mennecier and J. C. Dreyfus, Molecular aging of fructose-biphosphate aldolase in tissues of rabbit and man. *Biochim. Biophys. Acta, 364* (1974) 320–326.

22 J. C. Dreyfus, A. Kahn and F. Schapira, Post-translational modifications of enzymes. *Curr. Top. Cell. Regul., 14* (1978) 243–297.

23 N. V. Pushkina, Amidirorannost' belkor aristarenii organizma. *Ukr. Biokhim. Zh., 51* (1979) 680–683.

24 T. Flatmark, Multiple molecular forms of bovine heart cytochrome *c*: A comparative study of their physicochemical properties and their reactions in biological systems. *J. Biol. Chem., 242* (1967) 2454–2459.

25 C. Y. Lai, C. Chen and B. L. Horecker, Primary structure of two COOH-terminal hexa-peptides from rabbit muscle aldolase: A difference in the structure of the alpha and beta subunits. *Biochem. Biophys. Res. Commun., 40* (1970) 461–468.

26 F. S. M. Van Kleef, W. W. De Jong and H. J. Hoenders, Stepwise degradations and deamidation of the eye lens protein α-crystallin in aging. *Nature, 258* (1975) 264–266.

27 U. Reiss and M. Rothstein, Age-related changes in isocitrate lyase from the free living nematode, *Turbatrix aceti. J. Biol. Chem., 250* (1975) 826–830.

28 M. Rothstein, Aging and alteration of enzymes — a review. *Mech. Ageing Dev., 4* (1975) 325–338.

18

Reprinted from *Mech. Age. Devel.* **20**:93–101 (1982)

INCREASED LABILITY OF TRIOSEPHOSPHATE ISOMERASE IN PROGERIA AND WERNER'S SYNDROME FIBROBLASTS

TRYGVE O. TOLLEFSBOL, M. REBECCA ZAUN and ROBERT W. GRACY*

Departments of Chemistry and Biochemistry, North Texas State University/Texas, College of Osteopathic Medicine, Denton, Texas 76203 (U.S.A.)

(Received December 22, 1981; in revised form April 30, 1982)

SUMMARY

Triosephosphate isomerase was found to have an increased thermolabile component in skin fibroblasts from patients with progeria and Werner's syndrome when compared with normal fibroblasts. Mixtures of cell extracts from progeria or Werner's syndrome with normal fibroblasts gave intermediate levels of the heat-labile triosephosphate isomerase suggesting the absence of cytosolic destabilizing factors. The incorporation of the protease inhibitors 1-tosylamide-2-phenylethyl chloromethyl ketone, N-α-p-tosyl-L-lysine chloromethyl ketone, phenylmethyl-sulfonyl fluoride, and pepstatin A in cell extracts failed to affect the level of the labile form of triosephosphate isomerase. The labile component also accumulates in normal fibroblasts in late passage in tissue culture and appears to be identical to the deamidated form of the enzyme which accumulates in other aging cells.

INTRODUCTION

Several pathological syndromes characterized by accelerated aging processes are known. Progeria, or the Hutchinson-Gilford syndrome, is one such disorder. The precise genetic mode of transmission of this disease is still unknown [1] although an autosomal recessive transmittance has been proposed [2–4]. Werner's syndrome is considered to be the adult form for progeria and its etiology results from a rare autosomal recessive allele [5]. The features of each disease include dwarfism, aged appearance, graying of the hair with alopecia, wide-spread tissue degeneration, and insulin resistance [1, 5]. Severe atherosclerosis also occurs frequently leading to death from coronary or cerebrovascular occlusion [1, 6]. Overt diabetes is common in Werner's syndrome as are the occurrences of sarcomatous tumors [5, 6]. This syndrome also frequently manifests with skin ulcerations, lenticular cataracts, and osteoporosis [7].

---

*To whom correspondence should be directed.

Skin fibroblasts from persons with progeria and Werner's syndrome have a reduced replicative life span in comparison to normals [8]. The mean population doubling (MPD) of 50.3 for normal fibroblasts [9] is significantly reduced in fibroblasts from progeria (MPD = 32.6) and Werner's syndrome (MPD = 8.6) [5, 10]. Several studies have suggested enzyme abnormalities in fibroblasts from progeria and Werner's syndrome [11–17]. Goldstein and Moerman [14, 16] showed that glucose-6-phosphate dehydrogenase (G6PDH) and 6-phosphogluconate dehydrogenase (6PGDH) activities in skin fibroblasts decrease in the premature aging syndromes. They also observed biphasic thermal inactivation rates for these enzymes from the premature aging syndromes in contrast to monophasic inactivation rates from young normal individuals. Houben *et al.* [17] have recently measured the thermolability of certain cytosolic, lysosomal and mitochondrial enzymes in Werner's syndrome skin fibroblasts. Although the lysosomal or mitochondrial enzymes in these cells did not exhibit biphasic inactivation patterns, there was a thermolabile component of the cytosolic G6PDH. The authors postulated that a cytoplasmic (but not lysosomal or mitochondrial) post-translational mechanism may account for the increased heat-labile forms of the enzyme found in premature aging fibroblast cells. Unfortunately, the nature of the post-synthetic modifications leading to the more labile enzymes has not been elucidated. Clearly, if the nature of such post-synthetic structural alterations were known, it could provide a great deal of information relating to the mechanism for the accumulation of labile enzymes in aging cells.

We have recently demonstrated that triosephosphate isomerase (TPI; EC 5.3.1.1) undergoes two specific, spontaneous deamidations during aging of erythrocytes and other human tissues [18]. The deamidations of TPI are in the subunit—subunit contact sites and cause the dimeric enzyme to dissociate. In normal young cells the deamidations seem to initiate the catabolism of the protein and make the dissociated enzyme more susceptible to proteolysis. In old cells the deamidated forms accumulate, perhaps due to defects in proteolytic degradation [18–20]. It was thus of interest to determine if the deamidated forms of TPI accumulate in the cells from individuals with premature aging syndromes.

MATERIALS AND METHODS

Cell cultures

Normal, Werner's syndrome and progeroid skin fibroblast cultures were obtained from the Human Genetic Mutant Cell Repository. Cell lines GM 2938, AG 4054, AG 2602, AG 2261 and AG 4059 (ages 3, 29, 35, 61 and 96 years, respectively) were from normal males. Cultures AG 3593 and AG 780 were from males aged 28 and 54 years with Werner's syndrome. The progeroid cell lines used were AG 1177, AG 3199, and AG 3513 from a 9-year-old male, a 10-year-old female and a 13-year-old male, respectively.

All cell lines were grown in a humidified 5% CO_2 atmosphere at 37 °C in Eagle's minimum essential medium, Earle's salt base supplemented with 2 mM L-glutamine,

50 μg/ml gentamycin sulfate and fetal calf serum (10–20%). Maintenance of cells consisted of three feedings per week and subculturing monolayers 1–2 days post confluence at a split ratio of 1:2 for cells with premature aging and 1:3 for normal cells. Cultures were discontinued if confluency was not reached in two weeks. Cells were harvested with 0.05% trypsin, 0.02% EDTA, washed twice with cold phosphate buffered saline (PBS), resuspended in cold PBS and placed in ice. The cells were counted in a hemacytometer and assessed for viability using 0.4% trypan blue. The cells were disrupted in a Dounce homogenizer in 10 mM triethanolamine (TEA), 1 mM EDTA, 0.05% 2-mercaptoethanol, pH 7.6. The suspension was then centrifuged at 10 000 g for 3 min, and the cell extract supernatant solution removed and immediately assayed.

Triosephosphate isomerase assay

Triosephosphate isomerase catalytic activity was measured by a continuous coupled enzyme spectrophotometric assay [21]. The assays were monitored at 30 °C in a total volume of 1.0 ml containing 50 mM TEA, 0.15 mM NADH, 1.5 mM glyceraldehyde-3-phosphate, 1.0 mM EDTA, and 1.0 unit of α-glycerolphosphate dehydrogenase, pH 7.6.

Thermolability studies

Enzyme levels of extracts were adjusted to equal concentrations and incubated at 57 °C. For each experiment, normal (control) cell extracts were simultaneously incubated with the extracts from the premature aging syndromes. Aliquots of each sample were removed at timed intervals and assayed for TPI catalytic activity. Initial activities were measured in triplicate and subsequent assays were measured in duplicate. In cases utilizing protease inhibitors the cell extracts contained 0.5 mM phenylmethyl-sulfonyl fluoride (PMSF), 0.5 mM 1-tosylamide-2-phenylethyl chloromethyl ketone (TPCK), 0.5 mM N-α-p-tosyl-L-lysine chloromethyl ketone (TLCK), or 5×10^{-4} mM pepstatin A. All incubations were in 10 mM triethanolamine (hydrochloride), 1 mM EDTA, 0.05% 2-mercaptoethanol, and 1% 2-propanol, pH 7.6. Linear regression, correlation coefficient, best fit, slope, half-life, and Y-intercept (percentage heat-labile enzyme) were calculated for all experiments.

RESULTS

Table I shows the mean population doublings (MPD) and percentage viability of the skin fibroblast cultures of progeroid, Werner's syndrome, and normal cells. The absolute initial yield of enzyme (expressed as units per mg of protein) from the different cell types ($n = 5$) were: normal = 9.47 ± 1.19 units/mg; Werner's = 7.05 ± 0.67 units/mg; and progeria = 5.81 ± 0.92 units/mg. The MPD for both the progeria and Werner's cells were significantly below the normal. This is consistent with previous reports [8–12, 22]. It is interesting to note, however, that the cell viability remained at a fairly high level for all three cell types.

TABLE I

MEAN POPULATION DOUBLING AND CELL VIABILITY OF FIBROBLASTS FROM NORMAL, PROGERIA AND WERNER'S SYNDROME

Cell type	n[a]	Mean population doubling	Percentage viability
Normal	6	50.3 ± 2.1[b]	94.7 ± 3.8
Progeria	6	10.8 ± 4.6	94.4 ± 4.3
Werner's syndrome	3	8.8 ± 3.5	88.8 ± 7.3

All values represent the mean ± standard deviation.
[a]n = number of cell cultures.
[b]From ref. 10.

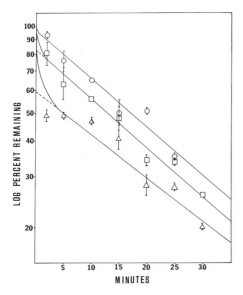

Fig. 1. Stability of progeria and normal fibroblast TPI at 57 °C. The log percentage remaining TPI is indicated at time intervals. Dashed lines represent the linear regression extrapolation to the Y-axis. Progeria extract TPI is represented by triangles (△) and normal by circles (○). An equal mixture of the two is represented by squares (□). The range of the mean is shown by vertical bars. Initial assays at zero time were measured in triplicate and subsequent assays were measured in duplicate. All correlation coefficients are better than −0.95. Each extract contained 0.5 mM PMSF. The normal extract was obtained at passage 8 and was from a 35-year-old male. The progeria extract was at passage 6 and was from a 13-year-old male.

    Stability studies on TPI from fibroblasts of progeria and normal individuals are shown in Fig. 1. The enzyme from normal fibroblasts exhibited a slow linear loss of activity with time. In contrast, the enzyme from progeroid fibroblasts exhibited a very labile component in addition to a component with identical stability to the normal cells. Mixtures of these two extracts gave intermediate values. Table II summarizes the percentages of the thermolabile TPI.

185

TABLE II

CONTENT OF LABILE TRIOSEPHOSPHATE ISOMERASE FROM NORMAL, PROGEROID, AND WERNER'S SYNDROME FIBROBLASTS*

Extract	Percentage heat-labile enzyme	Linear regression correlation coefficient
Normal	1.6	−0.9714
50% normal + 50% progeria	18.2	−0.9844
Progeria	41.4	−0.9560
50% normal + 50% Werner's syndrome	12.9	−0.9846
Werner's syndrome	20.1	−0.9891
Young normal (early passage)	1.1	−0.9538
Young normal (late passage)	24.2	−0.9900
50% young normal + 50% old normal	9.0	−0.9047
Old normal	9.8	−0.9117

*Incubations were at 57.0 °C. Percentage heat-labile enzyme was determined by linear regression and extrapolation to the Y-axis (log percentage remaining TPI). All samples contained 0.5 mM PMSF. The "normal" extracts were from a 35-year-old male at passage 8 and a 29-year-old male at passage 7. The progeria extract was from a 13-year-old male at passage 6 and the Werner's sample was from a 54-year-old male at passage 5. The "young normal" was from a 3-year-old male at passage 6 and the "old normal" was from a 61-year-old male at passage 10. The "young normal-late passage" was from a 13-year-old male at passage 27 (38.5 MPD).

Figure 2 depicts the lability of TPI from Werner's syndrome fibroblasts. A similar pattern emerged with TPI from normal cells being monophasic and the most stable whereas the enzyme from Werner's cells was biphasic and less stable (see also Table II).

All skin fibroblasts represent early passage, but since the ages of the normals varied, a comparison of early passage "young normal" with early passage "old normal" was conducted. Table II shows that the young normal cells exhibited only 1.1% heat-labile TPI as compared to a 9.8% fraction in the "old normal" sample. However, even this value is significantly lower than that found for either progeria or Werner's syndrome. It is also interesting to note that the content of deamidated TPI increased in fibroblasts from young donors as the mean population doubling increased (Table II).

In order to determine the extent to which proteolysis might account for the heat-labile TPI components, a comparison of the enzyme activity in progeria versus normal cells was conducted in the presence of various protease inhibitors. Figure 3 compares TPI from progeria and normal cells in the presence and absence of TLCK. The enzyme from normal fibroblasts was much more stable than that from progeria, as in the previous experiments. The presence of the tryptic inhibitor had no effect on the stability of the normal or progeroid cells' enzymes. Table III summarizes the protease inhibitor studies. There was essentially no labile component of TPI from normal skin fibroblasts in the presence or absence of the protease inhibitors. In all cases, the enzyme from progeria fibroblasts exhibited a major labile component. Moreover, in all cases (both normal and progeria) no significant difference was found in the presence or absence of any of the protease inhibitors.

186

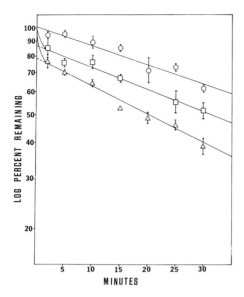

Fig. 2. Stability of Werner's syndrome and normal fibroblast TPI at 57 °C. Werner's extract is represented by triangles ($\triangle$), a normal extract by circles ($\circ$), and an equal mixture of the two by squares ($\square$). The range for the two determinations is shown by vertical bars. Correlation coefficients were better than -0.95. All extracts contained 0.5 mM PMSF. See Fig. 1 for further explanation. The normal extract was from a 29-year-old male at passage 7. The Werner's extract was from a 54-year-old male at passage 5.

DISCUSSION

This study indicates that triosephosphate isomerase has a substantial portion of thermolabile enzyme present in cultured skin fibroblasts from progeroid and Werner's syndrome individuals and late-passage normal cells. The increased lability does not appear to be the result of an increase in general proteolysis of these cells. The increased enzyme lability cannot be attributed to increased cell death since the cell viability was very similar to normal controls.

Several studies [11–17] have indicated that other enzymes, including glucose-6-phosphate dehydrogenase, 6-phosphogluconate dehydrogenase, and hypoxanthine-guaninephosphoribosyl transferase, also contain an increased heat-labile enzyme fraction in fibroblasts from these premature aging syndromes. It has been postulated [23–26] that an age-related increase in proteolysis might account for the accumulation of defective or unstable enzymes. However, our results indicate that acid proteases, chymotrypsin-like proteases, and trypsin-like proteases do not appear to have an effect on the appearance of the heat-labile fractions of TPI. The presence of pepstatin A, TPCK, TLCK, and PMSF had no effect on the level of the labile TPI. A more likely mechanism seems to be consistent with that proposed by Goldstein *et al.* [27] and our laboratory [18, 19], in which a decrease in proteolysis may occur with age. Reznick *et al.* [28] have also

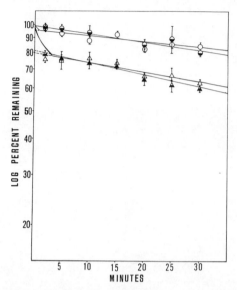

Fig. 3. Stability of progeria and normal fibroblast TPI at 57 °C. Normal extract without TLCK is represented by open circles (○) and with 0.5 mM TLCK by half-closed circles (◑). Progeria extract without TLCK is depicted by open triangles (△) and with 0.5 mM TLCK by half-closed triangles (◬). Vertical bars represent the range for the two determinations. All correlation coefficients were better than -0.90 except the normal without inhibitor which had a correlation of better than -0.80. See Fig. 1 for further explanation. The normal extract was at passage 8 from a 3-year-old male. The progeria extract was at passage 8 from a 13-year-old male.

TABLE III

CONTENT OF LABILE TRIOSEPHOSPHATE ISOMERASE FROM NORMAL AND PROGERIA FIBROBLASTS IN THE PRESENCE OF PROTEASE INHIBITORS

Cells	Protease inhibitor	Percentage heat-labile enzyme	Linear regression correlation coefficient
Normal	None	0–4.7	-0.802
Normal	0.5 mM TLCK	0.11	-0.916
Normal	0.5 mM TPCK	0.00	-0.962
Normal	5×10^{-4} mM pepstatin A	0.00	-0.976
Progeria	None	20.2	-0.922
Progeria	0.5 mM TLCK	18.9	-0.978
Progeria	0.5 mM TPCK	24.4	-0.974
Progeria	5×10^{-4} mM pepstatin A	25.6	-0.966

Incubations were at 57.0 °C. The normal extract was from a 3-year-old male at passage 8 and the progeria extract was from a 13-year-old male at passage 8. See Table II for further explanation.

observed that the degradation enzymes may be less effective in aging animals. In studying the accumulation of "abnormal" forms of fructose 1,6-bisphosphatase in mice liver cells, Reznick *et al.* [28] also came to the conclusion that protein catabolism may become defective in aging.

The study of TPI has been particularly informative in this respect. Deamidation of Asn15 and Asn71 appears to be the first step in the normal degradation of the enzyme, resulting in dissociation of the subunits and subsequent proteolysis [18, 19]. In the case of these premature aging syndromes the labile enzyme that accumulated appears to be identical to the deamidated forms of TPI with respect to stability and electrophoretic and isoelectric focusing properties.

Several major mechanisms have been proposed to account for the observed accumulation of altered enzymes during aging. A defect in protein synthesis was proposed [11–13], but strong evidence against this mechanism has arisen from several laboratories. Post-synthetic defects occurring in aged cells have also been proposed [27]. The data presented here support the concept that, for these two premature aging syndromes, an increase in proteolytic activity does not appear to play a significant role. Rather it seems more likely that a decrease in proteolysis with age may result in the accumulation of labile normal catabolic intermediates. Deamidation appears to be a rather common post-synthetic step involved in the initial stages of protein catabolism [26]. Moreover, the accumulation of such catabolic intermediates might be expected to inhibit synthesis of new enzymes, and thus such cells, when presented with an environmental stress requiring the synthesis of new enzyme, may not be able to respond to this biochemical challenge as effectively as young cells.

ACKNOWLEDGEMENTS

The authors express their appreciation to the Human Genetic Mutant Cell Repository for the supply of fibroblasts and to M. Lynne Chapman for her technical assistance. This research was supported in part by grants from the National Institutes of Health (AM14638, AG01274), a Biomedical Research Support Grant (BRSG-S07 RR07195), the R.A. Welch Foundation (B-502), and the Burroughs Wellcome Osteopathic Research Fellowship Program.

REFERENCES

1 F. L. DeBush, The Hutchinson-Gilford progeroid syndrome: Report of four cases and review of the literature. J. Pediatr., 80 (1972) 695–724.
2 A. H. Mostafa and M. Gabr, Hereditary progeria with follow-up of two affected sisters. Arch. Pediatr., 71 (1954) 163–172.
3 V. A. McKusick, Mendelian Inheritance in Man. Catalogs of Autosomal Dominant, Autosomal Recessive and X-Linked Phenotypes, Johns Hopkins University Press, Baltimore, 1975, p. 548.
4 S. Goldstein and E. J. Moerman, Heat-labile enzymes in circulating erythrocytes of a progeroid family. Am. J. Hum. Genet., 30 (1978) 167–173.
5 C. J. Epstein, G. M. Martin, A. L. Schultz and A. G. Motulsky, Werner's syndrome: A review of its symptomatology, natural history, pathologic features, genetics and relationship to the natural aging process. Medicine (Baltimore), 45 (1966) 177–221.
6 S. Goldstein, Human genetic disorders which feature accelerated aging. In E. L. Schneider (ed.), The Genetics of Aging, Plenum, New York, 1978, pp. 171–224.

7 G. F. Beadle, I. R. Mackay, S. Whittingham, G. Taggart, A. W. Harris and L. C. Harrison, Werner's syndrome: A model of premature aging. *J. Med., 9* (1978) 377–404.

8 G. M. Martin, C. A. Sprague and E. J. Epstein, Replicative life-span of cultivated human cells: Effects of donor's age, tissue, and genotype. *Lab. Invest., 23* (1970) 86–92.

9 L. Hayflick, The limited *in vitro* lifetime of human diploid cell strains. *Exp. Cell Res., 37* (1965) 614–636.

10 S. Goldstein, Studies on age-related diseases in cultured skin fibroblasts. *J. Invest. Dermatol., 73* (1979) 19–23.

11 R. Holliday and G. M. Tarrant, Altered enzymes in aging human fibroblasts. *Nature, 238* (1972) 26–30.

12 R. Holliday, J. S. Porterfield and D. D. Gibbs, Premature aging and occurrence of altered enzyme in Werner's syndrome fibroblasts. *Nature, 248* (1974) 762.

13 S. Goldstein and D. P. Singal, Alteration of fibroblast gene products *in vitro* from a subject with Werner's syndrome. *Nature, 251* (1974) 719.

14 S. Goldstein and E. J. Moerman, Heat-labile enzymes in Werner's syndrome fibroblasts. *Nature, 255* (1975) 159.

15 S. Goldstein and E. J. Moerman, Defective proteins in normal and abnormal human fibroblasts during aging *in vitro. Interdiscipl. Top. Gerontol., 10* (1976) 34–43.

16 S. Goldstein and E. J. Moerman, Heat-labile enzyme in skin fibroblasts from subjects with progeria. *N. Engl. J. Med., 292* (1975) 1305–1309.

17 A. Houben, A. Houbion and J. Remacle, Lysosomal and mitochondrial heat-labile enzymes in Werner's syndrome fibroblasts. *Exp. Gerontol., 15* (1980) 629–631.

18 P. M. Yuan, J. M. Talent and R. W. Gracy, Molecular basis for the accumulation of acidic isozymes of triosephosphate isomerase on aging. *Mech. Ageing Dev., 17* (1981) 151–162.

19 R. W. Gracy, H. S. Lu, J. M. Talent and P. M. Yuan, Structural analysis of altered proteins. In R. C. Adelman and G. S. Roth (eds.), *Methods in Aging Research*, CRC Press, 1980, p. 15.

20 P. M. Yuan, J. M. Talent and R. W. Gracy, A tentative elucidation of the sequence of human triosephosphate isomerase by homology peptide mapping. *Biochim. Biophys. Acta, 671* (1980) 211–218.

21 R. W. Gracy, Triosephosphate isomerase from human erythrocytes. *Methods Enzymol., 41* (1975) 442–447.

22 S. Goldstein, Life-span of cultured cells in progeria. *Lancet, i* (1969) 424.

23 A. L. Goldberg and J. F. Dice, Intracellular protein degradation in mammalian and bacterial cells. *Annu. Rev. Biochem., 43* (1974) 835–869.

24 R. T. Schimke, Intracellular protein turnover. In R. T. Schimke and N. Katanuma (eds.), Academic Press, New York, 1975, pp. 173–186.

25 N. Katanuma, E. Kominami, Y. Banno, K. Kits, Y. Aoki and G. Urata, Concept on mechansim and regulation of intracellular enzyme degradation in mammalian tissues. *Adv. Enzyme Regul., 14* (1976) 325–345.

26 J. Dreyfus, A. Kahn and F. Schapira, Posttranslational modifications of enzymes. *Curr. Top. Cell. Regul., 14* (1978) 243–297.

27 S. Goldstein, D. Stotland and R. A. J. Cordeiro, Decreased proteolysis and increased amino acid efflux in aging human fibroblasts. *Mech. Ageing Dev., 5* (1976) 221–233.

28 A. Reznick, Z. Lavie, H. E. Gershon and D. Gershon, Age-associated accumulation of altered FDP aldolase B in mice. *FEBS Lett., 128* (1981) 221–224.

19

Reprinted from *Proc. Natl. Acad. Sci. (U.S.A.)* **73**:2818–2822 (1976)

Decreased fidelity of DNA polymerase activity isolated from aging human fibroblasts

(DNA nucleotidyltransferase/error-prone replication/cell culture senescence)

Stuart Linn*, Michael Kairis, and Robin Holliday

Genetics Division, National Institute for Medical Research, Mill Hill, London NW7 1AA, England

Communicated by Bruce N. Ames, June 7, 1976

ABSTRACT   DNA polymerase (deoxynucleosidetriphosphate:DNA nucleotidyltransferase, EC 2.7.7.7 or DNA nucleotidyltransferase) activity, isolated from late and early passage cells of the diploid human fibroblast line, MRC-5, was compared. The level of activity dropped with increasing passage. In addition, when the fidelity of polymerization was monitored with four synthetic templates under a variety of conditions, it was observed that the enzyme from late passage cells was more error-prone. The possible relation of these observations to "senescence" of the fibroblasts is discussed.

Diploid human fibroblasts have a defined lifespan in culture that is somewhat related to the potential life expectancy of the individual from whom the cells were obtained (1, 2). This system has therefore been widely adopted as a model for studying the molecular bases of aging, physiological senescence, and cell death. Orgel has proposed that the fibroblast senescence results from an "error catastrophe" in macromolecular synthesis, which will inevitably lead to faulty and unregulated cellular metabolism (3, 4), and evidence has been obtained that defective enzyme molecules accumulate at increasingly higher levels in senescent fibroblasts, as well as in tissues from aging mice (5–10). Since the appearance of defective DNA polymerase could be an important component in any general breakdown of information transfer between macromolecules, we have compared the fidelity of DNA polymerase from cells in the early and late stages of cell culture during the replication of defined synthetic templates.

MATERIALS AND METHODS

Growth of Cells. MRC-5 human male fetal lung fibroblasts (5) were grown in Eagle's basal medium containing 10% fetal calf serum, 100 units/ml of penicillin, 100 μg/ml of streptomycin, and 50 μg/ml of aureomycin. Cells were monitored to assure the absence of mycoplasmic contamination (11). For collection, cells that had not reached confluence, as judged by microscopic examination, were rinsed with 5 mM Tris·HCl (pH 7.5), 0.15 M NaCl, scraped off of the glass with a piece of soft plastic, then harvested and washed by centrifugation in the same buffer. The pellet could be stored at −70° or used immediately with no effect upon the subsequently fractionated activity.

Assays for Enzyme Activity. Reaction mixtures (0.1 ml) contained 50 mM Tris·HCl (pH 8.5), 7.5 mM $MgCl_2$, 0.1 M KCl, 0.5 mM dithiothreitol, 0.5 mg/ml of bovine serum albumin (Sigma, Fraction V), 0.13 μmol of "activated" salmon sperm DNA (12), and 5 nmol each of dATP, dGTP, dCTP, and dTTP. One of the triphosphates was labeled with ^3H at 50–100 cpm/pmol. After 30 min at 37°, the reaction mixtures were chilled, then mixed with 0.2 ml of 0.1 M $Na_4P_2O_7$ and 0.7 ml

of 10% trichloroacetic acid. After at least 5 min, 3 ml of 1 M HCl, 0.1 M $Na_4P_2O_7$ were added. The contents were filtered through a Whatman GF/C glass filter that had been soaked in 0.1 M $Na_4P_2O_7$, then washed with the HCl-$Na_4P_2O_7$. The filter was then successively washed 12 times with 3 ml of the HCl-$Na_4P_2O_7$ and then with ethanol and dried under a heat lamp. Radioactivity was measured in 2 ml of toluene scintillator fluid. One unit of enzyme incorporates 1 nmol of total nucleotide in 30 min at 37°. The assay was linear with fraction I up to about 1 unit, whereas it was linear to about 0.2 unit of fraction II due to the presence of inhibitor(s). Early fractions give a nonlinear response, and less than 0.1 unit is generally used.

Other Materials. Synthetic polymers and unlabeled nucleotides were from P-L Biochemicals. Radioactive nucleotides were from Amersham.

RESULTS

Levels of DNA polymerase activity

DNA polymerase (deoxynucleosidetriphosphate:DNA nucleotidyltransferase, EC 2.7.7.7) was purified from whole cell sonicates through the DEAE-cellulose gradient step, essentially as described for HeLa cells (13) (Table 1). Two peaks of activity were obtained on the DEAE-cellulose, as reported for HeLa cells, but they both sedimented at 5.5–6 S and both had catalytic properties similar to those reported for the "cytoplasmic" enzyme, α (13). The reason for the apparent absence of "nuclear" DNA polymerase, β, is unknown, but it was not due to the preparation methods, since 3.5S material was found in peak I when the same procedure was used with human placental material (A. Divor and S. Linn, unpublished). For comparison, both fractions were used in the studies below, since each presumably contained a different complement of contaminating proteins.

The level of DNA polymerase activity was highly dependent upon the time after subculture of the cells; for instance, at passage 31 after a 1:4 splitting, activity reached a maximum at day 3, then dropped to about one-fourth that value and remained at that level for at least 9 days. At this passage, cells became confluent 4 days after subculture.

Activity also depended upon the passage number. The activity of DNA polymerase in cells prior to reaching confluence was monitored in a subline of MRC-5 until its death at passage 70. These cells first showed morphological changes associated with aging between passage 40 and 50. The activity in cells of passage 40–56 averaged roughly one-fifth that of earlier passage cells. At passage 56, activity began dropping off more rapidly, and at passage 62 it was barely detectable. All of the changes in activity affected both peaks of enzyme though peak I activity was usually more reduced (Table 1).

With several other sublines polymerase activity generally decreased prior to the appearance of altered cell morphology,

---

* Permanent address: Department of Biochemistry, University of California, Berkeley, Calif. 94720, USA.

Table 1. Fractionation of the DNA polymerases

Fraction	Volume (ml)	Protein (mg)	Activity (units)
Passage 19			
Cell extract	6.5	45	982
First DEAE-cellulose	13	18	705
Second DEAE-cellulose			
Fraction I	15		507
Fraction II	9		92
Dialyzed concentrate			
Fraction I	2.8	3.2	314
Fraction II	2.1	3.7	53
Passage 56			
Cell extract	8.6	27	58
First DEAE-cellulose	14	11	41
Second DEAE-cellulose			
Fraction I	10		4.9
Fraction II	10		10.0
Dialyzed concentrate			
Fraction I	2.2	1.6	2.0
Fraction II	2.4	3.0	6.9

For the passage 19 preparation, approximately 1.5 g (wet weight) of cells were suspended into 6 ml of 0.4 M potassium phosphate (pH 7.5), 10 mM 2-mercaptoethanol, 0.5 mM dithiothreitol, and disrupted by sonication ("cell extract"). After centrifugation at 40,000 × g, the solution was passed through a 7.5-ml DEAE-cellulose column (Whatman DE-32) equilibrated with the above buffer. Flow-through fractions containing the majority of enzyme and protein were then dialyzed for 4 hr against 20 mM potassium phosphate (pH 7.5), 1 mM 2-mercaptoethanol, 0.5 mM dithiothreitol. Material at this stage could be stored at −70° without loss of activity, but this and subsequent fractions were very prone to loss by adsorption to glass or dialysis tubing. Therefore, it was important to keep dialysis times to a minimum and to use nonwettable plastic in place of glass containers. The dialyzed material was passed through a 4-ml DEAE-cellulose column equilibrated with 20 mM potassium phosphate (pH 7.5), 10 mM 2-mercaptoethanol, 0.5 mM dithiothreitol, and the column was washed with 20 ml of the same buffer. A linear gradient (30 ml total volume) between 0.02 M and 0.50 M potassium phosphate (pH 7.5) was then applied. Fraction I appeared in the flow-through, whereas fraction II eluted at approximately 0.15 M. The individual fractions were concentrated by dialysis against 30% polyethyleneglycol, 50 mM potassium phosphate (pH 7.5), 10 mM 2-mercaptoethanol, 0.5 mM dithiothreitol, then dialyzed extensively against 20 mM potassium phosphate (pH 7.5), 1 mM 2-mercaptoethanol, 0.5 mM dithiothreitol. These fractions are stable for at least several months over liquid nitrogen, but lose roughly half of this activity after several weeks at −20°. Fractions from passage 56 cells were prepared in a similar manner from 1.1 g of cells. Enzyme was assayed as described in *Materials and Methods*. Protein was measured by the method of Lowry *et al.* (14), after precipitation with trichloroacetic acid.

then became difficult to detect when cell growth slowed. Mixed extracts from early and late passage cells indicated that there were no inhibitors peculiar to the later passage cells. However, the enzyme present in these extracts did show increased lability during storage and purification. A similar reduction of DNA polymerase activity has been reported in spleens of old BALB/c mice (15).

Fidelity of the DNA polymerase

We tested polymerization fidelity of enzyme from early and late passage cells by measuring the incorporation of nonhomologous nucleotides into synthetic templates (6, 17). Reactions were carried out with Mg^{++} or Mn^{++}; Mn^{++} has been reported to exaggerate the misincorporation frequency of DNA polymerases of bacteriophage T4 (17), whereas Mg^{++} might be

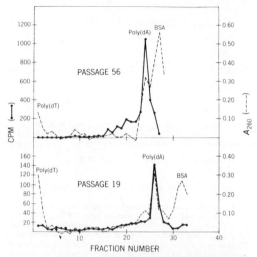

FIG. 1. Isopycnic centrifugation in alkaline CsCl of deoxyguanylate incorporated during the replication of poly(dA)·poly(dT). Reaction mixtures (230 μl) contained 45 mM Tris·HCl (pH 7.7), 0.1 M KCl, 0.1 mM dithiothreitol, 0.22 mM $MnCl_2$, 0.5 mg/ml of bovine serum albumin (BSA), 10 nmol of dATP, 3 nmol of [³H]dGTP (15 cpm/fmol), 15.3 nmol each of poly(dA) and poly(dT) (heated and cooled slowly together before use), and 0.67 unit (6.9 μg) of passage 19 fraction I or 0.067 unit (55 μg) of passage 56 fraction 1. After incubation for 2 hr at 37°, both EDTA and additional poly(dA)·poly(dT) were added. Mixtures were dialyzed extensively, first against 1 M NaCl, 10 mM Tris·HCl (pH 9.2), 1 mM EDTA, then against 1 mM Tris·HCl. (Parallel reactions in which the dATP was labeled showed 120 pmol and 105 pmol of polymer synthesized in the passage 19 and 56 reactions, respectively.) To each sample was then added 3 μmol of EDTA, 160 μmol of NaOH, and distilled water to a final solution weight of 2.2 g; then 2.73 g of CsCl was added. The solutions (2.8 ml final volume) were transferred to polyallomer tubes, topped off with liquid paraffin, and centrifuged for 43 hr at 38,000 rpm in the Spinco SW50.1 rotor. Drops were collected from the tube bottom, diluted to 0.4 ml to determine UV absorbance, then mixed with 50 μl of 1 mg/ml of bovine serum albumin, 0.2 ml of 0.1 M $Na_4P_2O_7$, and 0.4 ml of 25% trichloroacetic acid. Samples were filtered and radioactivity was measured. Poly(dA) peaks contained 359 cpm (24 fmol) and 2855 cpm (190 fmol) for passage 19 and 56 reactions, respectively. Misincorporation frequencies would therefore be 1/5000 and 1/550 for the respective reactions if the recovery of synthesized polymer was complete.

considered "natural." The ³H-labeled deoxyribonucleoside triphosphates were purified on Dowex 1 (18), but the enzymes were purposely purified to a level sufficient only to remove endogenous DNA (the "dialyzed concentrate," Table 1), in order to avoid the removal of any subpopulation of polymerase which might contribute to the error frequency. Consequently, preparations contained factors (e.g., nuclease activity) that limited the extent of the polymerization reactions. Therefore, each reaction was carried out to the furthest extent possible by adjustment of ionic strength, pH, time of incubation, and enzyme level. Blank values for misincorporated nucleotide (no enzyme, no template, or zero time) varied with the time of storage of the triphosphate from 0.001 to 0.01% of the input counts, but on a given day these were reproducible to within 20%. Therefore, levels of misincorporation of greater than 50% above the blank were considered significant, and values of up to five times the blank were not uncommon. In addition, blank values with template omitted never exceeded values with enzyme omitted or zero-time values, thus eliminating the possi-

Table 2.   Incorporation of dGMP or dCMP into polymers containing deoxyadenylate and/or deoxythymidylate

Polymer	Polymerase fraction	Passage number	Divalent cation and nonhomologous triphosphate		Polymer synthesized (fmol)	Nonhomologous nucleotide incorporated (fmol)	Level of misincorporation
Poly(dA-dT)	I	19	Mg^{++}	dGTP	17,300	9.5	1/1820
		56			5,400	11.4	1/470
		19	Mn^{++}	dGTP	26,800	14.5	1/1850
		56			18,800	33.8	1/560
Poly(dA)·poly(dT)	I	19	Mn^{++}	dGTP	61,600	<2.7	<1/24,000
		56			26,800	86.1	1/310
	II	19			32,300	44.5	1/730
		56			8,200	95.0	1/86
	II	19	Mn^{++}	dCTP	3,700	<2.0	<1/1900
		56			1,700	7.2	1/240
		19	Mg^{++}	dCTP	8,400	<1.5	<1/5600
		56			3,000	4.3	1/700
Poly(dA)	I	19	Mn^{++}	dGTP	<500	<8.7	
		56			<500	<8.7	
Poly(dA)·poly(rU)	II	19	Mn^{++}	dGTP	<500	<9.5	
		56			<500	<9.5	
Poly(rA)·poly(dT)	I	19	Mn^{++}	dGTP	75,300	54.1	1/1400
		56			75,200	88.6	1/750
	II	19			32,800	43.3	1/760
		56			32,300	229	1/140

Each misincorporation ratio is obtained from a parallel pair of reactions (0.1 ml) with either [^{3}H]dATP (55 cpm/pmol) or [^{3}H]dTTP (42 cpm/pmol) in one tube to measure total polymer synthesis, and [^{3}H]dCTP (1 μM, 27 cpm/fmol) or [^{3}H]dGTP (10 μM, 15 cpm/fmol) in the other tube to measure non-faithful synthesis. All incubations were at 37° under conditions that maximize polymer synthesis. Each pair of tubes had identical concentrations of all reagents; they differed only in the nucleotide labeled. Reaction mixtures had 50 mM Tris·HCl (pH 8.5), 0.5 mM dithiothreitol, 0.1 M KCl as indicated below, 0.5 mg/ml of bovine serum albumin, MnCl$_2$ at 0.25 mM when present, 50 μM complementary triphosphate(s), and "dialyzed concentrate" enzyme as follows: passage 19 fraction I, 0.22 unit (2.3 μg of protein); passage 19 fraction II, 0.13 unit (8.8 μg); passage 56 fraction I, 0.022 unit (18 μg); and passage 56 fraction II, 0.06 unit (25 μg). After incubation for 2 hr at 37°, reaction mixtures were filtered and radioactivity was measured (see *Materials and Methods*). Several individual reaction conditions were as follows. For poly(dA-dT): 4.7 nmol of polymer was present; KCl was added with MnCl$_2$, not MgCl$_2$; MgCl$_2$ was 2 mM when present; [^{3}H]dTTP was used to measure polymer synthesis. The dTMP incorporated has been multiplied by 2 to give total polymer synthesized. For homopolymers: 5.1 nmol of each homopolymer (10.2 nmol total for homopolymer pairs) were present; KCl was added throughout; MgCl$_2$ was 0.75 mM when present; [^{3}H]dATP was used to measure polymer synthesis (no dTTP was added). Homopolymer pairs were heated together in buffer and slowly cooled prior to the reaction to assure pairing.

bility of differences of error frequency being due to contamination of the two enzyme preparations with varying levels of endogenous template.

Table 2 presents typical data using primer-templates containing deoxyadenylate and/or deoxythymidylate. With the alternating polymer, poly(dA-dT), significantly more incorporation of dGMP occurred with fraction I from senescent cells than from the younger cells. (No significant difference in dCMP incorporation was detected with this polymer; fraction II could not be used because it degraded the polymer during the reaction.)

With the homopolymer pair, poly(dA)·poly(dT), fractions I and II synthesized only poly(dA); the reaction was unaffected by dTTP. In the presence of Mn^{++}, fractions I and II from senescent cells both incorporated significantly more dGMP than their counterparts from young cells. (A difference was not detectable in the presence of Mg^{++}.) Fraction II from senescent cells also showed higher incorporation of dCMP with either Mn^{++} or Mg^{++} present. No detectable incorporation of dCMP (<1/7000) was detectable with fraction I.

To verify that the incorporation of dGMP required the synthesis of polymer in a template-dependent reaction, we tested several analogous polymers (Table 2). In the presence of poly(dA) or poly(dA)·poly(rU), no polymer synthesis occurred, and no dGMP incorporation could be detected. On the other hand, poly(rA)·poly(dT) was quite active for the synthesis of

poly(dA), and also showed higher levels of infidelity with enzyme fractions from old cells.

Two lines of evidence were used to show that the [^{3}H]dGMP was incorporated into poly(dA). First, polymerization was carried out with poly(dA)·poly(dT), then the reaction mixtures were extensively dialyzed and subjected to isopycnic centrifugation in alkaline CsCl. The nondialyzable, acid-precipitable counts banded with the poly(dA) (Fig. 1). [From the counts recovered with the poly(dA) peak, the misincorporation levels were calculated to be 1/550 for enzyme from senescent cells and 1/5000 for enzymes from young cells.] Second, it was also verified that the dGMP incorporated by enzyme from senescent cells was sensitive to pancreatic DNase (Table 3).

In a complementary set of experiments, polymers containing deoxycytidylate and deoxyinosinate were used (Table 4). In some cases data for enzyme from passage 28 and from pooled cells of passages 58 and 60 are also included. (By passage 58, the level of fraction II enzyme is too low for purification, so experiments were with fraction I only.)

With the alternating polymer, poly(dI-dC), in the presence of Mn^{++}, significantly higher incorporation of dAMP was noted with fraction II of senescent cells than with that of young cells. The additional dAMP incorporated was similarly sensitive to pancreatic DNase (Table 3). With identical reagents and under identical conditions, DNA polymerase I from *Escherichia coli* showed no significant incorporation of dAMP (Table 4), thus

193

Table 3.   Sensitivity of incorporated nonhomologous nucleotides to DNase

Polymer present	Passage number	Divalent cation and nonhomologous triphosphate present		Pancreatic DNase present	Polymer found (fmol)	Nonhomologous incorporated nucleotide (fmol)
Poly(dA)·poly(dT)	19	Mn^{++}	dGTP	−	14,900	<10
				+	<300	<10
	56			−	13,900	54
				+	<300	<10
Poly(dI-dC)	19	Mn^{++}	dATP	−	17,200	<20
				+	<700	<20
	56			−	31,400	53
				+	<700	<20

Reactions were with fraction II enzyme and were identical to the respective reactions in Tables 2 or 4. After the 2 hr at 37°, 7.5 µl of 0.1 M $MgCl_2$ and 5 µl of 1 mg/ml of bovine serum albumin, which contained 2 mg/ml of pancreatic DNase where indicated, were added to each tube, and incubation was continued for 15 min at 37°. Reaction mixtures were then treated as in *Materials and Methods.*

verifying the purity of the dATP. Fraction I also showed somewhat higher incorporation of dAMP when isolated from senescent cells (Table 4), but the difference in this case was relatively slight. In the presence of Mg^{++} no detectable incorporation of dAMP ($<\frac{1}{5000}$) occurred with any enzyme fraction. Likewise, detectable incorporation of dTMP ($<\frac{1}{5000}$) occurred only with fraction I in the presence of Mn^{++}, where misincorporation successively increased with the passage number of four extracts.

With the homopolymer pair, poly(dI)·poly(dC), fractions I and II synthesized only poly(dG) [or potentially poly(dI)]; the reaction was unaffected by dCTP. In the presence of Mn^{++}, fractions I and II each incorporated significantly more dAMP

when isolated from the senescent cells. This incorporation of dAMP did not occur when poly(dG) synthesis was prevented by omission of dGTP, and was not inhibited when fractions from young and senescent cells were incubated together (Table 4). In addition, when the level of fraction II from the young cells was increased 3-fold so as to provide equal protein to that used for the senescent cell fractions, dAMP incorporation was still not detected. In the presence of Mg^{++}, the polymerase fractions incorporated dAMP at a level barely detectable in the assay system ($\frac{1}{10,000}-\frac{1}{5000}$), and we could detect no consistent difference between values for young versus senescent cells. Likewise, detectable dTMP incorporation was noted only with fraction I in the presence of Mn^{++}, where, as with the po-

Table 4.   Incorporation of dATP and dTTP into polymers containing deoxyinosinate and deoxycytidylate

Polymer	Polymerase fraction	Passage number	Divalent cation and nonhomologous triphosphate		Polymer synthesized (fmol)	Nonhomologous nucleotide incorporated (fmol)	Level of misincorporation
Poly(dI-dC)	II	19	Mn^{++}	dATP	10,800	11.5	1/940
		56			22,800	125	1/180
	I	19			66,200	11.5	1/5800
		56			66,600	18.6	1/3600
	E. coli I	—			27,200	<6	<1/5000
	I	19	Mn^{++}	dTTP	62,900	2.9	1/22,000
		28			59,400	4.3	1/13,300
		56			48,300	7.7	1/6300
		58/60			12,100	3.8	1/3200
Poly(dI)·poly(dC)	II	19	Mn^{++}	dATP	124,000	<14	<1/9000
		19 (3×)			81,200	<14	<1/6000
		56			233,000	111	1/2100
		19 + 56			200,000	87	1/2300
	II (-dGTP)	19			—	<14	—
		56			—	<14	—
	I	19			217,000	<14	<1/16,000
		56			157,000	94	1/1700
	I	19	Mn^{++}	dTTP	149,000	13.8	1/11,000
		56			91,000	24.7	1/3700

Reactions were at pH 7.7 and used 50 µM [³H]dGTP (55 cpm/pmol) to measure polymer synthesis, and [³H]dATP (8.5 µM, 24 cpm/fmol) or [³H]dTTP (4 µM, 47 cpm/fmol) to measure non-faithful synthesis. General conditions were as described in Table 2. Individual reaction conditions were as follows: For poly(dI-dC): 5 nmol of polymer was present; KCl was added for fraction I and *E. coli* polymerase I only; 50 µM unlabeled dCTP was present; incubations were 3 hr with fraction I, 2 hr with fraction II, and 30 min with *E. coli* polymerase. The dGMP incorporated has been multiplied by 2 to give total polymer synthesized. *E. coli* DNA polymerase I was the phosphocellulose fraction (19). For poly(dI)·poly(dC): 10 nmol of total polymer was present; KCl was added throughout; [³H]dGTP was used to measure polymer synthesis (no dCTP was added); incubations were for 3 hr. In the experiment labeled "3×", 0.39 unit (26 µg of protein) of passage 19 Fraction II was added so as to add protein equal to that of passage 56 fraction II. The polymer synthesized in this case was reduced by nuclease(s) in the extract.

ly(dI-dC) polymer, the senescent cell sample showed a higher level of incorporation.

DISCUSSION

Under a variety of conditions, enzyme fractions I and II from aged fibroblasts show a propensity to incorporate incorrect nucleotides at a rate that is higher than that found with enzyme from early passage cells. Representative data have been presented with four different synthetic polymers, each of the triphosphates, four different enzyme preparations, and Mg^{++} or Mn^{++} present. The cases where differences did not exist are likely to be due, at least in part, to limitations in the sensitivity of our assay, in which we may not have used saturating levels of noncomplementary triphosphates (20), and which utilized relatively crude enzyme fractions in which inhibitors limited the extent of synthesis. We can rule out the possibility that erroneous nucleotides are incorporated by a terminal deoxynucleotidyltransferase activity (21), since we observed no misincorporation when a complementary template was not provided [i.e., with poly(dA) or poly(dA)·poly(rU)] or when complementary triphosphate was omitted.

As generally observed for mammalian DNA polymerases, the levels of misincorporation we see are relatively high. It is, of course, unrealistic to believe that so many incorrect nucleotides are inserted into DNA *in vivo*. The overall machinery for DNA replication is certainly far more complex than our *in vitro* system—for example, we still await description of the "editing" function for mammalian enzymes. In addition, we do not know whether the enzyme fractions we have studied are actually involved in chromosome replication, or, indeed, whether the changes we noted are due to primary structural changes of a protein or a new distribution of polypeptide components involved in DNA replication (e.g., a corrective "editor" polypeptide). What we do feel is significant, however, is that our experiments show that at least some aspects of an *in vitro* system capable of DNA synthesis have become less specific with increasing passage number. In addition, these studies provide an alternative to those which have relied upon the detection of heat-labile or inactive forms of a particular enzyme and which have yielded conflicting positive (5–8) and negative (22) evidence for the presence of defective enzymes in aging cells.

Our results are consistent with the predictions made by a number of authors that alterations in the fidelity of DNA polymerase could be an important factor in cell aging (4, 5, 23–25), at least for MRC-5 fibroblasts. In this system it was already shown by DNA fiber autoradiography that the rate of replicon elongation is significantly reduced in the senescent phase, suggesting that faulty DNA polymerase molecules are present (26). Furthermore, the frequency of somatic mutations (27) and chromosome abnormalities (28) rises rapidly at the end of the lifespan, which would be expected if fidelity in replication or repair was impaired. Of course, further studies will be needed to correlate the loss of fidelity of DNA polymerase with the appearance of these genetic changes.

The general error theory of cellular aging proposes that once errors in proteins or DNA begin to accumulate, there will be a progressive breakdown in information transfer between macromolecules, which will eventually affect a wide range of cellular components. By this process, the accumulation of error-prone DNA polymerase would be due to mistakes in transcription, translation, or protein processing, and the altered molecules would then have a crucial role by introducing a high frequency of genetic errors. An alternative possibility is that a new, error-prone DNA polymerase appears prior to aging that causes cellular senescence. Although we detected no new enzyme fractions in senescent cultures, it is possible that the reduction in activity and concomitant loss of fidelity we observed during senescence is due to a modification of the normal polymerase.

A final question is how observations for cell culture senescence might relate to aging of an organism. It is obviously essential to verify these observations with cells obtained from a living organism before attempting to answer this question. Should similar error-prone DNA polymerase activity be found in aged tissues, as a preliminary report suggests (29), the molecular basis of cancer and other age-related degenerative diseases may be better understood.

S.L. was supported by a Guggenheim Memorial Fellowship and an EMBO grant during the tenure of this work. We thank G. M. Tarrant for aiding us in culturing cells.

1. Hayflick, L. (1965) *Exp. Cell Res.* **37**, 614–636.
2. Martin, G., Sprague, C. & Epstein, C. (1970) *Lab. Invest.* **23**, 86–92.
3. Orgel, L. E. (1963) *Proc. Natl. Acad. Sci. USA* **49**, 517–521.
4. Orgel, L. E. (1973) *Nature* **243**, 441–445.
5. Holliday, R. & Tarrant, G. M. (1972) *Nature* **238**, 26–30.
6. Lewis, C. M. & Tarrant, G. M. (1972) *Nature* **239**, 316–318.
7. Holliday, R., Porterfield, J. S. & Gibbs, D. D. (1974) *Nature* **248**, 762–763.
8. Goldstein, S. & Moerman, E. (1975) *N. Engl. J. Med.* **292**, 1306–1309.
9. Gershon, H. & Gershon, D. (1973) *Proc. Natl. Acad. Sci. USA* **70**, 909–913.
10. Wulf, J. H. & Cutler, R. G. (1975) *Exp. Gerontol.* **10**, 101–117.
11. Russell, W. C., Newman, C. & Williamson, D. H. (1975) *Nature* **253**, 461–462.
12. Schlabach, A., Fridlender, B., Bolden, A. & Weissbach, A. (1971) *Biochem. Biophys. Res. Commun.* **44**, 879–885.
13. Weissbach, A., Schlabach, A., Fridlender, B. & Bolden, A. (1971) *Nature New Biol.* **231**, 167–170.
14. Lowry, O. H., Rosebrough, N. J., Farr, A. L. & Randall, R. J. (1951) *J. Biol. Chem.* **193**, 265–275.
15. Barton, R. W. & Yang, W.-K. (1975) *Mech. Ageing Dev.* **4**, 123–136.
16. Trautner, T. A., Swartz, M. N. & Kornberg, A. (1962) *Proc. Natl. Acad. Sci. USA* **48**, 449–455.
17. Hall, Z. W. & Lehman, I. R. (1968) *J. Mol. Biol.* **36**, 321–333.
18. Lehman, I. R., Bessman, M. J., Simms, E. S. & Kornberg, A. (1958) *J. Biol. Chem.* **233**, 163–170.
19. Richardson, C. C., Schildkraut, C. L., Aposhian, H. V. & Kornberg, A. (1964) *J. Biol. Chem.* **239**, 222–232.
20. Gillins, F. D. & Nossal, N. G. (1975) *Biochem. Biophys. Res. Commun.* **64**, 457–464.
21. Chang, L. M. S. & Bollum, F. J. (1971) *Biochemistry* **10**, 536–542.
22. Penbergrass, W. R., Martin, G. M. & Bornstein, P. (1976) *J. Cell. Physiol.* **87**, 3–14.
23. Woolhouse, H. W. (1969) *Int. Congr. Gerontol. Proc. 8th.*, Vol. I, pp. 162–166.
24. Lewis, C. M. & Holliday, R. (1970) *Nature* **228**, 877–880.
25. Burnet, F. M. (1974) *Intrinsic Mutagenesis: A Genetic Approach to Aging* (Wiley, New York).
26. Petes, T. D., Farber, R. A., Tarrant, G. M. & Holliday, R. (1974) *Nature* **251**, 434–436.
27. Fulder, S. J. & Holliday, R. (1975) *Cell* **6**, 67–75.
28. Thompson, K. V. A. & Holliday, R. (1976) *Exp. Cell Res.*, in press.
29. Barton, F. W., Waters, L. C. & Yang, W. K. (1974) *Fed. Proc.* **33**, 1419.

20

Reprinted from *J. Mol. Biol.* **146:**55–76 (1981)

Increased Error Frequency of DNA Polymerases from Senescent Human Fibroblasts

VINCENT MURRAY† AND ROBIN HOLLIDAY

Genetics Division
National Institute for Medical Research
Mill Hill, London, NW7 1AA
England

(Received 23 April 1980, and in revised form 9 October 1980)

The error catastrophe theory of ageing predicts that the fidelity of DNA polymerase should be reduced in extracts from senescent cells. This prediction has been experimentally verified with MRC-5 human diploid fibroblasts. Using "cytoplasmic" DNA polymerase α the old/young ratio of error frequencies was 3·4 with the poly[d(A-T)]/dGTP/Mg^{2+} system, 1·9 with poly[d(A-T)]/dGTP/Mn^{2+} and 2·0 with poly[d(I-C)]/dTTP/Mg^{2+}. With DNA polymerase γ the old/young ratio was 3·8. The fidelity of DNA polymerase was examined at seven points during the life span of MRC-5 fibroblasts and the increase in error frequency with cell age was found to be statistically highly significant ($P<0·001$).

By means of control experiments, artifactual explanations of the results can be eliminated. The close correlation between homologous and non-homologous DNA synthesis was demonstrated by following the time-course of the reaction, varying the enzyme concentration and by other means. The non-homologous dNMP incorporation was sensitive to DNase, but not RNase or protease treatment. No terminal transferase activity could be detected. The discrimination of the DNA polymerase from young and old extracts was constant over a wide range of homologous and non-homologous dNTP concentrations. Degradation of the products of the misincorporation assay to 3′-dNMPs revealed that a guanine–thymine base mispairing was the main method of non-homologous synthesis in young and old cells while degradation to 5′-dNMPs revealed that dGTP was not contaminated by other homologous dNTPs.

By use of DNA polymerase with a lower error frequency than MRC-5 polymerase, the contamination of poly[d(A-T)] by cytosine could be ruled out. Mixing experiments could not detect a diffusible agent in young or old extracts which was capable of modifying the error frequency of a DNA polymerase from another extract. Since the DNA polymerase extract was not pure, an alternative explanation is possible for these results. Nevertheless, these control experiments do strongly suggest that the misincorporation assay measures the frequency of DNA polymerase-directed errors which are present as single base substitutions.

† Present address: Department of Biochemical Sciences, Princeton University, Princeton, N.J. 08544, U.S.A.

1. Introduction

Human foetal lung fibroblasts have a finite lifespan in culture (Hayflick & Moorhead, 1961; Hayflick, 1965) and they display several of the characteristics of ageing in multicellular organisms. The error catastrophe theory of ageing (Orgel, 1963,1973) proposes that senescence may be the result of an exponential increase of error frequency in macromolecular synthesis, which is caused by the positive feedback of errors in protein and RNA synthesis. Indirect evidence supporting this hypothesis has been obtained from both fibroblasts and mammals by the investigation of heat labile enzymes (Holliday & Tarrant, 1972; Goldstein & Moerman, 1976; Fry & Weisman-Shomer, 1976; Houben & Remacle, 1978), immunologically cross-reacting material (Lewis & Tarrant, 1972; Gershon & Gershon, 1973), loss of enzyme specificity (Lewis & Tarrant, 1972; Holliday & Tarrant, 1972), protein turnover (Shakespeare & Buchanan, 1976; Bradley et al., 1976), the effect of amino acid, divalent cation and base analogues (Holliday, 1975; Holliday & Stevens, 1978; R. Holliday and L. I. Huschtscha, unpublished observations), and variant frequency (Fulder & Holliday, 1975; Fulder, 1979).

The approach of this study was to test the prediction of the error theory that the *in vitro* fidelity of macromolecular synthesis decreases in senescent cells. This was accomplished using the enzyme DNA polymerase from MRC-5 human diploid fibroblasts. Linn et al. (1976) found a two- to tenfold increase in the *in vitro* error frequency of DNA polymerase in senescent MRC-5 fibroblasts, but in most of their experiments the non-physiological divalent cation manganese was used.

In this study the misincorporation assay technique was improved so that the DNA polymerase error frequency could be accurately measured using the physiological divalent cation magnesium. Also the DNA polymerase fidelity was determined throughout the fibroblast lifespan, thus enabling the kinetics of error accumulation to be investigated. In addition, numerous controls (including nearest neighbour analysis) were carried out to ensure the validity of the misincorporation assay in measuring the *in vitro* error frequency of DNA polymerase.

2. Materials and Methods

(a) *Materials*

Micrococcal nuclease from *Staphylococcus aureus* (19,682 units/mg), bovine spleen phosphodiesterase (24 units/ml), pancreatic DNase 1 (2029 units/mg) and snake venom phosphodiesterase (22 units/mg) were obtained from the Worthington Corporation. *Micrococcus luteus* DNA polymerase was from P-L Biochemicals. *Escherichia coli* DNA polymerase I was a gift from G. T. Yarranton and was the hydroxyapatite fraction of the purification procedure of Richardson et al. (1964). *Ustilago maydis* DNA polymerase was a kind gift from A. Spanos and was prepared by the method described by Holliday et al. (1979).

Bovine serum albumin fraction V was obtained from Miles Laboratories. Calf thymus DNA was from the Worthington Corporation. DEAE-cellulose (DE32) was supplied by Whatman Biochemicals. Homopolymers and alternating copolymers of ribo- and deoxyribonucleotides, deoxyribonucleoside mono- and triphosphates were obtained from P-L Biochemicals. Dithiothreitol, phenylmethane sulphonyl fluoride and N-ethyl maleimide were from Sigma. All radiochemicals were obtained from the Radiochemical Centre, Amersham. Cell culture materials were from Gibco-Biocult.

(b) Cell culture

MRC-5 human foetal lung diploid fibroblasts (obtained from the National Institute for Biological Standards and Control, Hampstead, London, U.K.) were grown at 37°C in 50 ml of Eagle's basal medium containing 10% (v/v) foetal calf serum, 10% (v/v) tryptose phosphate broth, 5000 units of penicillin and 5 mg of streptomycin in 500 ml glass bottles. When the cells reached confluence, they were subcultured using trypsin, at either a 1:2 or 1:4 split ratio. Contamination by mycoplasma was regularly monitored by the method of Chen (1977).

Large scale cell cultures were grown at intervals throughout the life-span of one ampule of MRC-5. Approximately sixty 500 ml bottles were harvested at passage 28, 32, 40 and 48; and fifteen 500 ml bottles at passage 54, 55, 56, 57, 58 and 59. (The cell line ceased to divide at passage 59.) In addition, a cell culture of sixty 500 ml bottles was prepared at passage 22 from a different ampule of MRC-5. For collection, cells that had almost reached confluence, as judged by microscopic examination, were rinsed with phosphate buffered saline (Dulbecco "A") and scraped from the glass with a piece of soft plastic. The cells were pooled, centrifuged (500 g), washed with phosphate buffered saline, centrifuged and stored frozen at -70°C. Cells could be stored at -70°C without loss of DNA polymerase activity.

(c) Fractionation of DNA polymerase

DNA polymerases were not purified to homogeneity because (1) during purification error-prone DNA polymerases would be lost through instability and differing purification properties (even with the degree of purification performed error-prone polymerases were purified from the preparation; V. Murray, unpublished results); (2) it would not be feasible to obtain sufficient quantities of DNA polymerase (especially from senescent cells) to be able to purify the polymerases to homogeneity.

All operations were carried out at 4°C. Cells were suspended in 0·3 M-sucrose, 20 mM-KPO_4 (pH 7·8), 1 mM-$MgCl_2$, 1 mM-dithiothreitol, 0·1 mM-PMSF† (buffer A) to an approximate protein concentration of 2 mg/ml and broken with 20 strokes of a close fitting Dounce homogenizer (this and subsequent steps were monitored by microscopy using the stain acridine orange). This extract was centrifuged (500 g) for 10 min and the supernatant ("cytoplasmic" fraction) retained. The pellet was suspended in buffer A, further subjected to 12 strokes of the Dounce homogenizer and centrifuged (500 g for 10 min). The pellet was suspended in 0·3% Triton X100, 20 mM-KPO_4 (pH 7·8), 20 mM-NaEDTA, 1 mM-dithiothreitol, 0·1 mM-PMSF, 2 M-NaCl, left to extract for 45 min with occasional vortexing, centrifuged (500 g for 10 min) and the supernatant ("nuclear" fraction) kept. The supernatants were dialysed overnight against 20 mM-KPO_4 (pH 7·8), 0·5 mM-dithiothreitol, 0·1 mM-PMSF, 10% (v/v) glycerol, centrifuged 40,000 g and the supernatants retained. For the routine preparation of DNA polymerase, the fraction was chromatographed onto an 8 ml DEAE-cellulose (DE-32) column, washed with the 20 mM-KPO_4 buffer, washed with 80 mM-KPO_4 buffer and eluted with 200 mM-KPO_4 buffer. The fractions eluted by 200 mM-KPO_4, containing DNA polymerase activity were pooled, concentrated by vacuum dialysis and dialysed against the 20 mM PO_4 buffer. The above method produced 7 S DNA polymerase α. 5 S DNA polymerase α (which has been proteolytically degraded by the isolation procedure) was prepared by the method of Linn et al. (1976).

(d) Assay for DNA polymerase activity

The "activated" calf thymus assay measures DNA polymerase α and β activity. The reaction mixture (0·1 ml) contained 50 mM-Tris·HCl (pH 8·5), 0·5 mM-dithiothreitol, 0·5 mg bovine serum albumin/ml, 7·5 mM-$MgCl_2$, 70 nmol "activated" calf thymus DNA (Aposhian & Kornberg, 1962), 2·9 μM each of dATP, dTTP, dGTP and dCTP, 1·25 μCi [³H]dTTP (680 cts/min per pmol) and an appropriate quantity of DNA polymerase extract. The reaction mixture was incubated for 30 min at 37°C and terminated in ice. A "batch" assay procedure

---

† Abbreviations used: PMSF, phenylmethane sulphonyl fluoride; NEM, N-ethyl maleimide; CG, cell generation; KPO_4 denotes potassium phosphate.

was employed (Bollum, 1968). 80 μl of the reaction mixture was spotted onto Whatman 3 MM filter discs, dipped in 5% (w/v) trichloracetic acid, 1% (w/v) tetrasodium pyrophosphate (PPi), placed on a "looped" piece of zinc-coated copper wire, suspended (along with the other discs in the assay) in a beaker containing 5% trichloroacetic acid/1% PPi (at least 20 ml per disc). This trichloroacetic acid/PPi wash was repeated twice and was followed by a wash with 96% ethanol. The discs were dried under an infrared lamp and radioactivity measured in 3 ml of toluene-based scintillation fluid. One unit of DNA polymerase activity incorporates 1 nmol of total nucleotide in 30 min at 37°C. The specific activity is defined as the units of DNA polymerase activity divided by the total cellular protein present. Protein concentration was determined by the procedure of Lowry et al. (1951).

(e) Misincorporation assays

The misincorporation assay was carried out with 2 parallel sets of tubes. Each parallel reaction contained the same quantity of reactants and differed only in the deoxynucleoside triphosphate labelled.

(i) Misincorporation system 1 A (poly [d(A-T)]/dGTP/Mg^{2+})

The reaction mixture (0·1 ml) contained 50 mM-Tris·HCl (pH 8·5), 0·5 mM-dithiothreitol, 0·2 mg bovine serum albumin/ml, 2 mM-MgCl$_2$, 34 nmol poly[d(A-T)], 2·5 μM dATP, 2·5 μM-dTTP, 20 μM-dGTP, 1·25 μCi [^{3}H]dTTP (1200 cts/min per pmol) to measure correct synthesis or 13·3 μCi [^{3}H]dGTP (3·9 cts/min per fmol) to measure incorrect synthesis. "Cytoplasmic" DNA polymerase extracts were used with this system and systems 1 B, 2 and 3 and except where stated, there was 20 μg protein from passage 28 in each assay, 32 μg for passage 32, 29 μg for passage 40, 25 μg for passage 48, 51 μg for passage 54/5, 57 μg for passage 56/7 and 24 μg for passage 58/9. The assay for 5 S DNA polymerase contained 100 mM-KCl. (The [^{3}H]dGTP is stored in 50% ethanol which was removed immediately prior to the assay by blowing a gentle stream of nitrogen onto its surface.) Each parallel reaction was carried out in duplicate and 2 blanks, minus poly[d(A-T)], were included in every misincorporation assay. After the addition of the DNA polymerase extract, the assay was incubated for 50 min at 37°C. At the end of this period 15 μl of 6·7 mg/ml subtilisin, 167 mM-EDTA and 6·7 mM cold dGTP was added to the reaction mixture and further incubated at 37°C for 20 min. The tubes were placed on ice and 2 ml of 12% (w/v) trichloroacetic acid/0·1 M-PPi, followed by 50 μl lysozyme (10 mg/ml) were added. The precipitate was centrifuged (10,000 g), washed with 2 ml 12% trichloroacetic acid/0·1 M-PPi, centrifuged, resuspended in 0·1 ml 1 M-NaOH, spotted onto 1 inch square Whatman 3 MM filter paper and processed by the "batch" method as described before. The error frequency (or misincorporation ratio) is defined as incorrect synthesis divided by correct synthesis. A property of DNA polymerase can be defined called the discrimination (Bernardi & Ninio, 1978).

$$\text{Discrimination} = \frac{\text{Correct (incorporated)}}{\text{Incorrect (incorporated)}} \times \frac{\text{Incorrect concentration}}{\text{Correct concentration}}$$

Thus it is apparent that 1/discrimination is equivalent to a measure of the misincorporation ratio corrected for substrate concentrations.

(ii) Misincorporation system 1B (poly[d(A-T)]/dGTP/Mg^{2+})

This system is exactly the same as system 1A except that different concentrations of dNTPs were used. This assay contained dATP, dTTP and dGTP each at 10 μM, 1·25 μCi [^{3}H]dTTP (300 cts/min per pmol) to measure correct synthesis or 12·4 μCi [^{3}H]dGTP (5·7 cts/min per fmol) to measure incorrect synthesis. The assay for U. maydis DNA polymerase also contained 100 mM-KCl and the pH of the Tris·HCl was 7·5.

(iii) *Misincorporation system 2* (poly[d(A-T)]/dGTP/Mn^{2+})

This system is exactly the same as system 1A except that 0·075 mM-MnCl$_2$ is present instead of MgCl$_2$ and the concentrations of dATP and dTTP are both 10 μM.

(iv) *Misincorporation system 3* (poly[d(I-C)]/dTTP/Mg^{2+})

This system differs from system 1A in that 26·4 nmol of poly[d(I-C)] are present instead of poly[d(A-T)]; 2·5 μM-dGTP, 2·5 μM-dCTP, 8 μM-dTTP are the substrate concentrations and 1·25 μCi [^{3}H]dGTP (1140 cts/min per pmol) measures correct synthesis and 24 μCi [^{3}H]dTTP (17·8 cts/min per fmol) measures incorrect synthesis.

(v) *Misincorporation system 4* (poly(A)·oligo(dT)/dGTP/Mn^{2+})

The altered conditions in this system from that of 1A are: 28·2 nmol of poly(A) and 2·6 nmol of oligo(dT)$_{12-18}$ (mixed together and annealed at 37°C for 15 min) in place of poly[d(A-T)], 0·5 mM-MnCl$_2$ instead of MgCl$_2$, 100 mM-KCl, 5 μM-dTTP and 15 μM-dGTP, 1·25 μCi [^{3}H]dTTP (1000 cts/min per pmol) to measure correct synthesis and 10 μCi [^{3}H]dGTP (3·9 cts/min per fmol) to measure incorrect synthesis. "Nuclear" DNA polymerase extracts were used for this system.

When the effects of various agents on the misincorporation assay were investigated, the following procedures were adopted. With the sulphydryl reagent NEM a portion of the enzyme was pre-incubated for 15 min at 0°C with the appropriate NEM concentration and then the reaction mixture, which contains no dithiothreitol, was added and the assay proceeded as normal. The DNase and RNase treatments occurred after the 50-min incubation but before the protease step. A 5 μl portion of 2 mg/ml DNase in 0·5 mg/ml bovine serum albumin was added to the reaction mixture for 15 min. For RNase 10 μl of 0·5 mg/ml RNase A in 77 mM-NaEDTA were added for 15 min. In the mixing experiments the young and the old extracts were mixed and incubated for 30 min at 0°C prior to the beginning of the misincorporation assay.

(f) *Nucleotide base analysis of the products of the misincorporation assay*

By using dGTP labelled with ^{32}P at the α-phosphate position, a nearest neighbour base analysis of the products of the misincorporation assay can be obtained by degradation to 3'-deoxynucleoside monophosphates. Alternatively, if the degradation is to 5'-deoxynucleoside monophosphates, the identity of the incorporated base can be elucidated. For both types of analysis, misincorporation system 1A was used except that 9 μCi [α-^{32}P]dGTP (8·3 cts/min per fmol) were present instead of [^{3}H]dGTP and no protease treatment was employed. Four centrifugation steps were used instead of 2 and the final precipitate was washed 3 times with 3 ml diethylether (to remove trichloroacetic acid). For degradation to 3' dNMPs (nearest neighbour analysis) the final precipitate was dissolved in 100 μl of 0·2 M-Tris·HCl (pH 7·5) and 0·11 units of spleen phosphodiesterase, 150 units of micrococcal nuclease and CaCl$_2$ (final concentration 2 mM) was added to give a final volume of 200 μl. This mixture was incubated for 3 h at 37°C. The products were spotted onto Whatmann 3 MM paper and developed for 19 h with a solvent containing isobutyric acid/concentrated NH$_4$OH/0·1 M-EDTA/water (100:2·3:1·6:50, by vol). The chromatogram was cut into either 1, 1·5 or 2 cm strips and counted in a toluene-based scintillation fluid. Cold 3' dNMPs were also run as standards on the chromatogram and visualised with a u.v. light source. For degradation to 5' dNMPs the final precipitate was dissolved in 100 μl of 0·2 M-Tris·HCl (pH 8·1) and 10·2 units of pancreatic DNase and MgCl$_2$ (final concentration 5 mM) were added to give a volume of 190 μl. This was incubated for 3 h at 37°C. After a further addition of 1·1 units of snake venom phosphodiesterase (10 μl), the mixture was incubated for 2 h at 37°C. The products were analysed with the same chromatography system as for 3' dNMPs but 5' dNMPs were used as chromatography standards. In both cases the radioactivity was compared to a blank that lacked poly[d(A-T)].

200

3. Results

(a) Levels of DNA polymerase activity during the lifespan of MRC-5 fibroblasts

In senescent cells there was a reduction in the specific activity of DNA polymerase. At passage 28 the specific activity was 0·92 units/mg; passage 32, 0·74; passage 40, 0·60; passage 48, 0·67; passage 54/5, 0·23; passage 56/7, 0·29; passage 58/9, 0·26. This drop in DNA polymerase activity with age (approximately 3-fold) is related to population growth rate, since from passage 44 the cells were significantly slower in reaching confluency and by passage 50 they had to be subcultured at a split ratio of 1:2 instead of 1:4.

(b) Use of protease in the misincorporation assay

In the course of developing the misincorporation assay, it was noted that commercial preparations of [^{3}H]dGTP contained a contaminant (or contaminants) that were insoluble in trichloroacetic acid in the presence of protein. This artifact was removed by treating the products of the misincorporation assay with a protease (see Materials and Methods).

The protease digestion step does not affect the error frequency of DNA polymerase. This can be shown with *E. coli* DNA polymerase I since this enzyme preparation does not contain significant quantities of protein and an error frequency can therefore be determined without the protease digestion step. Without protease treatment the error frequency was 118×10^{-6}; with protease treatment it was 141×10^{-6}. These values are within the known variation of the misincorporation assay and are therefore taken as being the same.

(c) Misincorporation of dGTP into poly[d(A-T)] with Mg^{2+}

Table 1 shows a typical experiment involving the misincorporation of dGTP into poly[d(A-T)] with Mg^{2+} (system 1A). The incorrect synthesis is about 100% above blank values and, since 25% above background is taken as the limit of detection in this type of experiment, this incorporation is significant. The variation within an experiment can be judged from the duplicates and is seen to be quite small. In this experiment a 4·5-fold increase in error frequency is observed between cell generation 28 and 73. (The relationship between passage number and number of cell generations (CG) is given in the footnote to Table 1.) The data in Table 1 are shown graphically in Figure 1(a) (triangular symbols) together with the limits of detection.

A summary of all the data obtained on separate occasions with this system 1A is shown in Figure 1(a). This reveals that the variation from experiment to experiment is generally quite small and is of the order of ±10% (standard deviation) (see Fig. 1(b)). The reproducibility of the measurements from one cell extract to another of similar age can be compared and no significant differences can be perceived.

Figure 1(a) reveals that the error frequency is lowest at CG 28, rises to a plateau from 32 to 52·5 and then increases sharply to give an average 3·8-fold increase in error frequency at the termination of the life-span. An alternative interpretation is that up to CG 52·5 the error level is constant then rises steeply in the senescent phase to give a biphasic curve. Whatever the shape of the curve may be, statistical analysis shows that the increase in error frequency is highly significant ($P < 0.001$).

TABLE 1

Misincorporation of $[^3H]dGTP$ into $poly[d(A-T)]$ with Mg^{2+}

"Cytoplasmic" DNA polymerase (cell generation)	Misincorporation (cts/min)	Blank − poly[d(A-T)] (cts/min)	Net (cts/min)	Incorrect synthesis (fmol)	Correct synthesis (fmol)	Error frequency (×10⁶)	Discrimination 1
28	120	70 } 68	52	13·9 } 13·5	33,400	414 } 404	1/18,300
	117	67	49	13·1		392	
32	182	80 } 75	107	28·7 } 30·6	58,700	489 } 522	1/14,200
	197	70	122	32·6		555	
40	144	75 } 79	65	17·4 } 19·7	30,500	571 } 646	1/11,400
	161	83	82	22·0		722	
52·5	151	90 } 84	66	17·8 } 18·7	32,500	548 } 577	1/12,800
	158	79	74	19·7		607	
64	122	63 } 64	58	15·5 } 15·4	12,200	1270 } 1260	1/5900
	121	65	57	15·4		1260	
67·4	149	63 } 64	85	22·8 } 22·6	15,200	1500 } 1490	1/5000
	148	65	84	22·4		1470	
73	114	63 } 64	50	13·4 } 13·1	7100	1890 } 1840	1/4000
	112	65	47	12·7		1790	

Passage number is related to the number of cell generations (CG) in the following manner: passage 28- CG 28, 32-32, 40-40, 48-52·5, 54/5-64, 56/7-67·4, 58/9-73 (Kirkwood & Holliday, 1975; Fulder, 1979; T. B. L. Kirkwood, unpublished results).

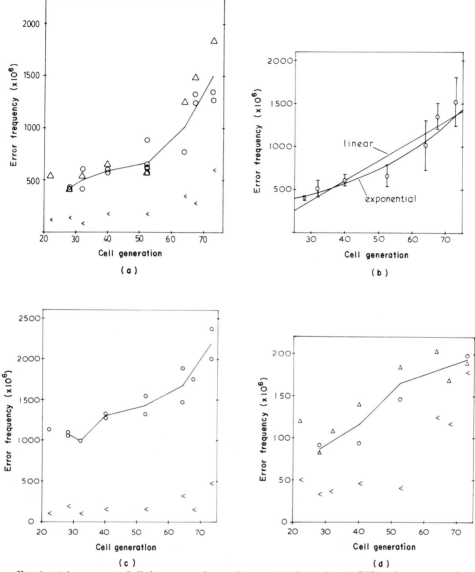

FIG. 1. (a) A summary of all the measured error frequencies of cytoplasmic DNA polymerase α using the poly[d(A-T)]/dGP/Mg^{2+} misincorporation assay system 1A during the lifespan of MRC-5. (△) Data from Table 1 and (<) the limits of detection also for this data. (○) Data obtained on other occasions.

(b) The average error frequency of cytoplasmic DNA polymerase α with the poly[d(A-T)]/dGTP/Mg^{2+} system is plotted ± one standard deviation. The best fitting linear and exponential curves, as calculated by linear regression, are also shown.

(c) Summary of all the error frequencies of cytoplasmic DNA polymerase α obtained with the poly[d(A-T)]/dGTP/Mn^{2+} system during the lifespan of MRC-5. The limits of detection (<) are from the same experiment as in Table 2.

(d) The error frequency of cytoplasmic DNA polymerase α with the poly[d(I-C)]/dTTP/Mg^{2+} system. The results of 2 experiments are shown (○, △) and the limits of detection (<) are for the (△) experiment whose results are also shown in Table 2.

203

The misincorporation frequency of DNA polymerase extracts from CG32 and 67·4 has also been measured in the poly[d(A-T)]/Mg^{2+} system 1A using [^{32}P]dGTP instead of [^{3}H]dGTP. The error frequencies with [^{32}P]dGTP are very similar to that obtained with [^{3}H]dGTP. For CG 32 the error frequency was 652×10^{-6} and for CG 73 was 1740×10^{-6}, giving an old/young ratio of 2·67.

Earlier misincorporation experiments used a different batch of [^{3}H]dGTP. Misincorporation system 1B used this earlier batch of [^{3}H]dGTP. This batch tended to give higher blank values and so the error frequency measurements are less accurate than with system 1A. However, very similar results were obtained and the old/young ratios for two experiments are 5·06 and 4·78.

(d) Misincorporation of dGTP into poly[d(A-T)] with Mn^{2+}

Using the divalent cation manganese instead of magnesium in the misincorporation assay, the error frequency is expected to be amplified. In these experiments with MRC-5 DNA polymerase a Mn^{2+}/Mg^{2+} error frequency ratio of between 4 and 10 is obtained. The young cell extracts tend to have a higher error frequency amplification than old extracts: CG 28 to 52·5 have a seven to tenfold increase; CG 64 to 73 have a four to fivefold increase.

Representative data from a typical experiment are shown in Table 2. It can be seen that the error frequencies can be measured more accurately in this system than with Mg^{2+} since the misincorporation is much higher. (In some cases, it is 300% above blank values.)

A summary of data obtained on two separate occasions is shown graphically in Figure 1(c) and the limits of detection from the results of the experiment in Table 2 are also depicted. The same shape of curve is found for Mn^{2+} as for Mg^{2+}. The initial low value at CG 28 and 32 increases slowly up to CG 52·5, then rises steeply after CG 63. The ratio of CG 73 : CG 28 is 2·17 which is lower than the Mg^{2+} value of 3·8. Nevertheless the increase in error frequency is statistically highly significant ($P < 0.001$).

(e) Misincorporation of dTTP into poly[d(I-C)] with Mg^{2+}

The rate of misincorporation of dTTP into poly[d(I-C)] with Mg^{2+} is approximately fivefold less than with the dGTP/poly[d(A-T)]/Mg^{2+} system. From the data in Table 2 and Figure 1(d) (triangular symbols) it can be seen that up to CG 52·5 the incorrect synthesis is reasonably above the blank values but for CG 64 to 73 the misincorporation values are not much higher than the limits of detection. Thus, these misincorporation frequencies should be treated with caution since they are probably not very accurate.

A summary of all the data obtained with this system is shown in Figure 1(d). The shape of the curve is approximately linear. The average ratio of error frequencies from young and old extracts is 2·3. The increase in error frequency in senescent cells is statistically highly significant ($P < 0.001$).

(f) Misincorporation of dGTP into poly(A)·oligo(dT) with Mn^{2+}

"Nuclear" extracts are used in this system, as opposed to "cytoplasmic" extracts which were used in the last three systems, and therefore the error frequency of DNA

TABLE 2

Misincorporation of $[^3H]dGTP$ into $poly[d(A\text{-}T)]$ with Mn^{2+}, and $[^3H]dTTP$ into $poly[d(I\text{-}C)]$ with Mg^{2+}

"Cytoplasmic" DNA polymerase (cell generation)	Misincorporation (cts/min)	Blank (– DNA template) (cts/min)	Net (cts/min)	Incorrect synthesis (fmol)	Correct synthesis (fmol)	Error frequency ($\times 10^6$)	1/Discrimination
poly[d(A-T)]/dGTP/Mn²⁺							
32	298 334	83 88 } 85	213 249	59·3	60,000	990	1/2020
52·5	309 278	92 84 } 88	221 190	52·8	40,000	1320	1/1520
73	178 187	74 87 } 80	98 101	25·6	10,800	2370	1/840
poly[d(I-C)]/dTTP/Mg²⁺							
32	60 61	36 33 } 35	25 26	1·4	13,400	108	1/27,000
52·5	64 58	28 30 } 29	35 29	1·8	9700	184	1/16,000
67·4	48 47	37 33 } 35	13 12	0·7	4200	168	1/18,000

205

polymerase can be measured in a different environment. With CG 32 the error frequency was $1·24 \times 10^{-3}$; CG 40, $1·79 \times 10^{-3}$; CG 52·5, $4·45 \times 10^{-3}$; CG 64, $1·08 \times 10^{-3}$; CG 73, $4·68 \times 10^{-3}$. At CG 32, the error rate was sixfold higher than with the poly[d(A-T)]/dGTP/Mg^{2+} system. The incorrect synthesis was in most cases much higher than the blank values (for CG 52·5 it was 400% higher). The ratio of CG 73:CG 32 was 3·77. The increase in error frequency at CG 52·5 and 73 compared to CG 32 and 40 was statistically highly significant ($P < 0·001$). However, the error frequency of CG 64 was approximately the same as CG 32. This effect was reproducible since a preliminary experiment also showed a fourfold decrease in error frequency of CG 64 compared with CG 73 (data not shown).

(g) Conditions reducing the level of homologous synthesis

In this and the subsequent four sections substantial evidence is presented that the misincorporation assay measures DNA polymerase-directed errors. Most of the control experiments were carried out with the poly[d(A-T)]/dGTP/Mg^{2+} system, since the major experiments used this system.

Manipulations which reduce the quantity of homologous synthesis should also reduce the quantity of non-homologous synthesis by a similar proportion. DNA polymerase requires a divalent cation for activity. Removing Mg^{2+} from the misincorporation assay completely abolishes correct and incorrect synthesis to a background level (Table 3). DNA polymerase α and γ are sensitive to the sulphydryl reagent N-ethyl maleimide. A preincubation with 10 mM-NEM reduces correct synthesis by 94% and incorrect synthesis to below the limit of detection (i.e. by greater than 80%).

TABLE 3

Controls with poly[d(A-T)]/dGTP/Mg^{2+} and "cytoplasmic" passage 48 DNA polymerase

Conditions	Incorrect synthesis (fmol)	Correct synthesis (fmol)
Complete	25·8	40,600
$-Mg^{2+}$	<5·6	<74
+NEM	<5·6	2500
$-$poly[d(A-T)]$+$poly(dT)	<5·6	<74
$-$poly[d(A-T)]$+$oligo(dT)$_{12-18}$	<5·6	<74
+RNase A	23·0	36,100
+DNase 1	<5·2	<65

DNA polymerase require a 3′ OH primer for activity and therefore the removal of poly[d(A-T)] and its replacement by either poly(dT) or oligo(dT) should abolish DNA polymerization. This is indeed found to be the case for both correct and incorrect synthesis (Table 3). This eliminates the possibility that terminal transferase is responsible for incorrect synthesis. It also eliminates the possibility that contaminating endogenous DNA or RNA is causing the misincorporation. This is because synthesis with endogenous DNA or RNA might not be observed in the "minus-DNA" blank because of nuclease degradation. However, when poly(dT) or oligo(dT) are added, a nuclease activity would be saturated, as is the case when poly[d(A-T)] is present, and any misincorporation caused by endogenous DNA or RNA would be observed if it occurred. The non-homologous synthesis is dependent on the DNA polymerase extract since the "minus-template" blank is the same as the "minus-enzyme" blank.

Treatment of the product of the misincorporation assay with DNase completely abolishes correct synthesis and Table 3 shows that incorrect synthesis is also DNase-sensitive. Treatment with RNase has no significant effect on correct or incorrect synthesis. Since the products of the reaction are routinely treated with a protease, the misincorporation is also not susceptible to proteolytic degradation.

(h) *Variation of deoxynucleoside triphosphate concentrations*

As shown in Figure 2(a), on increasing the complementary dNTP concentration while keeping the non-complementary dGTP constant, the incorrect synthesis falls steeply. The decline in error frequency is inversely proportional to the correct/incorrect substrate concentration ratio and hence the discrimination is constant over this range of correct dNTP concentration.

In Table 4 the discrimination at various correct dNTP concentrations is shown for three enzymes. The discrimination is approximately constant with an average of 15,400 for passage 32, 11,900 for passage 48 (from Fig. 2(a)), 5310 for passage 56/7 and 54,000 for *E. coli* DNA polymerase I. The only deviations occur (at 50 μM correct dNTP) where the measurements are the least accurate because the quantity of incorrect synthesis is so low. Therefore, this deviation is not considered to be significant.

When the non-complementary dGTP concentration is varied, the incorrect synthesis increases with increasing dGTP concentration (Fig. 2(b)). The discrimination (for passage 48) is approximately constant with an average of 12,900. The measurement at 5 μM-dGTP is inaccurate because of the small quantity of incorrect synthesis.

Removal of dATP and/or dTTP completely from the poly[d(A-T)]/dGTP/Mg^{2+} assay system gives some information on the mechanism of misincorporation. In the absence of dATP the misincorporation ratio was 1/50 for passage 32 and 1/53 for passage 56/7. Without dTTP the misincorporation ratio was 1/1100 for passage 32 and 1/450 for passage 56/7. Thus, it is apparent that dGTP is more likely to be incorporated in place of dATP than dTTP (by a factor of between 9 and 22). Nearest neighbour analysis confirms this result (see section (j), below). In the absence of both dATP and dTTP, a level of incorrect synthesis occurs which is approximately similar to that measured in the absence of dATP.

(i) *Time-course and variation in enzyme concentration*

The time-course for the misincorporation assay is shown in Figure 3(a). Correct and incorrect syntheses are both linear for up to 75 minutes. The error frequency is approximately constant for all stages of the time-course: for passage 32 it is 516×10^{-6}, and for passage 56/7 it is 1540×10^{-6}. The same result has been

TABLE 4

Effect of varying correct dNTP concentrations with the poly[d(A-T)]/dGTP/Mg^{2+} system

Concentration each of dATP and dTTP (μM)	Incorrect synthesis (fmol)	Correct synthesis (fmol)	Error frequency ($\times 10^6$)	Discrimination
"Cytoplasmic" passage 32 enzyme				
2·5	14·1	33,700	417	17,700
5	13·3	53,000	250	15,300
10	11·1	70,000	159	12,300
20	5·9	97,000	61	16,300
50	5·9	128,000	46	8620
"Cytoplasmic" passage 56/7 enzyme				
2·5	22·9	20,000	1130	6560
5	17·0	28,000	600	6360
10	17·7	40,000	450	4380
20	15·3	61,000	250	3920
50	16	100,000	160	2480
E. coli Pol 1				
2·5	99·2	616,000	159	47,000
5	61·5	762,000	80	48,000
7·5	41·4	708,000	58	45,000
10	25·2	976,000	26	76,000
20	19·2	936,000	21	48,000

obtained with passage 48 enzyme (data not shown). This control demonstrates that the kinetics of both incorrect and correct syntheses are linear with time for both young and old enzymes.

The effect of varying the quantity of DNA polymerase extract is shown in Figure 3(b). For passage 32 the correct synthesis is linear for up to 24 μg protein then gradually tails off. The incorrect synthesis duplicates this trend. For passage 56/7 correct synthesis is linear up to 57 μg protein then the rate of increase falls slowly. The incorrect synthesis closely parallels this pattern. The point at 19 μg protein from passage 56/7 is inaccurate because of the small quantity of correct synthesis. For passage 48 the correct and incorrect syntheses are linear up to 20 μg protein and then remain approximately constant (data not shown). The error frequency is approximately constant for all three enzymes at all concentrations of enzyme (for passage 32 the average is 650×10^{-6}, for passage 48 it is 970×10^{-6} and for passage 56/7 it is 1190×10^{-6}).

In both time-course and variation in enzyme concentration it can be shown that at similar levels of correct synthesis, the incorrect synthesis is greater in old extracts than in young. For example, in Figure 3(b), 16 μg protein from passage 32 cells and

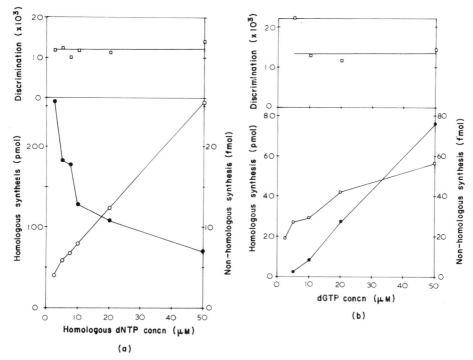

FIG. 2. (a) Effect of varying the homologous dNTP concentration on homologous and non-homologous synthesis in the poly[d(A-T)]/dGTP/Mg^{2+} system. Cytoplasmic DNA polymerase α from passage 48 was used. Lower Figure: homologous synthesis (○) and non-homologous synthesis (●). Upper Figure: discrimination (□) at each homologous dNTP concentration.

(b) Effect of varying the non-homologous dGTP concentration on homologous and non-homologous synthesis in the poly[d(A-T)]/dGTP/Mg^{2+} system. Cytoplasmic DNA polymerase α from passage 48 was used. Lower Figure: homologous synthesis (○) and non-homologous synthesis (●). Upper Figure: discrimination (□) at each non-homologous dGTP concentration.

57 μg from passage 56/7 cells give similar correct synthesis, but the incorrect synthesis is 12·5 and 22·4 fmoles, respectively. Similarly in Figure 3(a) after 15 minutes (passage 32) and 30 minutes (passage 56/7) incubation the correct synthesis is similar, but the incorrect synthesis is 5·7 and 13·2 fmoles, respectively. In addition correct synthesis is similar after 30 minutes (passage 32) and 60 minutes (passage 56/7) and incorrect synthesis is 14·5 and 32·8 fmoles, respectively.

(j) Nearest neighbour analysis and degradation to 5' deoxynucleoside monophosphates

Nearest neighbour analysis of the products of the misincorporation of [^{32}P]dGTP into poly[d(A-T)] with Mg^{2+} for both young and old extracts is shown in Figure 4(a). For passage 32, 84% of the radioactivity was present as 3'-deoxynucleoside monophosphates. Of these counts 80% was present as dTMP, 11% as dGMP, 9% as dAMP and 0% as dCMP. In the case of passage 56/7, 72% of the radioactivity was present as 3' dNMPs. 74% of this radioactivity was present as dTMP, 8% as dGMP,

209

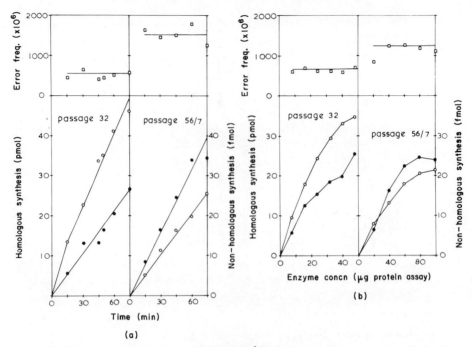

FIG. 3. (a) Time-course of the poly[d(A-T)]/dGTP/Mg^{2+} misincorporation assay system. Left, passage 32 and right, passage 56/7 cytoplasmic DNA polymerase α. Lower Figure: homologous synthesis (○) and non-homologous synthesis (●). Upper figure: error frequency (□) at each time point.
(b) Effect of varying the enzyme concentration on the poly[d(A-T)]/dGTP/Mg^{2+} misincorporation assay system. Left: passage 32 and right: passage 56/7 cytoplasmic DNA polymerase α. Lower Figure: homologous synthesis (○) and non-homologous synthesis (●). Upper Figure: error frequency (□) at each enzyme concentration.

16% as dAMP and 1% as dCMP. The percentage of counts as dGMP is probably an overestimate due to incomplete separation from dTMP. The remaining 16% and 28% of the counts not present as 3′ dNMPs were either undegraded oligonucleotides (4 to 12 cm from origin) or phosphate (28 to 34 cm from origin).

The products of the misincorporation of [^{32}P]dGTP into poly[d(A-T)] with Mg^{2+} were degraded to 5′-deoxynucleoside monophosphates for both young and old cell extracts and the results are shown in Figure 4(b). The snake venom phosphodiesterase was contaminated with a 5′ nucleotidase activity which degraded 5′dNM^{32}P to deoxynucleosides and [^{32}P]phosphate. Thus a significant proportion of the recovered radioactivity is present as [^{32}P]phosphate, which migrates at 30 cm from the origin. However, in the region where the 5′ dNMPs migrate, the counts are sufficiently close to zero (except for 5′-dGMP) that the possibility that significant contamination of dGTP by dATP, dTTP or dCTP can be excluded. For passage 32, 53% of the radioactivity migrates as 5′ deoxynucleoside monophosphates and 44% as phosphate. 91% of the radioactivity in dNMPs was present as dGMP, 0% as dTMP, 3% as dCMP and 2% as dAMP. With passage 56/7, 64% of the counts migrate as 5′ dNMP and 28% as phosphate. dGMP accounted for 93% of the radioactivity in 5′ dNMPs, 0% as dTMP, 5% as dCMP and 2% as dAMP.

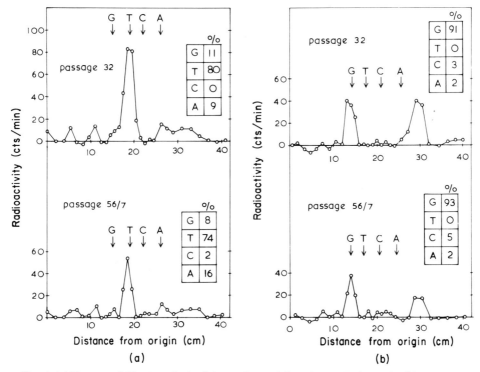

FIG. 4. (a) Nearest neighbour analysis of the products of the poly[d(A-T)]/dGTP/Mg^{2+} misincorporation assay using passage 32 (top) and passage 56/7 (bottom) cytoplasmic DNA polymerase α. The 3' dNMPs were chromatographed and counted as described in the Materials and Methods. The 3' dNMP markers were visualised with a u.v. light source. The numbers in the rectangular boxes represent the percentage of each nucleotide present.

(b) Degradation of the products of the poly[d(A-T)]/dGTP/Mg^{2+} misincorporation assay to 5' dNMPs. Passage 32 (top) and passage 56/7 (bottom) cytoplasmic DNA polymerase α were used. The 5' dNMPs were chromatographed and counted as described in the Materials and Methods. The 5' dNMP markers were visualised with a u.v. light source. The numbers in the rectangular boxes represent the percentages of each nucleotide present.

These results indicate that more than 90% of the radioactivity in dNMPs migrated as 5'-dGMP. This radioactivity could not be in the form of non-incorporated dGTP (which migrates at 7 cm from origin) or dGDP (which migrates at 10 cm). The possibility that a base analogue, which migrates at the same position as dGMP, is responsible for the misincorporation cannot be completely excluded, but is an unlikely explanation of the results.

The purity of the dGTP has been established but a cytosine impurity in the poly[d(A-T)] could explain the misincorporation of dGTP. This possibility can be excluded by using DNA polymerases that are more accurate during *in vitro* synthesis. Thus, if a cytosine impurity was present in poly[d(A-T)], these polymerases would accurately copy the cytosine and the error frequency of these polymerases would be the same as MRC-5 DNA polymerases. The discrimination of *E. coli* DNA polymerase I was approximately 54,000 (Table 4), for *Ustilago maydis* DNA polymerase 59,000, and *M. luteus* DNA polymerase 65,000 (data not shown), which is approximately threefold greater than MRC-5 DNA polymerase from young cells. Assuming that MRC-5 and the above DNA polymerases copy base analogues in a similar manner, then base analogue contamination of poly[d(A-T)] can also be excluded. This control also confirms the result obtained by degrading the misincorporation assay products to 5' dNMPs since significant contamination of dGTP with dATP can be excluded by the experiments with the three DNA polymerases with lower error frequencies.

(k) *Mixing experiments*

These experiments can eliminate the possibility that a diffusible agent is present in old or young cells which is capable of modifying the error frequency of DNA polymerase. If the accuracy of DNA polymerase is a property solely of the enzyme molecules themselves, then on mixing two DNA polymerase extracts with different error frequencies it is possible to predict the error frequency of the mixture assuming independent operation of the DNA polymerase molecules.

Old (passage 56/7) and young (passage 32) extracts were mixed in three different proportions (Table 5). The calculation of the predicted error frequency is complicated by the fact that the correct synthesis is not additive. Therefore to calculate

TABLE 5

Mixing experiments with old and young extracts with poly[d(A-T)]/dGTP/Mg^{2+}

Mixture	Quantity of extract added (μg protein)		Incorrect synthesis (fmol)	Correct synthesis (fmol)	Observed error frequency ($\times 10^6$)	Predicted error frequency ($\times 10^6$)
	Passage 32	Passage 56/7				
	8	—	5·7	9500	605	—
	32	—	18·4	29,400	628	—
	—	19	10·0†	8000	1250	—
	—	57	22·4	18,000	1250	—
1	8	57	25·5	21,200	1200	1030
2	32	57	28·1	32,200	870	860
3	32	19	24·4	35,800	680	760

† Estimated value since the actual measurement (6·5) is too inaccurate. This value was calculated from an error frequency of 1250×10^{-6}.

the predicted error frequency, the total incorrect synthesis of the young plus the old extracts was divided by the total correct synthesis of young plus old. In all three cases the experimentally observed values were very close to the predicted values. In mixture 1, the predicted error frequency is biased towards that of the old cell extract and a slightly higher error frequency than predicted is observed. In mixture 2 an intermediate error frequency is predicted and found. In mixture 3 equal amounts of protein are present from young and old extracts and the expected error frequency is biased towards that of the young cell extract and this is experimentally observed.

(1) *Kinetics of error accumulation*

It is difficult to distinguish between linear and exponential kinetics when the increase in error frequency is only two- to fivefold. This is shown in Figure 1(b) where the best-fitting linear and exponential curves are not very different. However, statistical analysis of all the data with systems 1, 2, 3 and 4 reveals that there is a better fit to exponential rather than linear kinetics.

4. Discussion

The fall in DNA polymerase specific activity with age is probably due to the decreased mitotic capacity of the cell population, since DNA polymerase α is induced in mitotic cells (Weissbach, 1977; Falaschi & Spadari, 1978), and the α-polymerase is responsible for 80 to 95% of the DNA polymerase activity in the cell extract. However, the presence of inactive or highly labile DNA polymerase molecules in senescent cells could be partly responsible for the reduced polymerase activity.

It is very important to demonstrate that the misincorporation assay measures the frequency of DNA polymerase-directed errors and not an artifact of the system. This can be accomplished by numerous control experiments. It was shown that any manipulation which reduced the quantity of homologous synthesis (removal of Mg^{2+}, reaction with NEM, removal of a 3'-OH primer, treatment with DNase) also similarly reduced the quantity of non-homologous synthesis. The time-course and variation in enzyme concentration experiments also clearly demonstrated the close relationship of correct and incorrect synthesis. Both homologous and non-homologous synthesis were unaffected by RNase or protease treatments. No contaminating terminal transferase activity could be detected. Variation of the homologous and non-homologous dNTP concentrations showed that the discrimination of DNA polymerase was approximately constant over a wide range of dNTP concentrations. Nearest neighbour base analysis of the products of the misincorporation assay revealed that errors occurred mainly as single base substitutions and a guanine–thymine base mis-pair was the main form of misincorporation for both young and old enzymes. On theoretical grounds the most likely base mis-pair is guanine–thymine and this result has also been obtained by Agarwal *et al.* (1979) and Seal *et al.* (1979). The purity of the labelled dGTP was examined by degradation of the products of the misincorporation assay to 5' dNMPs and found to be free from significant contamination. The purity of the template was checked by using DNA polymerase with a lower error frequency. The presence of a diffusible, error frequency-modifying agent in young or old cells could not be detected by mixing experiments.

There is an alternative explanation for the results of the mixing and other experiments. Since the DNA polymerase extracts employed in the experiments were impure, an agent that modifies the error frequency of DNA polymerase could be present in the extract. Examples of this error frequency-modifying agent include: DNA nucleases; binding proteins; a hypothetical enzymic activity that can insert a non-homologous dNTP at the 3'-OH terminus of primer template-DNA; etc. If the activity of this agent is present in *limiting* amounts, then this could account for the observed error frequencies after mixing the old and young extracts.

In the time-course, variation in enzyme concentration and variation in dNTP concentration experiments, the possible error frequency-modifying agent could have affected the results. This alternative possibility could not be distinguished from the hypothesis that DNA polymerase was solely responsible. The impure nature of the DNA polymerase extracts could also influence the error frequency values obtained in this study. However, the error frequency of young cell DNA polymerase α was similar to that found by other workers for highly purified DNA polymerase α (see below). This would appear to suggest that the possible error frequency-modifying agent would have to be present in senescent cell extracts. In addition, there is an unlikely alternative explanation for the control experiment where the products of the misincorporation assay were degraded to 5' dNMPs. The presence of a base analogue nucleoside triphosphate contamination of dGTP whose monophosphate has the same chromatographic properties as dGMP, would be indistinguishable from the case when pure dGTP was present. Nevertheless, the control experiments do strongly suggest that the measured misincorporation is directed by DNA polymerase and that the misincorporation assay provides an accurate estimate of the *in vitro* error frequency of DNA polymerase.

Four misincorporation assay systems were used to measure the *in vitro* error frequency of DNA polymerase. In the systems 1A and 1B, dGTP into poly[d(A-T)] with Mg^{2+}; system 2, dGTP into poly[d(A-T)] with Mn^{2+}; and system 3, dTTP into poly[(I-C)] with Mg^{2+}, "cytoplasmic" extracts were used. The "cytoplasmic" extracts contain greater than 99% DNA polymerase α. DNA polymerase γ is present in cytoplasmic extracts at less than 1% (assayed with poly(A)·oligo(dT) plus or minus 10 mM-NEM (data not shown)) and no DNA polymerase β is present (removed by DEAE-cellulose chromatography). The DNA polymerase activity in the cytoplasmic extracts possesses all of the characteristics of DNA polymerase α (Weissbach, 1977; Falaschi & Spadari, 1978): sedimentation coefficient of 7·5 S, inhibition to greater than 95% by 10 mM-NEM, inhibition by KCl, non-utilization of poly(A)·oligo(dT) as template-primer, correct elution properties from DEAE-cellulose. Thus it is assumed that the error frequency of only DNA polymerase α is being measured in systems 1A, 1B, 2 and 3 since DNA polymerase γ has tenfold more activity with a ribo-template than a deoxy-template (Weissbach, 1977; Knopf *et al.*, 1976). In system 4, dGTP into poly(A)·oligo(dT) with Mn^{2+}, "nuclear" extracts were employed. DNA polymerase α cannot copy a ribo-template to a significant extent (Weissbach, 1977; Knopf *et al.*, 1976) and since poly(A)·oligo(dT) is the preferred template for the γ polymerase, it is assumed that this is a misincorporation assay solely for DNA polymerase γ. DNA polymerase β is removed by DEAE-cellulose chromatography.

The error frequency of DNA polymerase α from young cells was similar to that found in other experimentally rigorous studies. Dube *et al.* (1979) examined the fidelity of DNA polymerase α from five sources with the poly[d(A-T)]/dGTP/Mg^{2+} system and found values from 1/3360 to 1/11,300. The average value of 1/18,000 found in this study, is slightly lower and probably represents a difference in the cellular origin of the DNA polymerase. This view is confirmed, since the fidelity of

DNA polymerase α from HeLa cells was 1/5480 (data not shown) which is very similar to the value obtained by Dube *et al.* (1979). The fidelity of DNA polymerase γ observed in this study, 1/3000, is approximately the same as found by Krause & Linn (1980) when the effect of dNTP concentrations is taken into account.

The major conclusion that can be drawn from this study is that the *in vitro* error frequency of DNA polymerase is increased in senescent fibroblasts. This result has been obtained with four different template/dNTP/divalent cation systems, both [^{3}H]- and [^{32}P]dNTPs, with DNA polymerase α and γ, numerous DNA polymerase extracts and with several batches of DNA templates and [^{3}H]dNTPs. A histogram (Fig. 5) summarizes all the data comparing old and young DNA polymerase error frequencies carried out on 19 separate occasions.

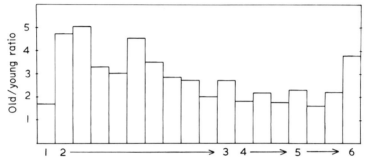

FIG. 5. A histogram showing the ratio of error frequencies from young and old extracts. The various conditions are as follows: (1) 5 S DNA polymerase α with [^{3}H]dGTP/poly[d(A-T)]/Mg^{2+}. (2) 7 S cytoplasmic DNA polymerase α with [^{3}H]dGTP/poly[d(A-T)]/Mg^{2+}. (3) 7 S cytoplasmic DNA polymerase α with [^{32}P]dGTP/poly[d(A-T)]/Mg^{2+}. (4) 7 S cytoplasmic DNA polymerase α with [^{3}H]dGTP/poly[d(A-T)]/Mn^{2+}. (5) 7 S cytoplasmic DNA polymerase α with [^{3}H]dTTP/poly[d(I-C)]/Mg^{2+}. (6) Nuclear DNA polymerase γ with [^{3}H]dGTP/poly[d(A-T)]/Mn^{2+}.

Almost all the misincorporation experiments were carried out with DNA polymerase α which is thought to be the main replicative polymerase in mammalian cells. DNA polymerase α from old fibroblasts has an average of 3·4-fold higher error frequency than young cells when measured by the misincorporation of dGTP into poly[d(A-T)] with Mg^{2+}. The increase is 1·9-fold with dGTP/poly[d(A-T)]/Mn^{2+} and 2·0-fold with dTTP/poly[d(I-C)]/Mg^{2+}. The error frequency of DNA polymerase γ increased 3·8 times. DNA polymerase extracts were prepared at seven points during the lifespan of MRC-5 and the reproducibility of the *in vitro* fidelity of these extracts relative to one another gives further confidence that the method is measuring a real biological phenomenon. Statistical analysis shows that the increase in error frequency with age is highly significant ($P < 0·001$) for all four misincorporation systems. The results presented here are consistent with the decrease in the fidelity of DNA polymerase from senescent MRC-5 fibroblasts reported by Linn *et al.* (1976). They found a two- to tenfold increase in error frequency and used mainly Mn^{2+} as divalent cation. Barton *et al.* (1974) have reported a twofold decrease in the fidelity of DNA polymerase β from aged mouse liver. However, Agarwal *et al.* (1978) could detect no increase in the error frequency of DNA polymerase α derived from lymphocytes of old people.

The error frequency of DNA polymerase *in vitro* is several orders of magnitude greater than that of DNA replication *in vivo*. Thus, it is not being suggested that the error frequencies of DNA polymerase found in this study are the same as those of DNA replication in young and old cells. We are instead suggesting that an important enzymic component of the DNA replication machinery, which is intimately associated with the accuracy of base selection, has a decreased fidelity *in vitro*. It is thought that *in vivo* DNA polymerase exists in a multi-enzyme complex for DNA replication, which is disrupted on isolation (Weissbach, 1977). The decrease in fidelity of DNA polymerase could be representative of the decrease in specificity of the other enzymes and proteins involved in the accuracy of DNA replication. Thus, the increase in the error frequency of DNA replication might be greatly amplified above the two- to fivefold DNA polymerase increase. This idea would correlate with the results of Fulder & Holliday (1975) and Fulder (1979) who found evidence with MRC-5 fibroblasts that the mutation frequency increases exponentially during the lifespan of MRC-5. Their observed increase in cell variants was 20 to 50-fold between young and old cells. Using the technique of DNA fibre-autoradiography, Petes *et al.* (1974) have shown with MRC-5 fibroblasts that the rate of replicon elongation is significantly reduced in senescent cells which suggests that the DNA replication complex contains faulty protein molecules. Morley, Cox & Holliday, unpublished, have recently shown that there is an exponential rise in somatic mutations in human lymphocytes from donors of increasing age. A statistical analysis of the kinetics relating the increase in error frequency with cell age revealed that the data were slightly better expressed by an exponential relationship than a linear.

We can conclude that the data presented are fully consistent with one of the major predictions of Orgel's error catastrophe theory of ageing. The demonstration of decreased fidelity of DNA polymerase extracts for senescent cells in this study and by Linn *et al.* (1976), provide evidence that an error feedback mechanism is responsible for the senescence of fibroblasts in culture. It is significant that the reductions in specificity we see are in enzymes that play a crucial role in information transfer. Once DNA polymerase is affected, a number of other changes in macromolecules will follow.

The results are possibly compatible with other theories of ageing. For instance, according to the programme theory, two new species of DNA polymerases, with a similar high error frequency, would be induced as part of the developmental programme for ageing. Alternatively, the alterations in DNA polymerase fidelity could be caused by post-translational modification of the enzyme. Further experiments would be required to rule out these possibilities.

One of the authors (V. M.) acknowledges financial support from the Medical Research Council. We thank T. B. L. Kirkwood for carrying out the statistical calculations.

REFERENCES

Agarwal, S. S., Tuffner, M. & Loeb, L. A. (1978). *J. Cell. Physiol.* **96**, 235–244.
Agarwal, S. S., Dube, E. K. & Loeb, L. A. (1979). *J. Biol. Chem.* **254**, 101–106.
Aposhian, H. V. & Kornberg, A. (1962). *J. Biol. Chem.* **237**, 519–525.

Barton, F. W., Waters, L. C. & Yang, W. K. (1974). *Fed. Proc. Fed. Amer. Soc. Exp. Biol.* **33**, 1419.

Bernardi, F. & Ninio, J. (1978). *Biochimie*, **60**, 1083–1095.

Bollum, F. J. (1968). *Methods Enzymol.* **12B**, 591–611.

Bradley, M. O., Hayflick, L. & Schimke, R. T. (1976). *J. Biol. Chem.* **251**, 3521–3529.

Chen, T. R. (1977). *Exp. Cell Res.* **104**, 255–262.

Dube, D. K., Kunkel, T. A., Seal, G. & Loeb, L. A. (1979). *Biochim. Biophys. Acta*, **561**, 369–382.

Falaschi, A. & Spadari, S. (1978). In *DNA synthesis: present and future*, pp. 487–516, Plenum Press, New York.

Fry, M. & Weisman-Shomer, P. (1976). *Biochemistry*, **15**, 4319–4328.

Fulder, S. J. (1979). *Mech. Age Devel.* **10**, 101–115.

Fulder, S. J. & Holliday, R. (1975). *Cell*, **6**, 67–73.

Gershon, H. & Gershon, D. (1973). *Proc. Nat. Acad. Sci., U.S.A.* **70**, 909–913.

Goldstein, S. & Moerman, E. J. (1976). *Interdiscip. Top. Geront.* **10**, 24–43.

Hayflick, L. (1965). *Exp. Cell Res.* **37**, 614–636.

Hayflick, L. & Moorhead, P. S. (1961). *Exp. Cell Res.* **25**, 585–621.

Holliday, R. (1975). *Fed. Proc. Fed. Amer. Soc. Exp. Biol.* **34**, 51–55.

Holliday, R. & Stevens, A. (1978). *Gerontology*, **24**, 417–425.

Holliday, R. & Tarrant, G. M. (1972). *Nature (London)*, **238**, 26–30.

Holliday, R., Pukkila, P. J., Dickson, J. M., Spanos, A. & Murray, V. (1979). *Cold Spring Harbor Symp. Quant. Biol.* **43**, 1317–1323.

Houben, A. & Remacle, J. (1978). *Nature (London)*, **275**, 59–60.

Kirkwood, T. B. L. & Holliday, R. (1975). *J. Theoret. Biol.* **53**, 481–496.

Knopf, K. W., Yamada, M. & Weissbach, A. (1976). *Biochemistry*, **15**, 4540–4548.

Krause, S. W. & Linn, S. (1980). *Biochemistry*, **19**, 220–228.

Lewis, C. M. & Tarrant, G. M. (1972). *Nature (London)*, **239**, 316–318.

Linn, S., Kairis, M. & Holliday, R. (1976). *Proc. Nat. Acad. Sci., U.S.A.* **73**, 2818–2822.

Lowry, O. H., Rosenbrough, N. J., Farr, A. L. & Randall, R. J. (1951). *J. Biol. Chem.* **193**, 265–275.

Orgel, L. E. (1963). *Proc. Nat. Acad. Sci., U.S.A.* **49**, 517–521.

Orgel, L. E. (1973). *Nature (London)*, **243**, 441–445.

Petes, T. D., Farber, R. A., Tarrant, G. M. & Holliday, R. (1974). *Nature (London)*, **251**, 434–436.

Richardson, C. C., Schildkraut, C. L., Aposhian, H. V. & Kornberg, A. (1964). *J. Biol. Chem.* **239**, 222–232.

Seal, G., Shearman, C. W. & Loeb, L. A. (1979). *J. Biol. Chem.* **254**, 5229–5237.

Shakespeare, V. A. & Buchanan, J. H. (1976). *Exp. Cell Res.* **100**, 1–8.

Weissbach, A. (1977). *Annu. Rev. Biochem.* **46**, 25–47.

21

Reprinted from J. Biol. Chem. **251**:3521-3529 (1976)

Protein Degradation in Human Fibroblasts (WI-38)

EFFECTS OF AGING, VIRAL TRANSFORMATION, AND AMINO ACID ANALOGS*

(Received for publication, August 26, 1975)

Matthews O. Bradley,‡ Leonard Hayflick, and Robert T. Schimke

From the Department of Medical Microbiology and Biological Sciences, Stanford University, Stanford, California 94305

Protein degradation occurs more rapidly in senescent WI-38 cultures than in phase II cultures or in SV-40 transformed WI-38 cells (VA-13). The first differences are found in early phase III, when short lived but not long lived proteins are degraded more rapidly. At the end of phase III long lived proteins are also degraded more rapidly as shown by both intermittent perfusion and approach to equilibrium methods. By both methods the rates of protein degradation for the virally transformed derivative are the same as those for phase II WI-38, implying that transformation has not altered these characteristics of protein degradation.

WI-38 cells incorporate canavanine, an analog of arginine, into protein. This analog, as well as p-fluorophenylalanine and azetidine carboxylic acid, accelerates the degradation of proteins labeled with [³H]leucine in their presence but does not alter the degradation rates of proteins prelabeled with [¹⁴C]leucine in the absence of the analogs. These results imply that the analogs increase the intracellular degradation rates of proteins because they render them more susceptible to the degradative system.

Late phase III WI-38 cells may not selectively catabolize proteins containing canavanine as rapidly as do phase II and VA-13 cells. These results imply that the phase III protein degradative system becomes partially defective, thereby losing its ability to rapidly catabolize altered protein which leads to increased levels of abnormal proteins and decreased cell function.

Proteins are continuously synthesized and degraded in both prokaryotic and eukaryotic cells and this turnover process can consume a large share of a cell's metabolic energy (1, 2). Among the possible multiple functions of protein degradation is the selective removal of "abnormal" proteins resulting from transcriptional, translational, or post-translational errors and alterations (3-5). Most evidence supporting this concept comes from prokaryotes where it has been shown that structural mutations of the lac repressor (6) and of β-galactosidase (7) result in such rapid degradation of these proteins that their steady state levels are extremely low, even though the wild type proteins are stable. Other studies indicate that amino acid analogs (8-11), puromycin (8, 9), which causes premature release of incomplete polypeptides, and a ribosomal mutation causing increased errors in translation (9) also increase protein degradation in *Escherichia coli* cells.

Evidence also exists for eukaryotes that abnormal proteins are degraded rapidly. Human reticulocytes, for example,

rapidly degrade the β chain subunits of certain variant hemoglobins (12-14), even though the variant subunits differ from the normal ones by only one amino acid. Knowles *et al.* (15) have recently demonstrated that amino acid analogs increase protein degradation in hepatoma cells. Another example is the accelerated degradation of mutant forms of HGPRT in mouse L-cells (16). These findings all suggest that one of the major functions of protein degradation is the preferential elimination through degradation of abnormal proteins that contain transcriptional, translational, or post-translational errors or alterations. If this idea is correct, then functional failures in the protein degradative system could increase the cellular concentration of defective proteins so that cell function declines, eventually leading to cell senescence, transformation, or death (3-5).

The normal human embryonic fibroblast, WI-38 (17), provides a convenient and well characterized *in vitro* system for examining protein degradation in normal, senescent, and transformed cells. WI-38 cultures double approximately 50 times (phase II) before mitosis ceases and the cells die (phase III), a phenomenon that has been interpreted to represent aging at the cellular level (17). These same WI-38 cultures can be infected with SV-40 virus to yield morphologically altered, indefinitely propagable, and malignant cell lines (18).

Proteins with abnormal properties have been detected in a

* This investigation was supported by Research Grant HD04004 to L. Hayflick, and by Research Grant GM14931 to R. T. Schimke, both grants from the National Institutes of Health.

‡ A Postdoctoral Fellow of the Leukemia Society of America. Present address for correspondence and reprint requests, Laboratory of Molecular Pharmacology, National Cancer Institute, National Institutes of Health, Building 37, Room 5D27, Bethesda, Maryland 20014.

variety of senescent cells and organisms. Holliday's group (19), for instance, has demonstrated altered thermolability of two enzymes in phase III human fibroblast cultures. Gershon and co-workers (20, 21) have detected enzymes with decreased specific activity in both senescent mice and nematodes (22). In addition, we have shown that terminal phase III WI-38 cultures synthesize and accumulate proteins with an increased susceptibility to exogenous proteases (23). All of these studies suggested to us that protein degradation might be altered to senescent WI-38 cultures since increased amounts of abnormal proteins are probably present in these cultures and also since abnormal proteins are more rapidly catabolized.

We report here our initial studies on protein degradation in these cells with the eventual goal of defining what role protein degradation may play in the altered regulatory states of transformation and senescence. We describe two different methods of measuring protein degradation in monolayers of cultured cells. Both methods attempt to minimize isotopic precursor reutilization by first measuring the approach to equilibrium incorporation of isotopic leucine into proteins and secondly, by measuring the decay of prelabeled protein during intermittent perfusions with high leucine concentrations in shaking cultures. Using these methods, we have measured the rates of protein degradation in phase II, phase III, and SV-40 transformed WI-38 cells. Furthermore, we have shown that proteins containing any one of three different amino acid analogs are more rapidly degraded in WI-38 cells and we have assessed the relative ability of phase II, phase III, and SV-40 transformed WI-38 cells to degrade proteins that have incorporated canavanine, an analog of arginine.

EXPERIMENTAL PROCEDURES

Materials—Materials were purchased as follows: Aureomycin from Lederle; Hepes,[1] canavanine, *p*-fluorophenylalanine, azetidine carboxylic acid, phenylalanine, proline, arginine, and leucine from Calbiochem; powdered BME G-13 media, sodium bicarbonate, and glutamine from Grand Island Biological; calf serum from Pacific Biological; dialyzed calf serum from Microbiological Associates; L-[*U*-^{14}C]leucine (310 mCi/mmol), L-[4,5-^{3}H]leucine (40 Ci/mmol), D,L-[*guanido*-^{14}C]canavanine·2 HCl (33 mCi/mmol), and L-[4-^{3}H]arginine (25 Ci/mmol) from Schwarz/Mann; NCS from Nuclear Chicago; bovine serum albumin from Sigma; Aquasol, [^{3}H]toluene standard, and [^{14}C]toluene standard from New England Nuclear; and Instagel from Packard. All other chemicals were of reagent grade.

Cell Culture—The human embryonic lung fibroblast strain WI-38 (17) and its SV-40 transformed counterpart, VA-13 (18), were routinely subcultured in glass 32-ounce prescription bottles at 1/4 and 1/8 split ratios, respectively. For experiments the cells were seeded at 2×10^4 cells/cm^2 in either glass scintillation vials, 25 cm^2, or 75-cm^2 growth area flasks (Falcon Plastics, Inc.). The amount of medium used was 0.5 ml/cm^2 of growth area. Cell numbers were measured with a Coulter counter model B. The medium used was BME supplemented with 2.0 g/liter of NaHCO$_3$, 10% calf serum, 28 mM Hepes, 1 mM glutamine, and 50 μg/ml of Aureomycin or gentamycin. All cells used in these experiments were free of mycoplasma as determined by tests previously described (24).

In these experiments we express the finite replicative capacity of WI-38 cultures as the number of population doubling levels remaining before the final subculture of sister cultures when mitosis ceases and death ensues. Therefore, "young," phase II cultures are defined as those between 20 and 30 population doublings before death; whereas "old," phase III cultures are a few or zero population doublings before death. This information, while only useful in retrospect, provides a means of comparing different experiments with phase III cultures. Such a comparison cannot be made from the enumeration of population

doubling levels alone, since it is only known retrospectively at what population doubling levels the cultures reached terminal phase III.

Isotope Incorporation into Acid-soluble and -Insoluble Fractions—Radioactive amino acids were incorporated into cells by adding the appropriate isotopes to the culture media for various lengths of time as described for the individual experiments. Trichloroacetic acid-soluble and -insoluble radioactivity was determined as follows for media and for cells. Samples of media (100 μl of 1 ml) were added to test tubes (400-μl Beckman or 1.5-ml Eppendorf) containing enough 25% (w/v) trichloroacetic acid to make a 5% (w/v) final solution. After precipitating at 4° for at least 1 hour, the samples were centrifuged in a Beckman microcentrifuge for 1 min (10,000 × g maximum). One hundred microliters of the acid-soluble fraction were removed to a scintillation vial for determination of radioactivity. The remainder was removed by aspiration and the pellet rinsed once and broken up in 300 μl of 5% (w/v) trichloroacetic acid. The precipitate was centrifuged again for 15 s, the acid removed, and the tip of the test tube cut off and placed in a scintillation vial containing 1 ml of 0.1 N NaOH. The closed vials were incubated at 37° until the pellets were dissolved (at least 3 hours).

Isotope incorporation into cells was determined by rapidly aspirating away all radioactive medium with a vacuum line, by rinsing and aspirating away all rinse four times with ice-cold BME (without serum), and finally by adding 5% (w/v) trichloroacetic acid directly to the cultures (0.2 ml/cm^2 of growth area). After at least 2 hours of equilibration at 4°, a portion of the acid-soluble fraction was removed, centrifuged, and added to a scintillation vial. The remainder of the acid was removed and the cultures were rinsed three times with ice-cold 5% (w/v) trichloroacetic acid (0.4 ml/cm^2 of growth area) containing 1 mM the appropriate amino acid. The cells were dissolved in 0.1 N NaOH with 0.3% Sarkosyl (0.1 to 0.2 ml/cm^2 of growth area) by incubating them for at least 2 hours at 37°. An aliquot (usually 1.0 ml) of the NaOH digest was removed for scintillation counting and was considered the acid-insoluble fraction.

Leucine Flux—The uptake of [^{3}H]leucine into confluent cultures of WI-38 was determined by adding the isotope (2 μCi/ml) in sequence to duplicate cultures in glass scintillation vials. Uptake was stopped by rapidly cooling and washing the cultures as described above.

The rate of efflux of free leucine (acid-soluble) from the soluble pools of confluent WI-38 cultures was determined in cultures that had been previously incubated in [^{3}H]leucine for either 15 min or 50 hours. After the uptake period, the vials were rapidly chilled and rinsed 10 times with 5 ml of ice-cold BME until the rinse radioactivity was negligible. Then the vials were reincubated at 37° in 2 ml of BME containing 10% calf serum, and every 2 min the medium was removed and another 2 ml of medium added. The acid-soluble radioactivity in each rinse was determined as described above. At the end of each experiment, determinations were made of the amounts of acid-soluble and -insoluble radioactivity remaining in the cultures.

Approach to Equilibrium Measurements of Protein Degradation—The protocol was to first inoculate the required number of replicate 25-cm^2 Falcon flasks with 5×10^5 cells in 5 ml of BME for 6 days until the cultures were confluent. The medium was changed 2 days before the experiment was to begin in order to stimulate postconfluent growth. After stimulation a wave of mitoses began in phase II WI-38, but had subsided by 48 hours. Only a small wave of mitoses began in the phase III cells after stimulation, but this had also subsided by 48 hours. Therefore, these WI-38 cells are in G1 or G0 at the start of these experiments (40). The VA-13 cells are stimulated into division by a medium change and afterwards become quiescent again (40). Nevertheless, a small fraction of VA-13 cells continue to cycle and these cells will be in stages of the cell cycle other than G1. The majority of VA-13 cells, however, will be in G1 or G0 (40).

At the beginning of the experiment [^{3}H]leucine was added simultaneously to all cultures at approximately 1 μCi/ml. Periodically throughout the experiment, duplicate cultures of each cell type were harvested for determination of the amount of acid-soluble and -insoluble radioactivity as described above. The rate of protein synthesis was estimated by measuring the amount of [^{14}C]leucine (0.5 μCi/ml) incorporated in 1 hour. The amount of acid-soluble and -insoluble radioactivity was determined in the medium from each culture harvested.

Intermittent Perfusion Measurements of Protein Degradation—Confluent cultures, similar to those described in the preceding section, were incubated in 0.25 μCi/ml of [^{14}C]leucine for 2 or 3 days depending upon the experiment. This medium was removed and the cultures were rinsed four times with 10 ml of BME (without serum) containing 2 mM unlabeled leucine (Leu/BME). After each rinse, the

---

[1] The abbreviations used are: Hepes, Hydroxyethylpiperazine-*N'*-2-ethanesulfonic acid; BME, Basal Medium Eagle; Leu/BME, BME containing 2 mM extra leucine; PDBD, population doublings before death.

remaining medium was aspirated away with a sterile pipette attached to a vacuum line. The cultures were reincubated in fresh Leu/BME without isotope for periods between 2 and 17 hours in order to permit the degradation of the rapidly turning over fraction of cellular protein (25). The cultures were then relabeled for 25 or 30 min with between 10 and 40 μCi/ml of [³H]leucine. After this second isotope incorporation, the cultures were rinsed eight times with Leu/BME at 37° (without serum) within a period of 20 to 30 min. The cultures were rinsed one after another in sequence, so that each culture was kept in contact with the medium for as long as it took to rinse the total number of cultures in the experiment. In this way the intracellular pools were emptied of isotope as much as possible.

After the last rinse, 5 ml of Leu/BME (containing 10% serum) were added and the cultures were placed on a New Brunswick shaking water bath incubator at 37° (setting 3.5). At each experimental point, 5 ml of medium were removed and another 5 ml were added. The acid-soluble and -insoluble radioactivity in these samples was determined as described above. In most experiments the amount of acid-insoluble counts per min released to the medium was less than 5% of the total radioactivity in the system, but nevertheless was included in that total. At the end of the experiment the cultures were rinsed three times with 10 ml of ice-cold Leu/BME (without serum) and the amount of acid-soluble and -insoluble isotope remaining was determined. The protein fraction degraded during a certain period was calculated by dividing the amount of acid-soluble radioactivity contained in that sample by the total radioactivity in that flask. This total was calculated for each flask by summing the acid-soluble and -insoluble disintegrations per min in each sample of medium obtained from that culture during the experiment and from the total radioactivity remaining at the end of the experiment.

Scintillation Counting—Trichloroacetic acid-soluble fractions of 0.1 or 1.0 ml were added to distilled water to make a total volume of 3.0 ml. Ten milliliters of either Instagel or Aquasol were added to the 3 ml to make a stiff gel for scintillation counting.

Acid-insoluble fractions were dissolved in 0.1 N NaOH/1.0 ml of which was added to 2 ml of distilled water and 10 ml of Instagel or Aquasol.

In each experiment, standards of [¹⁴C]- and [³H]toluene were used to determine the counting efficiencies and overlap of ¹⁴C into the ³H channels. The tritium efficiency was usually 19% with 1.0 ml of trichloroacetic acid and 21% with 1.0 ml of 0.1 N NaOH; the ¹⁴C efficiency was usually 43% with 1.0 ml of acid-soluble BME, 47% with 1.0 ml of the acid-soluble fractions from the cultures, and 51% with 1.0 ml of 0.1 N NaOH. The ¹⁴C overlap into the tritium channel was usually 8.0% when calculated as counts per min of ³H channel/disintegrations per min of ¹⁴C channel.

Protein Determinations—Protein was measured by the Folin-Lowry (26) procedure at 540 nm, using crystalline bovine serum albumin as a standard.

RESULTS

Leucine Reutilization—Measurements of protein degradation are often complicated by the reutilization of isotopic precursor amino acids (27–29). In order to determine whether leucine reutilization occurs in these cultures, we determined the effect of higher leucine concentrations in the medium on the rate of release of leucine from prelabeled cells. WI-38 cultures in 25-cm² growth area flasks were prelabeled for 40 hours with [³H]leucine and [¹⁴C]arginine and then washed and reincubated either in normal medium containing 0.2 mM leucine or in medium supplemented with 2 mM leucine. Aliquots (100 μl from 6 ml total) of the medium were removed and assayed for acid-soluble radioactivity. The data in Fig. 1 show that the mean half-life of protein labeled with [³H]leucine and incubated in normal medium was 87 hours, while that incubated in medium supplemented with 2 mM leucine was 57 hours. The half-life of protein labeled with [¹⁴C]arginine was not altered by the additional leucine. This experiment indicates that reutilization of isotopic precursor can occur in cell monolayers, since increasing the leucine concentration of the medium significantly reduces the apparent half-life of protein

labeled with leucine but leaves unaltered the half-life of protein labeled with arginine.

Approach to Equilibrium—The problem of isotopic precursor reutilization can theoretically be overcome by measuring the rate of approach of cellular protein to a constant isotopic specific activity. For exponentially growing cells, labeled protein will replace unlabeled protein according to the equation:

$$P = P \propto (1 - e^{-\ln 2(1/T_{Dbl} + 1/T_{1/2})t}) \qquad (1)$$

where P is the counts per min per milligram of protein at time t; $P\propto$ is the counts per min per milligram of protein under steady state conditions at equilibrium labeling; T_{Dbl} is the doubling time of the culture in exponential growth; and $T_{1/2}$ is the protein half-life. This equation assumes: that the cells are in a steady state, that the rate of protein synthesis is constant, that proteins are degraded randomly (by first order kinetics) but synthesized with zero order kinetics, and that the specific activity of the precursor pool for protein synthesis is constant. Greenberg (30) and Brandhorst and McConkey (31) have derived equations similar to this one for describing mRNA degradation in growing cells.

In our experiments the cultures are postconfluent so that the cells are neither growing (see Figs. 2 and 3) nor mitosing and Equation 1 can be simplified to:

$$P = P \propto (1 - e^{-\ln 2t/T_{1/2}}) \qquad (2)$$

One of the assumptions underlying Equations 1 and 2 is that the specific radioactivity of the cellular pool from which leucine is withdrawn for protein synthesis remains constant during the time course of the experiment. There is at present no method for measuring the specific radioactivity of this pool if it is different from the total cellular leucine pool. For this reason our data may not measure the appropriate soluble pools and is subject to that limitation. Nevertheless, as shown below (see Figs. 7 and 8) leucine is transported into and out of the soluble pools of WI-38 with a $T_{1/2}$ of approximately 4 min. This

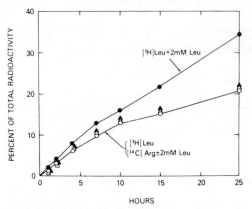

FIG. 1. Effect of excess leucine on release of acid-soluble radioactivity into the medium. WI-38 confluent monolayers were labeled with [³H]leucine and [¹⁴C]arginine; then they were washed and reincubated in normal medium or medium supplemented with 2 mM leucine. Aliquots (100 μl from 6 ml total) of the medium were removed during the experiment and assayed for acid-soluble radioactivity. O, [³H]leucine without additional leucine; ●, [³H]leucine plus 2 mM leucine; △, [¹⁴C]arginine without additional leucine; ▲, [¹⁴C]arginine plus 2 mM leucine.

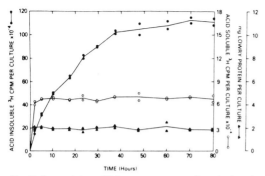

FIG. 2. Approach to equilibrium measurements of protein degradation in phase II (34 PDBD) WI-38 cultures. [³H]Leucine was added to confluent cultures of phase II WI-38 at the beginning of the experiment. Periodically, duplicate cultures were rinsed, harvested, and the amount of [³H]leucine incorporated into 5% acid-soluble (O——O) and insoluble (●——●) material was determined. The amount of protein per culture (▲——▲) was measured with the Lowry (26) procedure.

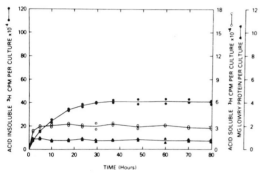

FIG. 3. Approach to equilibrium measurements of protein degradation in late phase III (0 PDBD) WI-38 cultures. The details are the same as in legend to Fig. 2.

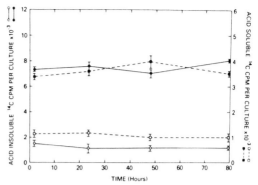

FIG. 4. Rates of protein synthesis in the experiments of Figs. 2 and 3. The rates of protein synthesis were estimated by measuring the incorporation of [¹⁴C]leucine into acid-soluble and insoluble material during 1 hour. Each point represents the mean of data from two flasks. ●——●, acid-insoluble incorporation per culture, phase II (34 PDBD); O——O, acid-insoluble incorporation per culture, phase III (0 PDBD); ■- - -■, acid-soluble incorporation per culture, phase II (34 PDBD); □- - -□, acid-soluble incorporation per culture, phase III (0 PDBD).

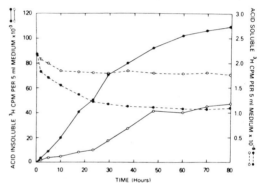

FIG. 5. Acid-soluble and insoluble radioactivity in the media from the experiments of Figs. 2 and 3. The amount of acid-soluble and insoluble counts per min in the media was determined throughout the experiments. Each point represents the mean of data from two flasks. ●——●, acid-insoluble, phase II (34 PDBD); O——O, acid-insoluble, phase III (0 PDBD); ●- - -●, acid-soluble, phase II (34 PDBD); O- - -O, acid-soluble, phase III (0 PDBD).

time is short relative to the time of equilibrium radioactive labeling of cellular protein. Furthermore, the data in Figs. 2 and 3 show that the counts per min of leucine in the acid-soluble pools per milligram of total protein in the cultures had reached an almost constant value within 2 hours after addition of radioactivity. In some other experiments this value increased slowly by 10 to 25% between 1 and 4 hours. Because the cells remain in a stationary phase during these experiments, all of these data imply that the pool leucine specific radioactivity reaches an essentially constant value within a time period that is short relative to the total accumulation curve of radioactivity into protein.

The total incorporation of leucine into acid-soluble and -insoluble material as a function of time is shown in Fig. 2 for early phase II WI-38 (34 population doublings before death) and in Fig. 3 for late phase III (0 population doublings before death). In order to estimate the rate of protein synthesis during the experiments, we measured the amount of [¹⁴C]leucine incorporated in 1 hour. These rates are shown in Fig. 4 and remain constant during the experiments (Figs. 2 and 3). In addition, the amount of protein per culture was also constant (Figs. 2 and 3). The invariance of all of these parameters implies that the cells remain in a steady state during the experiments and that the assumptions inherent in the deriva-

tions of Equations 1 and 2 are approximately valid.

The acid-soluble and -insoluble counts per min in the medium were determined throughout the experiments. The data in Fig. 5 show that trichloroacetic acid-insoluble material, presumably protein, is released continually to the medium, while the counts per min of soluble radioactivity declines as expected, reflecting its incorporation into protein. In order to determine the total accumulation of radioactive protein at time (t), the amount in the cell layer was added to the amount released into the medium. These totals were used to determine the half-life of protein in the different cultures according to Equation 2.

If the ln $[P\infty - P/P\infty]$ is plotted against time, one obtains a straight line with a slope equal to $-\ln 2/T_{1/2}$ (Fig. 6). The half-life determined in this way is between 15 and 25 hours for phase II (34 population doublings before death) cultures and between 11 and 14 hours for phase III (0 population doublings before death) cultures. By this analysis, the proteins in

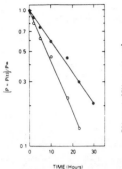

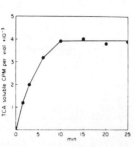

TIME (Hours)

FIG. 6 (*left*). Semilogarithmic plot of the data obtained from Figs. 2, 3, and 5. In order to determine the total accumulation of radioactive protein at time t, the amount in the cell layer was added to the amount released into the medium. The totals were used to determine the half-life of protein in the cultures according to Equation 2. ●, Phase II (34 PDBD); ○, phase III (0 PDBD). See the text for further details.

FIG. 7 (*right*). Leucine uptake by confluent cultures of WI-38. [³H]leucine was added in series to duplicate cultures in glass scintillation vials. Uptake was stopped by rapidly cooling and washing the cultures three times with 5 ml of ice-cold BME. After aspirating off the last of the wash, trichloroacetic acid (5%, 1 ml, ice-cold) was added for 2 hours. Acid-soluble radioactivity taken up by the cultures is plotted on the ordinate as a function of time after isotope addition.

terminal phase III WI-38 cultures are turning over faster than are the proteins in phase II cultures.

·Equations 1 and 2 are derived for a single class of protein decaying with a constant half-life. However, eukaryotic cells have many unique proteins, each turning over with their own half-life ranging from minutes to days (2, 3). The fact that we obtain a first order decay curve under these conditions is probably because this method will not detect turnover of proteins with half-lives less than 2 hours (the first time point) or of proteins with very slow half-lives (greater than 2 days). Within this restricted range protein half-lives may be similar. Nevertheless, if the half-lives of all proteins in the cell were to be increased or decreased proportionately, then these methods should detect that change.

Table I is a summation of all the data obtained with this method in phase II, phase III, and VA-13 cells. The general conclusions are first, that VA-13 and phase II WI-38 cells have nearly identical rates of protein degradation and second, that phase III cultures degrade protein detected by this method faster only at 0 population doublings before death, the very end of their *in vitro* lifespan.

Leucine Flux through Steady State WI-38 Cells—The experiments of Figs. 2 and 3 and of Table I imply that leucine uptake by monolayers of cultured cells might occur rapidly since the soluble pool was almost saturated with isotope within 2 hours. We therefore measured the rates of uptake and loss of free [³H]leucine by confluent cultures of WI-38 (25 population doublings before death) over shorter time periods. The data in Fig. 7 show that acid-soluble leucine radioactivity was taken up by the cultures to a maximum value within 7 min after addition of isotope and with a $T_{1/2}$ for uptake of 4-min. After this initially rapid uptake occurs, a slower rate of increase in the amount of acid-soluble leucine continues for approximately 2 hours, after which it remains almost constant (Figs. 2 and 3).

The rate of loss of free [³H]leucine from internal pools was determined in cells that had been labeled for either 15 min or

TABLE I

Approach to equilibrium

Rates of protein degradation in WI-38 and VA-13 cells are as determined by the approach to equilibrium method.

Cell	Population doublings before death	T^a	Mean
WI-38	0–1	11–14 (2)	13 ± 2.1
WI-38	2–10	17–24 (4)	20 ± 3.8
WI-38	>25	15–23 (6)	21 ± 3.6
VA-13		20–23 (3)	22 ± 1.5

[a] Range of half-lives; number of experiments given in parentheses. The half-life, T, is defined as: $T = \ln 2/k$, where k is the rate constant for the protein fraction degraded per hour. The means are ± the standard deviation.

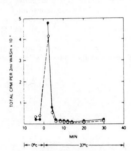

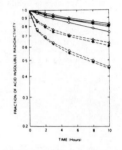

FIG. 8 (*left*). Leucine efflux for cultures of WI-38. Confluent cultures in glass scintillation vials were labeled with [³H]leucine for either 15 min or 50 hours. After labeling, the vials were chilled and washed 10 times with ice-cold BME until the wash radioactivity was negligible. Then the vials were reincubated at 37° in BME containing 10% calf serum; every 2 min the medium was removed from duplicate vials and another 2 ml of medium added. The mean radioactivity in each wash is plotted as a function of time. The kinetics of label efflux are the same, no matter whether the cells were labeled for 15 min or 50 hours. At the end of each experiment, 5% trichloroacetic acid was added to determine the amount of acid-soluble radioactivity remaining in the cultures.

Cultures pulsed for 15 min contained 4% of the initial acid-soluble label, while cultures labeled for 50 hours contained 25% of the initial acid-soluble label. ○- - -○, 15-min labeling; ●——●, 50-hour labeling.

FIG. 9 (*right*). Protein degradation as measured by the intermittent perfusion method. Proteins that turnover slowly were labeled by incubating the cultures with [¹⁴C]leucine for 40 hours and then by rinsing away the [¹⁴C]leucine and reincubating the cultures for 2 hours in isotope-free medium to allow the rapidly labeled proteins to degrade. Proteins that turnover rapidly were next labeled by incubating the same cultures with [³H]leucine for 25 min. After rinsing away free isotope, the rate of protein degradation was estimated by measuring, as a function of time, the percentage of the total initial acid-insoluble ³H and ¹⁴C disintegrations per min released from the cells into the medium in acid-soluble form. The data are plotted semilogarithmically as (1.0 – the acid-soluble fraction) to give the fraction of radioactivity remaining acid-insoluble. ▼——▼, WI-38, early phase III (2-10 PDBD), % ¹⁴C released; △——△, VA-13 cells, % ¹⁴C disintegrations per min released; ●——●, WI-38 phase II (>25 PDBD), % ¹⁴C released; ○——○, WI-38 terminal phase III (0 PDBD), % ¹⁴C disintegrations per min released; △- - -△, VA-13 cells, % ³H released; ●- - -●, WI-38 phase II (>25 PDBD), % ³H released; ○- - -○, WI-38 terminal phase III (0 PDBD), % ³H released; ▼- - -▼, WI-38 early (2-10 PDBD) phase III, % ³H released.

50 hours. Free leucine was lost by the internal pools within 5 to 7 min in both experiments (Fig. 8). Fifteen minutes after washing began, 4% of the initial trichloroacetic acid-soluble radioactivity remained within pulse-labeled cells, while 25%

remained within continuously labeled cells. In the latter case, the amount of acid-soluble radioactivity was 0.25% of the total acid-insoluble radioactivity. It is not yet known whether this retained radioactivity represents a contamination by acid-soluble protein, a nonexchangeable leucine pool, or a steady state flux through internal pools due to protein breakdown and leucine efflux. These data demonstrate that leucine fluxes in and out of the cells occur extremely rapidly. Furthermore, the rate of uptake is equal to the rate of loss as one would expect in steady state conditions.

Intermittent Perfusion—The approach to equilibrium method, although not subject to reutilization, has other limitations that are inherent in its mathematical derivation from certain steady state assumptions (see "Results" and "Discussion"). For this reason, we have designed a method to measure the release of acid-soluble radioactivity from cultures that had previously incorporated [³H] and [¹⁴C]leucine into protein. The method minimizes isotopic leucine reutilization by creating efficient traps for intra- and extracellular isotope by (*a*) washing the cultures at 37° eight times in Leu/BME (in order to empty intracellular leucine pools of isotope), (*b*) by incubating the cultures in Leu/BME on a rotary shaking water bath incubator during the degradation measurements (in order to break up boundary layers of trapped isotope around the cells), and (*c*) by changing the medium periodically throughout the experiment (to prevent extracellular isotope from re-entering the cells and being reincorporated into protein).

In these experiments we first labeled stationary phase cultures with 0.25 μCi/ml of [¹⁴C]leucine for 40 hours. Then the cells were rinsed four times and reincubated in fresh medium containing 2 mM unlabeled leucine. The ¹⁴C-protein was allowed to decay for 2 hours; this medium was discarded and the cells were labeled again with 40 μCi/ml of [³H]leucine for 25 min. After this labeling regime, the cultures were handled as described under "Experimental Procedures." Such a protocol will label long lived proteins with [¹⁴C]leucine and short lived proteins with [³H]leucine.

The results from a representative experiment are illustrated in Fig. 9, while all our results are collected together in Table II. These experiments show that short lived, but not long lived, protein is degraded at faster rates in WI-38 cells at the beginning of senescence (4 to 10 population doublings before death). At the very end of phase III (0 population doublings before death), however, both short lived and long lived protein turn over faster than do proteins in young, phase II cells. The rates of degradation in the transformed derivative, VA-13, are approximately the same as those of phase II WI-38 cells.

Effects of Amino Acid Analogs on Protein Degradation—In the following experiments phase II WI-38 cultures were first

labeled with L-[¹⁴C]leucine for 40 hours and then the [¹⁴C]leucine was removed for 2 hours and the cultures rinsed. Next, one of the amino acid analogs (5 $\times$ 10⁻³ M) or control media (5 $\times$ 10⁻³ M the corresponding amino acid) were added for 1 hour; finally, L-[³H]leucine was added to each culture containing either the natural amino acid or its analog for 30 min and after eight washes the release of acid-soluble radioactivity from the cells was measured. In this way we were able to measure the degradation of protein within the same cells that had been labeled with and without the amino acid analogs. This procedure provides an internal control for differential analog effects on both reutilization and on the degradative system itself. The experiments shown in Fig. 10 demonstrate that canavanine, azetidine carboxylic acid, and *p*-fluorophenylalanine each increase the degradation rates of proteins labeled simultaneously in their presence but do not effect the degradation rates of proteins prelabeled without them. Canavanine (5 $\times$ 10⁻³ M) appears to have the greatest effect, but this may be because the medium used in this experiment contained no arginine to compete with canavanine. Both *p*-fluorophenylalanine and azetidine carboxylic acid were also present at concentrations of 5 $\times$ 10⁻³ M; however, their corresponding natural amino acids, phenylalanine and proline, were present at concentrations of 10⁻⁴ M which may have reduced the amount of analog incorporated.

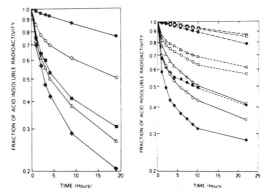

Fig. 10 (*left*). Effects of amino acid analogs on protein degradation. The cultures were labeled with [¹⁴C]leucine for 40 hours, washed, reincubated in isotope-free medium for 2 hours, and then relabeled with [³H]leucine in the presence of one of the amino acids or its corresponding amino acids. The data are expressed as in Fig. 9. ●——●, release of acid-soluble ¹⁴C radioactivity into the medium. There were no consistent differences in the release of ¹⁴C by cultures that were incubated with or without the amino acid analogs. ○——○, release of acid-soluble ³H radioactivity into the medium. There were no differences in the release of ³H for any of the controls that contained 5 $\times$ 10⁻³ M the natural amino acids corresponding to the analogs. Release of acid-soluble ³H radioactivity into the medium for cultures that had been incubated in either 5 $\times$ 10⁻³ M *p*-fluorophenylalanine (■——■), 5 $\times$ 10⁻³ M azetidine carboxylic acid (△——△), or 5 $\times$ 10⁻³ M canavanine (◆——◆).

Fig. 11 (*right*). Degradation of altered protein by phase II, phase III, and VA-13 cells. Stationary cultures were prelabeled with [¹⁴C]leucine for 72 hours, rinsed, and reincubated in medium for 18 hours. Canavanine (5 $\times$ 10⁻³ M) or arginine (5 $\times$ 10⁻³ M) was then added to the appropriate culture for 1 hour, followed by addition of [³H]leucine to the same media for another 30 min. The rate of protein degradation was measured by the intermittent perfusion method described in the text. The data are expressed as in Fig. 9. ●, WI-38 phase III (1 PDBD); ○, WI-38 phase II (28 PDBD); △, VA-13. - - -, [¹⁴C]leucine prelabeled control; ·····, [³H]leucine, arginine control; ——, [³H]leucine, canavanine.

TABLE II

Rates of protein degradation in WI-38 and VA-13 cells as determined by intermittent perfusion method

Cell	Population doublings before death	Phase	T, 25-min ³H labeling[a]	T, 40-hr ¹⁴C labeling[a]
WI-38	0–1	Late phase III	2.7 ± 0.6 (4)	38 ± 15 (4)
WI-38	2–10	Early phase III	4.8 ± 0.4 (6)	50 ± 17 (6)
WI-38	>25	Phase II	5.0 ± 0.4 (8)	58 ± 15 (8)
VA-13			5.2 ± 0.5 (8)	56 ± 16 (8)

[a] Values shown are the means ± the standard deviation with the number of determinations shown in parentheses. *T* is defined as in Table I.

None of the analogs alter the degradation rates of proteins that had been prelabeled with [¹⁴C]leucine in the absence of the analogs. This control emphasizes that the analogs do not stimulate the degradation of all proteins in the cell, nor do they differentially effect the reutilization of isotopic leucine; if they did, then the prelabeled protein degradation should have been altered as well.

Degradative Capacities for Altered Proteins—One explanation for the increased levels of abnormal proteins in phase III human fibroblast cultures (19, 23) is that the protein degradative system becomes defective, losing its ability to selectively catabolize altered proteins. In order to determine whether the degradative process in phase III cultures is defective, we have measured the relative abilities of phase II, phase III, and VA-13 cells to degrade protein that has been altered by canavanine incorporation (8–11, 23).

In these experiments, stationary cultures of phase II (>25 population doublings before death), phase III (1 and 2 population doublings before death) WI-38, and VA-13 were incubated in 0.25 μCi/ml of [¹⁴C]leucine for 72 hours and then rinsed and reincubated for 18 hours in Leu/BME. This medium was then removed, the cultures rinsed once in Leu/BME, and reincubated for 1 hour in medium containing either 5×10^{-3} M L-canavanine or 5×10^{-3} M L-arginine. After this hour, [³H]leucine (10 μCi/ml) was added to each of these cultures containing either canavanine or arginine for an additional 30 min. Finally, we measured the rates of protein degradation as described above for the intermittent perfusion method.

The results in Fig. 11 show that phase II (28 population doublings before death), phase III (1 population doubling before death), and VA-13 cultures can all degrade protein substituted with canavanine faster than protein containing

arginine. However, as shown in Table III, the ratios of the initial degradation rates (Δ canavanine/Δ arginine) for phase II and VA-13 cells are almost 2-fold greater than the ratio for phase III cells.

The amount of [¹⁴C]canavanine incorporated relative to the amount of [³H]leucine incorporated was determined for each of the cell types as shown in Table IV. Each culture was incubated in [³H]leucine and [¹⁴C]canavanine under conditions identical with those described for the experiments in Fig. 11. The results show that the ratios of disintegrations per min of canavanine incorporated to disintegrations per min of leucine incorporated are approximately the same for each of the cell types. This invariance indicates that the same amount of canavanine has been incorporated into the protein synthesized during the 1.5 hours. Although other explanations are possible, the differences between the slopes in Fig. 11 and Table III could be due to differences in the relative capacity to selectively degrade altered protein and not to differences in the incorporation of canavanine in phase II, phase III, and VA-13 cells.

DISCUSSION

Reutilization of isotopic precursor amino acids can bias measurements of protein half-lives toward larger values when they are determined by the decay of prelabeled protein in monolayer cell cultures (see Fig. 1). The fact that excess leucine (2 mM) in the medium did not alter the degradation rate of [¹⁴C]arginine protein implies that reutilization can occur in cell culture and furthermore that the excess leucine does not itself alter protein degradation. The approach to equilibrium method, theoretically at least, is not subject to reutilization; protein half-lives determined by this method are approximately 20 hours for phase II WI-38 and VA-13 cells. This value is much faster than previous estimates of protein degradation rates in cell culture (32) which were approximately 1% per hour (a half-life of 69 hours). Whether these discrepancies are due solely to reutilization is not known, although other authors (27, 29) have noted that reutilization can obscure accurate determinations of protein half-lives in cell cultures. An alternative explanation is that not all cellular proteins undergo turnover in stationary phase and that those proteins labeled to equilibrium within 3 days represent a more rapidly turning over fraction.

Whether or not the approach to equilibrium method measures the degradation rates of all cellular proteins, it is nevertheless experimentally limited because of its steady state

TABLE III

Comparison of altered protein degradation by phase II, phase III, and VA-13 cells

The initial rates of protein degradation with and without canavanine were determined from the data in Fig. 11 and from a second experiment. A ratio greater than 1.0 implies that protein containing canavanine is being degraded faster than protein containing arginine.

Cell	Population doublings before death	Ratio of initial slopes Δ canavanine/Δ arginine	
		Experiment 1	Experiment 2
WI-38	1–2	1.2	1.3
WI-38	>25	1.7	2.0
VA-13		1.8	2.2

TABLE IV

[¹⁴C]canavanine incorporation into phase II, phase III, and VA-13 cells

Confluent cultures of WI-38 (28 and 1 population doubling before death) and VA-13, identical with those in Fig. 11, were incubated in BME supplemented with 10% fetal calf serum, 5×10^{-3} M L-canavanine sulfate, 4 μCi/ml of D,L-[guanido-¹⁴C]canavanine·2 HCl (33 mCi/mM), and 1.0 μCi/ml of [³H]leucine (15 Ci/mM). After 1.5 hours the medium was removed and the amount of acid-soluble and -insoluble radioactive incorporation was measured.

	Acid-soluble			Acid-insoluble			Protein
	[¹⁴C]canavanine × 10⁻⁴	[³H]leucine × 10⁻⁴	Ratio ¹⁴C/³H	[¹⁴C]canavanine × 10⁻⁵	[³H]leucine × 10⁻⁵	Ratio ¹⁴C/³H	
	total dpm/culture			*total cpm/culture*			*total μg/ culture*
WI-38, 1 population doubling before death	3.3	2.3	1.4	1.8	0.5	3.6	566
WI-38, 28 population doublings before death	5.6	3.6	1.6	3.6	1.0	3.6	698
VA-13	19	5.4	3.8	9.7	2.8	3.5	1340

assumptions. For the measurements to be meaningful, the rates of protein synthesis and the specific activities of the appropriate precursor pools must remain constant throughout the experiment. These conditions cannot be met for many sorts of experiments where one wants to vary growth conditions or to introduce drugs. For these reasons we have attempted to minimize the amount of reutilization occurring in radioactive decay measurements of protein degradation.

The intermittent perfusion method generates shorter values for long labeled protein half-lives (35 to 60 hours) than does the standard aliquot removal method (60 to 90 hours). By this criterion, isotope reutilization may be significantly reduced by this method but still not completely eliminated, since half-lives are still longer than those measured by the approach to equilibrium method. Cell monolayers maintain a boundary layer of static medium around their periphery (33, 34) that may prevent the complete exchange of intracellular amino acids with the external medium. Rotary shaking of the cultures, as used here, should break up this boundary layer allowing a more complete mixing with the bulk medium. Nevertheless, the intermittent perfusion method is subject to whatever limitations shaking, frequent medium changes, and high leucine concentrations impose. For these reasons it is not suitable for all types of protein degradation experiments with monolayers of cultured cells, and in those cases appropriate modifications of the method may be unavoidable. However, a notable advantage of the method is that one can study in the same cells the degradation of both short and long lived proteins that may be degraded by different mechanisms (25).

As seen in Figs. 9, 10, and 11 the rate of degradation varies with time for the rapidly turning over fraction of protein, no doubt because each protein in that pool turns over at its own characteristic rate. For this reason one cannot obtain an accurate rate of degradation for the rapidly turning over protein as a whole, but one can estimate an initial rate of degradation for its fastest degrading component. The amount of rapidly turning over protein can be estimated by extrapolating the late part of the curve (the slowly turning over protein) back to the ordinate, however, this is an uncertain and imprecise approximation. Nevertheless, by extrapolating from Figs. 9, 10, and 11, one can draw the qualitative conclusion that some fraction of slowly turning over protein has been converted by amino acid analogs and senescence to rapidly turning over protein. Because of the heterogeneity in the rates of degradation, neither quantity (the amount of protein and its rate of degradation) needed to calculate a rate constant for degradation can be measured accurately. Therefore, we cannot determine whether the rate constants for degradation of senescent or analog proteins are faster or slower than the rate constant for the rapidly turning over protein in the controls. It is possible that there is a maximum rate constant for rapid protein degradation and that senescence and amino acid analogs cause slowly turning over protein to degrade with this fastest rate constant. However, our data with a heterogenous population of proteins cannot resolve this point.

Early phase III cells (4 to 10 population doublings before death) differ from phase II cells in that pulse-labeled protein is degraded more rapidly while long labeled proteins are degraded at the same rates (Table II). By the end of phase III (0 population doublings before death), both short and long lived proteins are being degraded faster as shown by both the intermittent perfusion method (Fig. 9) and by the approach to equilibrium method (Fig. 6). These findings, therefore, suggest that phase III cells contain proteins that are more rapidly degraded, perhaps because these proteins are in some way "abnormal." One alternative explanation is that a different set of "normal" proteins having shorter half-lives are synthesized in phase III cells. This explanation is unlikely since we have been unable to detect any consistent differences in the sodium dodecyl sulfate-acrylamide gel electrophoresis patterns of proteins from phase II and phase III cells (23).

If one assumes that lysosomal activity determines the rate of protein degradation (35), a second alternative explanation is that the increased rates of protein degradation are due to the increased number of residual lysosomes and lysosomal enzymes found in phase III (36, 37) and not necessarily because of abnormal proteins. However, as pointed out in the introduction to the text, the protein degradative system may function in a "quality control" capacity by preferentially degrading error-containing protein. If the error frequency of protein synthesis increases in phase III as postulated by Orgel (38) then preferential degradation of error-containing protein would increase the overall rate of protein degradation as we have demonstrated. Since error-containing protein should be rapidly degradable, it is reasonable that increased pulse-labeled protein degradation should be detectable first in early phase III. At the end of phase III more of the total protein may contain errors, so that the rate of long labeled protein degradation increases also. Terminal phase III might then result from a functional failure of the protein degradative system, due either to synthesis of an overwhelming amount of error-containing protein that cannot be degraded rapidly enough or to a defective degradative system that has lost its capacity to select "abnormal" protein or to some combination of the two possibilities.

Simian virus 40 transformation of WI-38 cells produces no consistent differences in the rates of either short or long lived protein degradation in confluent cultures (Tables I and II). Other characteristics of protein degradation, however, may vary between normal and virally transformed cells. For instance, SV-40 transformed 3T3 mouse fibroblasts do not increase the rate of protein degradation after serum withdrawal as do normal 3T3 cells (39), implying that whatever mechanism controls protein degradation during growth transitions has been altered by transformation.

The experiments of Fig. 10 demonstrate that human fibroblast proteins containing amino acid analogs are degraded intracellularly more rapidly than normal proteins. Thus, these data imply that "abnormal" proteins in WI-38 cells are more susceptible to catabolism and that such degradative selectivity may provide an important mechanism for maintaining low, steady-state protein error frequencies within these cells in order to allow normal cell function and survival.

The prelabeling of proteins with [^{14}C]leucine in the absence of the analogs provides an important control for the experiments of Fig. 10 and eliminates two alternative interpretations of these data. First, reutilization: if leucine reutilization was significantly altered by the analogs, then one would expect them to alter also the apparent rates of prelabeled protein degradation due to alteration in the rates of [^{14}C]leucine reutilization. The fact that these rates are identical in both amino acid analog and control cultures implies that reutilization is unaltered by the analogs so that the accelerated degradation cannot be due to unequal rates of leucine reutilization. Second, protease or lysosomal activation: if the degradative system was generally activated by the analogs so that all proteins were degraded faster, then the proteins prelabeled without the analogs should also have been degraded faster. However, since these rates were identical, we conclude that the analogs increase protein degradation because they render the proteins more susceptible to the action of an unaltered degradative system.

These findings provide strong evidence that diverse abnormal proteins are selectively degraded by human cells, and furthermore, that the rates of degradation are determined, in part, by the properties of the proteins as substrates for proteolysis. Selective degradation of abnormal protein may therefore provide a mechanism whereby cells can balance an increased production of protein errors by their increased degradation and maintain, under normal conditions, a low steady state level of abnormal proteins. Disturbance of this equilibrium, such as by defective protein degradation, could lead to a variety of pathological consequences for a cell (3, 5).

The evidence in Fig. 11 and Table III tends to support these ideas. Based on the ratios of the initial degradation rates, phase III cells may not be as efficient at degrading protein containing canavanine as are phase II or VA-13 cells. This loss could be due to a partially defective degradative system, it could be due to an inherent maximum rate constant for intracellular protein degradation, or it could be a consequence of a greater frequency of endogenous abnormal proteins in phase III cells that compete for degradation with the additional canavanine containing proteins. In any case the protein degradative system may provide the functional equivalent of "protein repair" by selectively eliminating those proteins with altered structures. If this is true, then senescent or malignant alterations may result, in part, from changes in either the rate of protein error production or the capacity of cells to recognize and degrade such altered proteins or from some combination of both processes.

REFERENCES

1. Swick, R. W. (1958) *J. Biol. Chem.* **231**, 751–764
2. Schimke, R. T. (1964) *J. Biol. Chem.* **239**, 3808–3817
3. Goldberg, A. L., and Dice, J. F. (1974) *Annu. Rev. Biochem.* **43**, 835–869
4. Bradley, M. O., and Schimke, R. T. (1975) in *Intracellular Protein Turnover* (Schimke, R. T., and Katanuma, H., eds) pp. 311–322, Academic Press, New York
5. Schimke, R. T., and Bradley, M. O. (1975) in *Proteases and Biological Control* (Reich, E., Rifkin, D., and Shaws, E., eds) Chap. 35, Cold Spring Harbor Laboratory, New York
6. Platt, T. J., Miller, J. H., and Weber, K. (1970) *Nature* **228**, 1154–1156
7. Goldschmidt, R. (1970) *Nature* **228**, 1151–1154
8. Pine, M. J. (1967) *J. Bacteriol.* **93**, 1527–1533
9. Goldberg, A. L. (1972) *Proc. Natl. Acad. Sci. U. S. A.* **69**, 422–426
10. Prouty, W. F., and Goldberg, A. L. (1972) *J. Biol. Chem.* **247**, 3341–3352
11. Prouty, W. F., and Goldberg, A. L. (1972) *Nature New Biol.* **240**, 147–150
12. Shaeffer, J. R. (1973) *J. Biol. Chem.* **248**, 7473–7480
13. Adams, J. G., III, Winter, W. P., Rucknagel, D. L., and Spencer, H. H. (1972) *Science* **176**, 1427–1429
14. DeSimone, J., Kelve, L., Longley, M., and Schaeffer, J. (1974) *Biochem. Biophys. Res. Commun.* **57**, 248–256
15. Knowles, S. E., Gunn, J. M., Hanson, R. W., and Ballard, F. J. (1975) *Biochem. J.* **146**, 595–600
16. Capecchi, M. R., Capecchi, N. E., Hughes, S. H., and Wahl, G. M. (1974) *Proc. Natl. Acad. Sci. U. S. A.* **71**, 4732–4736
17. Hayflick, L. (1965) *Exp. Cell Res.* **37**, 614–636
18. Girardi, A. J., Jensen, F. C., and Koprowski, H. (1965) *J. Cell. Comp. Physiol.* **65**, 69–84
19. Holliday, R., and Tarrant, G. M. (1972) *Nature* **238**, 26–30
20. Gershon, H., and Gershon, D. (1973) *Proc. Natl. Acad. Sci. U. S. A.* **70**, 909–913
21. Gershon, H., and Gershon, D. (1973) *Mech. Aging Dev.* **2**, 33–41
22. Zeelon, P., Gershon, H., and Gershon, D. *Biochemistry* **12**, 1743–1749
23. Bradley, M. O., Dice, J. F., Hayflick, L., and Schimke, R. T. (1975) *Exp. Cell Res.* **96**, 103–112
24. Hayflick, L. (1965) *Tex. Rep. Biol. Med.* **23**(Suppl. 1), 285–303
25. Poole, B., and Wibo, M. (1973) *J. Biol. Chem.* **248**, 6221–6226
26. Lowry, O. H., Rosebrough, N. J., Farr, A. L., and Randall, R. J. (1951) *J. Biol. Chem.* **193**, 265–275
27. Klevecz, R. (1971) *Biochem. Biophys. Res. Commun.* **43**, 76–81
28. Poole, B. (1971) *J. Biol. Chem.* **246**, 6587–6591
29. Righetti, P., Little, E. P., and Wolf, G. (1971) *J. Biol. Chem.* **246**, 5724–5732
30. Greenberg, J. R. (1972) *Nature* **240**, 102–104
31. Brandhorst, B. P., and McConkey, E. H. (1974) *J. Mol. Biol.* **85**, 451–463
32. Eagle, H., Piez, K. A., Fleischman, R., and Oyama, V. I. (1959) *J. Biol. Chem.* **234**, 592–597
33. Stoker, M. G. P. (1973) *Nature* **246**, 200–202
34. Stoker, M. (1974) *Cell* **3**, 207–215
35. DeDuve, C., and Wattiaux, R. (1966) *Annu. Rev. Physiol.* **28**, 435–439
36. Lipetz, J., and Cristofalo, V. J. (1972) *J. Ultrastruct. Res.* **39**, 43–56
37. Brunk, U., Ericsson, J., Ponten, J., and Westermark, B. (1973) *Exp. Cell Res.* **79**, 1–14
38. Orgel, L. E. (1963) *Proc. Natl. Acad. Sci. U. S. A.* **49**, 517–521
39. Hershko, A., Mamont, P., Shields, R., and Tomkins, G. M. (1971) *Nature New Biol.* **232**, 206–211
40. Costlow, M., and Baserga, R. (1973) *J. Cell. Physiol.* **82**, 411–420

22

Reprinted from *Exp. Cell Res.* **100**:1–8 (1976)

INCREASED DEGRADATION RATES OF PROTEIN IN AGING HUMAN FIBROBLASTS AND IN CELLS TREATED WITH AN AMINO ACID ANALOG

VALERIE SHAKESPEARE and J. H. BUCHANAN

Genetics Division, National Institute for Medical Research, Mill Hill, London NW7 1AA, UK

SUMMARY

The rates of degradation of total cellular protein in the diploid fetal lung fibroblast strain MRC-5 have been determined by measurement of acid-insoluble radioactivity remaining in protein during a non-radioactive chase period. We find that proteins in visibly senescent fibroblasts, and in fibroblasts grown in the presence of the amino acid analog *p*-fluorophenylalanine (*p*FPA) contain a fraction of molecules having an increased rate of breakdown relative to cells of earlier passage. The results suggest that altered or aberrant protein molecules may be synthesized in aging cells.

Human diploid fibroblasts have a finite life time in terms of their proliferative capacity in culture. Hayflick [7] showed that the number of population doublings (passages) undergone by these cells before death is inversely related to the age of the donor and independent of chronological time in culture. Martin et al. [17] have also demonstrated an inverse correlation between the in vitro lifespan of human fibroblasts and the age of the donor. To explain the loss of division potential and eventual cell death, Orgel [18, 19] suggested that cumulative errors in the protein synthetic mechanism of a cell may occur with age. Some evidence which suggests that cellular aging may be accompanied by the synthesis of altered or defective proteins has come from the work of Lewis & Tarrant [15] and Holliday & Tarrant [9], who showed alterations in the properties of some enzymes with age. Similarly, it has been shown that inactive en-zyme molecules accumulate in the tissues of aging mice [3, 4]. However, Holland et al. [10] were unable to demonstrate any alterations in viral nucleic acid or protein of viruses produced in aging human fibroblast cultures. No direct experimental demonstration that errors in transcription or translation occur during aging has yet been provided.

It is possible that altered protein molecules which may be synthesized by senescent fibroblasts are preferentially degraded by cellular mechanisms involved in the turnover of proteins. There is evidence that altered or abnormal protein molecules are degraded faster than the normal molecules in both bacterial and mammalian systems [2, 5, 14, 20]. We have, therefore, examined the rates of total protein degradation in the human fetal lung fibroblast strain MRC-5 to establish if changes in degradative rates occur as senescence is reached. These cells become visibly senescent after about 50–55 subcultures and cease growth after 60–65 subcultures.

We have measured the degradation rates of total cellular protein in fibroblasts of early and late passage by a radioisotope pulse–chase technique. Poole & Wibo [21] estimated protein degradative rates in cultured cells by measurement of radioactivity lost from pulse labeled cells into the culture medium. We find that it is necessary to measure acid-precipitable radioactivity to ensure that isotope lost from cells originates from degradation of labeled protein and not from unincorporated radioactive amino acid in the cellular pool.

We have also measured the protein degradative rate in fibroblasts grown in the presence of the amino acid analog p-fluoro-phenylalanine (pFPA) to demonstrate that abnormal (i.e. analog-containing) proteins can be shown to be degraded at a faster rate than normal proteins.

MATERIALS AND METHODS

Cell culture

The fetal lung fibroblast strain MRC-5 was cultured and harvested as previously described [9]. Cells were grown at 37°C in 150 ml glass bottles in Eagle's basal diploid medium containing 10% fetal calf serum, 10% tryptose-phosphate broth, 200 U/ml penicillin and 200 μg/ml streptomycin.

pFPA incorporation

DL-p-Fluorophenylalanine (pFPA) was obtained from Sigma Chemical Co. At high concentrations pFPA is an inhibitor of mitosis. A growth curve for MRC-5 in the presence of varying concentrations of DL-pFPA in the culture medium showed a suitable experimental concentration to be 0.1 mM DL-pFPA in the medium described above. No inhibition of cell growth could be detected at this concentration.

In order to show that pFPA is incorporated into the cellular protein of cultured fibroblasts, cells were grown in the presence of [3-^{14}C]DL-pFPA (52 mCi/mmol, Calatomic) at a concentration of 2.5 μCi/ml of culture medium for 72 h. A modified amino acid analysis using the Beckman 120B amino acid autoanalyzer, in which radioactivity in effluent from the autoanalyzer column was measured, showed that 2% of

radioactive counts were present in the labeled pFPA as phenylalanine contamination. Radioactive cellular protein was, therefore, precipitated by cold 10% trichloracetic acid (TCA) and hydrolysed for 16 h at 110°C in the presence of 6 N HCl. Amino acid analysis of the hydrolysed protein showed radioactive pFPA to be present.

Cell labeling

Fibroblasts of early and late passage and cells grown in the presence of pFPA for six passages were split 1:2 into culture medium free from leucine. The cells were then grown in the leucine-free medium with the addition of [4,5-^{3}H]L-leucine (38 Ci/mmol, Radiochemical Centre, Amersham) and 0.1 mM pFPA in the case of the analog-treated cells. The culture bottles contained 10 ml of medium plus 10 μCi of [^{3}H]leucine. The cells were grown in the radioactive medium for a period of 5 days when a confluent layer was formed. The labeling period of 5 days was used to ensure extensive isotopic labeling of cellular protein, and thus minimize errors in the measurement of acid-precipitable radioactivity.

Measurement of degradation rates of total cellular protein

After completion of the labeling period, the cell layer was rinsed several times with phosphate buffered saline (PBS) containing non-radioactive leucine (0.5 mg/ml). Eagle's basal diploid medium (10 ml) containing ten times the normal concentration of non-radioactive leucine (2 mM) was then added.

In order to try to minimize re-utilization of isotope during the chase period, the medium in all culture bottles was decanted and fresh medium added every 12 or 24 h.

At each time interval after change into non-radioactive medium (12 or 24 h intervals for 5 to 7 days) four separate culture bottles were taken, the medium decanted, and the cells rinsed twice with PBS. The cells were detached from the glass by treatment with 0.1% ethylene diamine tetra-acetic acid (EDTA) in saline, or by 0.12% trypsin/0.01% EDTA. The cell number was determined using a Coulter counter. The cells were washed twice by centrifugation at 1 000 g in PBS, resuspended in 10 mM TrisCl buffer pH 7.5 (1.0 ml/culture bottle) and lysed by sonication.

Cell membranes were removed by centrifugation at 12 000 g for 20 min. The supernatants were taken and protein precipitated by the addition of an equal volume of cold 10% TCA. The precipitates were collected by centrifugation and dissolved in 0.5 ml of NCS tissue solubilizer (Hopkin & Williams) at 37°C overnight. Acid-precipitable and acid-soluble radioactivity were measured by a Packard Liquid Scintillation Counter (Model no. 3380) using the toluene-based scintillant PCS (Hopkin & Williams) for aqueous samples containing TCA, and toluene/PPO (6 g/l)/POPOP (75 mg/l) for samples dissolved in NCS.

Radioactivity in detached cells

Two ml aliquots of medium from each culture bottle were filtered on 0.45 μM Millipore filters. Filters were

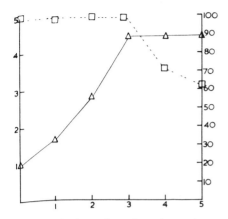

Fig. 1. *Abscissa:* time after change into non-radioactive culture medium (days); *ordinate: (left)* (△) rel. cell no.; *(right)* (□) % radioactivity remaining in protein.

Radioactivity remaining in cellular protein during a non-radioactive chase period for fibroblasts of passage 35 in rapid growth phase. TCA-precipitable radioactivity and cell numbers are expressed relative to the values at the time of change into non-radioactive culture medium. Each point represents the mean of determinations on four culture bottles.

rinsed three times with 2 ml of PBS. The filters were allowed to dry and the retained radioactivity was determined after the addition of toluene/PPO/POPOP scintillant.

RESULTS

Measurements of degradative rates in cultured fibroblasts are made more complex by the anchorage-dependence of the cells which does not allow sampling of cells at time intervals. Loss of radioisotope must therefore be investigated using separate culture bottles at each time interval. Knowles and associates [14] have used a similar experimental protocol to investigate degradation rates of protein synthesized in monolayer cultures. Corrections for cellular growth rate in parallel experiments must be made, since an increase in cell numbers during the non-radioactive chase period will cause dilution of specific radioactivity in protein.

We initially attempted to measure the loss of radioisotope from protein in early passage fibroblasts during rapid cell growth after subculture. Fig. 1 shows that total TCA-precipitable radioactivity remains constant over a period of 3–4 days until the cells reach confluence. Loss of radioactivity from cellular proteins only begins after the cells have formed a confluent layer.

Since little, or no measurable protein degradation could be detected in rapidly growing early passage cells, we carried out all further experiments with all cultures at confluence during the chase period. Fig. 2 shows that some increase in cell numbers is seen in response to the frequent addition of fresh medium. However, the relative increase in numbers is about the same with early passage, late passage and *p*FPA-treated cells. The results from three experiments to estimate loss of radioactivity from cellular protein in early passage, late passage, and *p*FPA-treated cells are presented in tables 1, 2 and 3.

The amount of [³H]leucine incorporated

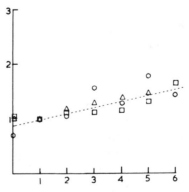

Fig. 2. *Abscissa:* time after change into non-radioactive culture medium (days); *ordinate:* rel. cell no.

Increase in cell numbers in confluent cultures in response to the addition of fresh medium: ○, cells of passage 26; □, cells of passage 26 plus *p*FPA; △, cells of passage 61. Cell numbers are expressed relative to the number at 24 h after change into non-radioactive medium. Each point represents the mean of four determinations.

Table 1. *Degradation of protein from fibroblasts of passage 26*

Time after change into non-radio-active medium (days)	Cell no. per culture bottle $\times 10^{-5}$	Total radio-active counts (TCA-pre-cipitable + TCA-soluble) $\times 10^{-5}$	TCA-pre-cipitable counts $\times 10^{-5}$	TCA-pre-cipitable counts expressed as % of total counts	% Radio-activity remaining in protein relative to 24 h time point
0	6.59±0.50	9.60±0.54	7.70±0.50	85	86
1	6.69±0.52	9.86±0.29	8.96±0.40	91	100
2	6.98±0.75	8.40±0.29	7.38±0.21	88	82±2
3	7.13±0.34	6.64±0.38	5.64±0.50	85	63±3
4	8.46±0.21	4.63±0.61	4.30±0.53	92	48±4
5	9.01±0.80	4.32±0.32	3.78±0.20	88	42±2

Values given for cell numbers and radioactivity are mean values ±S.D. for determinations on four culture bottles.
Values for radioactivity are corrected to the same counting efficiency.

Table 2. *Degradation of pFPA-containing proteins (fibroblasts of passage 26)*

Time after change into non-radio-active medium (days)	Cell no. per culture bottle $\times 10^{-5}$	Total radio-active counts (TCA-pre-cipitable + TCA-soluble) $\times 10^{-5}$	TCA-pre-cipitable counts $\times 10^{-5}$	TCA-pre-cipitable counts expressed as % of total counts	% Radio-activity remaining in protein relative to 24 h time point
0	5.44±0.51	10.62±1.04	8.91±0.46	85	92
1	6.25±0.48	10.69±0.54	9.65±0.58	90	100
2	6.26±0.90	7.66±0.64	6.65±0.47	87	69±4
3	6.72±0.24	5.88±0.27	5.06±0.24	86	52±2
4	8.34±1.07	4.60±0.34	4.17±0.33	90	43±3
5	9.08±0.03	3.75±0.11	3.19±0.29	85	33±2

Values given for cell numbers and radioactivity are mean values ±S.D for determinations on four culture bottles.
Values for radioactivity are corrected to the same counting efficiency.

into cellular protein after completion of the labeling period was of the same order of magnitude in cells of early and late passage and in *p*FPA-treated fibroblasts. However, in every case, we observed an increase in acid-precipitable radioactivity from zero time (the time of change into non-radio-active medium) to the first time point at 24 h. It appears that radioactivity continues to be incorporated into protein from free [^{3}H]leucine in cellular pools for some time after addition of non-radioactive medium. Measurement of acid-precipitable radioactivity at 12 h after zero time showed even higher incorporation in protein at this point. Thus, measurable loss of radioactivity due to degradation of protein begins between 12 and 24 h after addition of non-radioactive medium. Our results are therefore expressed relative to TCA-precipitable radioactivity present at 24 h after zero time.

Table 3. *Degradation of proteins from fibroblasts of passage 61*

Time after change into non-radio-active medium (days)	Cell no. per culture bottle $\times 10^{-5}$	Total radio-active counts (TCA-pre-cipitable + TCA-soluble) $\times 10^{-5}$	TCA-pre-cipitable counts $\times 10^{-5}$	TCA-pre-cipitable counts expressed as % of total counts	% Radio-activity remaining in protein relative to 24 h time point
0	6.48±0.70	10.08±0.90	8.45±1.15	84	98
1	7.40±0.39	10.43±0.34	8.61±0.30	83	100
2	7.96±0.57	6.69±0.45	5.23±0.50	78	61±5
3	8.58±0.52	5.67±0.46	3.80±0.27	70	44±3
4	9.98±0.58	3.70±0.64	2.75±0.63	74	32±6

Values given for cell numbers and radioactivity are mean values ±S.D. for determinations on four culture bottles. Values for radioactivity are corrected to the same counting efficiency.

Fig. 3 shows a logarithmic plot of the results obtained in tables 1, 2 and 3. From fig. 3, it can be seen that a straight line plot of radioactivity remaining in protein versus time is given in the case of early passage cells. Protein degradation in this case may be described by a single rate constant.

However, analog-containing and "senescent" proteins are initially more rapidly degraded. After this initial period, the rate of loss of radioactivity is diminished and approximates to that found for early passage proteins. From the initial slope of each line, the time taken for 50% of incorporated radioactivity to be lost from protein (half-time of degradation) can be estimated. The estimates of the half-times of degradation from thirteen such experiments are represented diagrammatically in fig. 4. In all experiments a linear plot of radioactivity remaining in protein versus time was given for early passage cells, with a more rapid initial loss of radioactivity in

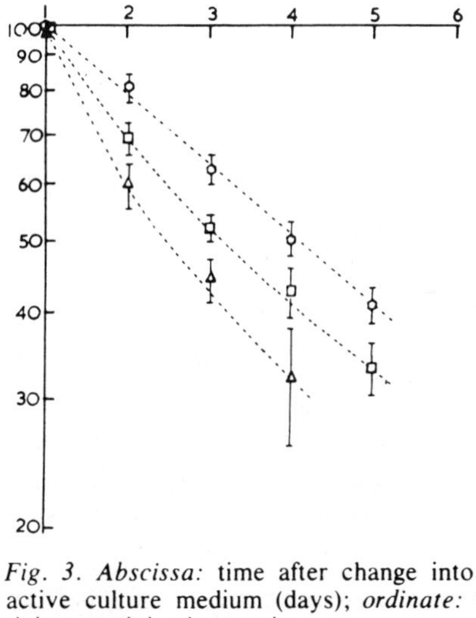

Fig. 3. Abscissa: time after change into non-radioactive culture medium (days); *ordinate:* % radioactivity remaining in protein.
Degradation of ^{3}H-labeled cellular protein; O, cells of passage 26; □, cells of passage 26 plus *p*FPA; △, cells of passage 61. TCA-precipitable radioactivity is expressed relative to radioactive counts present in protein at 24 h after change into non-radioactive medium. Each point represents the mean for determinations on four culture bottles. Error bars indicate one standard deviation from the mean.

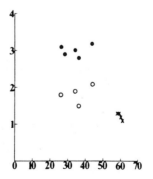

Fig. 4. Abscissa: passage no.; *ordinate:* half-time of degradation (days). ●, Early passage cells; O, *p*FPA-treated cells; ×, visibly senescent cells.

proteins from analog-treated or senescent cells.

In all experiments TCA-soluble radioactivity was found to be between 10 and 20% of the total radioactivity in the cell extracts over the chase period. A fraction of the TCA-soluble radioactivity may be due to the degradation of cellular protein by trace amounts of trypsin remaining in the cell extracts before the addition of cold TCA. When 0.1% EDTA was used in place of trypsin to harvest the cells, the amount of TCA-soluble radioactivity was reduced to 2% of the total radioactivity in the cell extract. The use of EDTA to harvest cells did not give rise to values for degradative half-life significantly different from those we obtained using trypsin.

We observed some detachment of cells from the glass culture bottles into the medium during the chase period. In order to establish that the shorter half-time of degradation observed in old cells was not due to greater detachment and loss of labeled cells, aliquots of medium containing detached cells were filtered as described in the Materials and Methods section. The percentage of radioactivity retained was found to be about 10% of the total radioactivity in the medium for both young and old cells. This level of radioactivity appears to be greater than would be expected from the small number of detached cells present in the medium. However, Herrmann [8] has observed that large molecular weight components of media (e.g. serum) will bind added radioactive compounds in the absence of cells. It is possible that such complexes are retained by Millipore filters and may explain our result.

DISCUSSION

Our initial finding was that, for early passage cells in the exponential phase of growth, TCA-precipitable radioactivity remained constant over the chase period until confluence was reached. It is possible that little or no degradation of protein occurs in growing cells. Spudich & Kornberg [22] have reported that bacteria in the exponential phase of growth do not degrade cellular protein. It is also possible that degradation of cellular protein and release of radioisotope does take place when cells are growing rapidly. However, the [3H]-leucine released may be immediately re-utilized during rapid cell division and is therefore not detectable by our methods.

Isotope re-utilization is a major limitation for obtaining accurate rates of protein degradation. We have attempted to reduce re-utilization by frequent changes of medium containing excess non-radioactive leucine during the chase period. The results demonstrate that protein degradation in pFPA-treated and senescent fibroblasts cannot be described by a single rate constant. These observations suggest that populations of proteins from such fibroblasts contain a fraction of molecules which are more susceptible to degradation than normal cell proteins. From the results shown diagrammatically in fig. 4, we estimate the mean half-time of protein degradation in early passage cells (passages 26 to 44) to be 3.0 ± 0.2 days. Estimates of the mean half-time of degradation of the more rapidly degraded fraction of protein from pFPA-treated and senescent cells are 1.8 ± 0.3 days and 1.2 ± 0.1 days, respectively.

Our interpretation of these results is that, for human fibroblasts in culture, cells approaching the end of the in vitro lifespan have an increased rate of protein breakdown and that the incorporation of an amino acid analog into cellular protein in earlier passage cells causes the resultant altered molecules to be degraded at a faster rate than normal proteins. The mechanisms responsible for protein degradation are complex, and alterations in conformation, subunit dissociation and interactions with other molecules may all play a part. However, we cannot rule out the possibility that the differences we observe in apparent rate constants of degradation for young, analog-treated and senescent cells are due to some factor other than altered breakdown rate. Alterations in the permeability of the cell membrane in the three cultures may lead to different intracellular pool sizes and hence rates of isotope re-utilization may be different in young, analog-treated and senescent cells.

The location of the precursor amino acid pool for protein synthesis is unknown. Ilan & Singer [11] have measured the specific activity of [³H]leucine immediately inserted into the nascent polypeptide chain in newt liver and suggest that leucine utilized for protein synthesis is not equilibrated with the intracellular pool of leucine. Other authors [6, 13] also favor the argument that amino acids for protein synthesis are used preferentially from an extracellular rather than an intracellular source. However, Li et al. [16] suggest that protein synthesis can proceed from one homogeneous intracellular pool.

It is still not clear why cellular proteins are continuously synthesized and degraded. One function of turnover may be to prevent the accumulation of abnormal or deleterious molecules. Pine [20] and Goldberg [5] have provided evidence that *E. coli* contains a mechanism for selective degradation of unfinished polypeptides containing puromycin, and proteins containing amino acid analogs. Knowles et al. [14] have shown that proteins synthesized in hepatoma cells in the presence of amino acid analogs have an increased degradation rate. Degradative mechanisms may be important in the regulation of enzyme activity in higher organisms, while the mechanisms of induction and repression, common in bacteria, are of less importance. Khairallah & Pitot [12] have shown that the half-lives of pyridoxal requiring enzymes are prolonged in rats by administration of the coenzyme in vivo and that an increase in the amount of enzyme is due to the protective ability of the coenzyme against degradation rather than an increase in the rate of synthesis.

Previous reports give estimates of half-life of cellular protein in cultured cells ranging from a few hours to several days, depending on the experimental method used. Bradley [1] has estimated protein degradative half-life in the human fibroblast strain WI38, by measurement of the rate of approach of cellular protein to a constant isotopic specific activity. The mean half-life given by this method is 20 h. Measurement of loss of radioisotope into the medium together with frequent changes of medium gave half-lives of 37 h in young cells, and 29 h in senescent cells. A mean half-life of about 3 days for cellular protein in several human cell lines has been determined by measurement of TCA-precipitable radioactivity incorporated in protein [23].

We suggest that the increased rate of protein degradation we observe in senescent fibroblasts is a cellular response to an increased rate of synthesis of aberrant or defective protein molecules in these cells, in accordance with Orgel's theory.

V. Shakespeare acknowledges student support from the Medical Research Council.

REFERENCES

1. Bradley, M O, Intracellular protein turnover (ed R T Schimke & N Katunuma) p. 311. Academic Press, New York (1975).

2. Capecchi, M R, Capecchi, N E, Hughes, S H & Wahl, G M, Proc natl acad sci US 71 (1974) 4732.
3. Gershon, H & Gershon, D, Mech age dev 2 (1973) 33.
4. — Proc natl acad sci US 70 (1973) 909.
5. Goldberg, A L, Proc natl acad sci US 69 (1972) 422.
6. Haider, R C, Fern, E B & London, D R, Biochem j 114 (1969) 171.
7. Hayflick, L, Exp cell res 37 (1965) 614.
8. Herrmann, H, Anal biochem 59 (1974) 293.
9. Holliday, R & Tarrant, G M, Nature 238 (1972) 26.
10. Holland, J J, Kohne, D & Doyle, M V, Nature 245 (1973) 316.
11. Ilan, J & Singer, M, J mol biol 91 (1975) 39.
12. Khairallah, E A & Pitot, H C, Symposium on pyridoxal enzymes (ed K Yamada, N Katunuma & H Wade) p. 159. Maruzen, Tokyo (1968).
13. Kipnis, D M, Reiss, E & Helmreich, E, Biochim biophys acta 51 (1961) 519.
14. Knowles, S E, Gunn, M J, Hanson, R W & Ballard, E J, Biochem j 146 (1975) 595.
15. Lewis, C M & Tarrant, G M, Nature 239 (1972) 316.
16. Li, J B, Fulks, R M & Goldberg, A L, J biol chem 248 (1973) 7272.
17. Martin, G, Sprague, C & Epstein, C, Lab invest 23 (1970) 86.
18. Orgel, L E, Proc natl acad sci US 49 (1963) 517.
19. — Nature 243 (1973) 441.
20. Pine, M J, J bacteriol 93 (1967) 1527.
21. Poole, B & Wibo, M, J biol chem 248 (1973) 6221.
22. Spudich, J A & Kornberg, A, J biol chem 243 (1968) 4600.
23. Eagle, H, Piez, K A, Fleischman, R & Oyama, V I, J biol chem 234 (1959) 592.

Received January 19, 1976
Accepted January 21, 1976

Part IV

CHANGES AT THE DNA LEVEL

Editor's Comments
on Papers 23 Through 30

It has often been argued that the process of aging may be largely due to accumulating genetic damage. Some of the evidence for this view comes from studies on the effect of ionizing radiation, which can both break chromosomes and cause gene mutations. Experiments with inbred mice clearly show that radiation significantly shortens the lifespan, and the animals that die prematurely appear to have most of the characteristic of normal aging (see, for example, Neary, 1960 and Lindop; Rotblat, 1961). An attempt by Szilard (1959) to put the somatic mutation theory of aging on a quantitative basis was only partially successful, since it was based on the restricting assumption that the target for genetic damage was the whole chromosome rather than the individual gene. Several persuasive arguments against the theory were put forward by Maynard Smith (1959, 1962); nevertheless, it continues to receive support (see, for example, Burnet, 1974; Morley, 1982).

In the context of error theories, it is acknowledged that genetic damage may be very important in aging, but this is regarded as a by-product of other defects in the cell. Most specifically, it has been suggested that abnormalities in DNA polymerase or other components of the machinery for DNA replication would reduce the fidelity of DNA synthesis and therefore increase mutation frequency or cause other genetic changes (Papers 4, 16, 19, and 20). Studies with human fibroblasts are generally consistent with this interpretation. Diploid cells maintain their karyotype for many generations in culture. When the cells reach the end of Phase II and enter Phase III, however, the proportion of metaphases with chromosome abnormalities becomes increasingly large. This was first documented by Saksela and Moorhead (Paper 23) using W1-38 and W1-26 and later confirmed for MRC-5 (Thompson and Holliday, 1975).

The direct measurement of mutations in human fibroblasts during in vitro aging poses serious difficulties, since mutations are usually measured by scoring colonies that are resistant to an inhibitory compound, such as 6-thioguanine (6TG). It is clearly impossible to do this when cells are running out of growth potential. Gupta (1980) measured resistance to 6TG and also diphtheria toxin during the first two-thirds of the in vitro growth of a long-lived strain of human fibroblasts and detected an approximate twofold increase. It is very important to measure mutation frequencies at the end of a lifespan, however, since the prediction of the error theory is that mutations should increase exponentially. This was attempted in Paper 24 using a histochemical staining procedure that detects rare cells with increased levels of G6PD. The evidence suggested that this variant phenotype is heritable, and the interpretation was that regulatory mutations were being scored. It was found that the frequency of variants increased dramatically during the senescent phase of fibroblast growth. The results are quite compatible with Gupta's, since the increase in variant frequency is only two- to threefold until the last ten or so population doublings. Another publication demonstrates that the increase in frequency of stained cells is close to exponential, when account is taken of the fact that each population doubling corresponds to more than one cell generation, when a proportion of the cells are no longer dividing (Fulder, 1978). It is interesting to note that the studies of fibroblast clones by Smith and his associates (Smith and Hayflick, 1974; Smith, Pereira-Smith, and Schneider, 1978; Paper 12) reveal that a subset of cells divides only a few times before the miniclone dies. It may well be that these cells (the proportion of which increases with age) have lost an indispensible gene function by mutation, but there is sufficient gene product remaining for a small number of residual divisions.

If senescent cells have lost accuracy in information transfer, the effect of this should be seen in the replication or assembly of infecting viruses. A widely cited study was carried out by Holland et al. (Paper 25) using three different viruses. Senescent or young W1-38 cells were infected with these viruses, but no differences were detected in virus yield, the heat stability of the virions produced, or mutation frequency. Of course, one problem with this type of experiment is that viruses overproduce their structural proteins and only a subset is assembled into intact virions. Any abnormal molecules may therefore be largely or entirely excluded. In the studies on mutation, poliovirus was used and this has an RNA genome, so the spontaneous mutation frequency would be very much higher than that of DNA viruses. There is no easy way of documenting the sensitivity of this viral-probe method for detecting errors in either RNA or protein synthesis, and it would have been valuable to have checked that the addition of an amino acid or RNA base analogue to normal cells produced measurable effects on the viruses. Other studies with viruses, using less sensitive techniques, have also failed to detect the differences between young and senescent fibroblasts (Tomkins, Stanbridge, and Hayflick, 1974; Pitha, Adams, and Pitha, 1974; Pitha, Stork, and Wimmer, 1975). Fulder (1977) showed in contrast that senescence had profound effects on mutation frequency in Herpes simplex viruses. This is a large DNA virus and Fulder used a better characterized genetic system for measuring mutations than that used in Paper 25. He employed fluctuation tests, which are essential for the detection of small changes in spontaneous mutation frequency, whereas only single populations were used in Paper 25. Fulder measured the frequency of reversion of three temperature-sensitive missense mutants, grown in young and senescent MRC-5 cells. In one case the mutation frequency increased in senescent cells, in another it decreased, and the third was unchanged. It is hard to interpret these results, but it is known that there are at least two pathways for mutagenesis (errors in DNA replication and error-prone repair), and it is possible that these are differentially affected in senescent cells. Further quantitative studies on the mutability of viruses infecting fibroblasts are needed.

Hayflick had first suggested (Paper 8) that the aging of cultured fibroblasts might be due to the steady accumulation of multiple genetic defects, or "hits." This can be tested by comparing the longevity of diploid and tetraploid cells. The study in Paper 26 is based on the discovery that a surprisingly high proportion of polyploid cells survives after treatment of cultures with colchicine. It was found that the proportion of diploid and tetraploid cells remained constant until the end of the lifespan. This showed that they grew at the same rate and had the same longevity. According to the somatic mutation theory of cellular aging, fibroblasts with more copies of the genome should survive longer. If mutations increase very rapidly in Phase III, however, very little, if any, difference would be expected. Hoehn et al. (1975) isolated clones of tetraploid fibroblasts by hybridizing diploids, and they also reported that the lifespan of the hybrids was similar to that of the diploid parents.

On the basis of reasonable assumptions about the number of indispensible genes in the human cell and also the number on the single active X-chromosome, a somatic mutation model for fibroblast aging is formulated in Paper 27. The model can explain the dying of populations of fibroblasts, but the mutation frequency per gene has to be unreasonably high (in the range of 10^{-3} to 10^{-4} per gene, per cell generation). A more serious difficulty is that the model predicts that the proportion of nonviable cells will be quite high even in early-passage cultures, but experimental determination of the proportion of dividing cells shows that this is not true. The authors conclude that the simple accumulation of recessive mutations during cell division cannot explain the so-called Hayflick limit to fibroblast growth. Another mortalization theory of fibroblast aging is based on the supposition that during serial passaging there is a progressive increase in the probability of producing nondividing cells (Shall and Stein, 1979). This model makes predictions that are quantitatively very similar to the somatic mutation model, and these are incompatible with experimental data.

Werner's syndrome cells have a very limited lifespan in culture (Paper 9; Thompson and Holliday, 1983). In Paper 28, Salk et al. investigated the chromosomes of these cells and made the major discovery that they were unstable. In particular, the frequency of translocations was very high, but other abnormalities were also seen. Although chromosome abnormalities in stimulated lymphocytes from Werner's syndrome cells have not yet been unequivocally demonstrated, there is little doubt that the results in Paper 18 establish the syndrome as another example of an inherited predisposition to genetic instability, at least at the chromosome level. In this and other cases, however (Fanconi's anaemia, ataxia telangiectasia), the biochemical defect has not been identified.

In spite of much discussion of the importance of somatic gene mutations in the aging of organisms, until recently no experimental data have been available. A new method has now been developed by Albertini, Castle, and Borcherding (1982) and Morley et al. (1983) that makes it possible to measure the frequency of mutation to 6-thioguanine resistance in human T lymphocytes. Resistant cells can either be scored by an autoradiographic technique or by clonal growth in the presence of the analogue. In Paper 29 the former method was used in the study of lymphocytes from 37 individuals aged 9 to 95 years. Although there is considerable scatter in mutation frequency, there is a highly significant increase with age and the data are more compatible with an exponential accumulation of mutations with time than with a linear accumulation. This technique is a very powerful one that I predict will be used in many future studies of somatic mutation in humans as well as in experimental animals.

Apart from gene mutations, which involve alteration in DNA sequence, other molecular changes in DNA are possible. Cytosine in DNA can be methylated at the 5-position (5-mC) through the activity of a methyl transferase. It was suggested independently by Riggs (1975) and Holliday and Pugh (1975) that the control of gene expression in higher organisms might be related to the heritable pattern of DNA methylation. Methylation would be maintained through cell division if the methyl transferase acted on hemi-methylated, but not unmethylated, DNA. Changes in DNA methylation might be mediated by sequence-specific "switch enzymes," which would be transient in activity. Evidence has subsequently been accumulated that the pattern of DNA methylation is indeed heritable and that transcription is associated with an absence of 5-mC at specific sites in DNA adjacent to structural genes (for reviews, see Doerfler, 1981, 1983; Riggs and Jones, 1983).

In Paper 30 5-mC levels are examined in human, hamster, and mouse diploid fibroblasts. In all cases, the amount of 5-mC decreases during serial subculture, but this is fastest for mouse, intermediate for hamster, and slowest for human cells. The rate of loss of methylation correlates with the cells' in vitro longevity, at least for these three species. Furthermore, it is shown that permanent lines have a constant level of DNA methylation, perhaps because they have the ability to replace methyl groups by a de novo methylase activity (Holliday, 1985). In any event, the loss of methylation in diploid cells may have progressively deleterious consequences, if the normal regulation of gene activity is disrupted. The results in Paper 30 provide evidence for a new type of error during aging: the appearance and subsequent inheritance of epigenetic defects. It will also stimulate many subsequent studies on DNA methylation and aging. For instance, it has now been shown that single treatment of early-passage cells with the base analogue 5-azacytidine, which is known to reduce the level of 5-mC in DNA, has a profound life-shortening effect on populations of human fibroblasts (Holliday, 1985).

REFERENCES

Albertini, R. J., K. L. Castle, and W. R. Borcherding, 1982, T-Cell Cloning to Detect the Mutant 6-Thioguanine-Resistant Lymphocytes Present in Human Peripheral Blood, *Proc. Natl. Acad. Sci. (U.S.A.)* **79:**6617-6621.

Burnet, F. M., 1974, *Intrinsic Mutagenesis: A Genetic Approach to Ageing,* Wiley, New York.

Doerfler, W., 1981, DNA Methylation—a Regulatory Signal in Eukaryotic Gene Expression, *J. Gen. Virol.* **57:**1-20.

Doerfler, W., 1983, DNA Methylation and Gene Activity, *Annu. Rev. Biochem.* **52:**93-123.

Fulder, S. J., 1977, Spontaneous Mutations in Ageing Human Cells: Studies Using a Herpes Virus Probe, *Mech. Ageing Dev.* **6:**271-282.

Fulder, S. J., 1978, Somatic Mutations and Ageing of Human Cells in Culture, *Mech. Ageing Dev.* **10:**101-115.

Gupta, R. S., 1980, Senescence of Cultured Human Diploid Fibroblasts—Are Mutations Responsible? *J. Cell. Physiol.* **103:**209-216.

Hoehn, H., E. M. Gryant, P. Johnston, T. H. Norwood, and G. M. Martin, 1975, Non-Selective Isolation, Stability and Longevity of Hybrids between Normal Human Somatic Cells, *Nature* **258:**608-609.

Holliday, R., 1985, The Significance of DNA Methylation in *The Molecular Biology of Ageing,* A. D. Woodhead, A. D. Blackett and A. Hollaender, eds., pp. 269-283, Plenum Press, New York.

Holliday, R., and J. E. Pugh, 1975, DNA Modification Mechanisms and Gene Activity during Development, *Science* **187:**226-232.

Lindop, P. J., and R. Rotblat, 1971, Shortening of Life and Causes of Death in Mice Exposed to a Single Whole-Body Dose of Radiation, *Nature* **189:**645-648.

Maynard-Smith, J., 1959, A Theory of Ageing, *Nature* **184:**956-957.

Maynard-Smith, J., 1962, The Causes of Ageing, *Proc. R. Soc.* **157B:**115-127.

Morley, A. A., 1982, Is Ageing the Result of Dominant and Co-dominant Mutations? *J. Theor. Biol.* **98:**469-474.

Morley, A. A., K. J. Trainor, R. Eshadri, and R. G. Ryall, 1983, Measurement of *in vivo* Mutations in Human Lymphocytes, *Nature* **302:**155-156.

Neary, G. J., 1960, Ageing and Radiation, *Nature* **187:**10-18.

Pitha, J., R. Adams, and P. M. Pitha, 1974, Viral Probe into the Events of Cellular (*in vitro*) Ageing, *J. Cell. Physiol.* **83:**211-218.

Pitha, J., E. Stork, and E. Wimmer, 1975, Protein Synthesis during Ageing of Human Cells in Culture, *Exp. Cell. Res.* **94:**310-314.

Riggs, A. D., 1975, X Inactivation, Differentiation and DNA Methylation, *Cytogenet. Cell. Genet.* **14:**9-25.

Riggs, A. D., and P. A. Jones, 1983, 5 Methyl Cytosine, Gene Regulation and Cancer, *Adv. Cancer Res.* **40:**1-30.

Shall, S., and W. D. Stein, 1979, A Mortalisation Theory for the Control of

Cell Proliferation and for the Origin of Immortal Cell Lines, *J. Theor. Biol.* **76:**219-231.,

Smith, J. R., and L. Hayflick, 1974, Variation in the Lifespan of Clones Derived from Human Diploid Cell Strains, *J. Cell. Biol.* **62:**48-53.

Smith, J. R., O. M. Pereira-Smith, and E. L. Schneider, 1978, Colony Size Distribution as a Measure of *in vivo* and *in vitro* Ageing, *Proc. Natl. Acad. Sci. (U.S.A.)* **75:**1353-1356.

Szilard, L., 1959, On the Nature of the Ageing Process, *Proc. Natl. Acad. Sci. (U.S.A.)* **45:**35-45.

Thompson, K. V. A., and R. Holliday, 1975, Chromosome Changes during the *in vitro* Ageing of MRC-5 Human Fibroblasts, *Exp. Cell Res.* **96:**1-6.

Thompson, K. V. A., and R. Holliday, 1983, Genetic Effects on the Longevity of Cultured Human Fibroblasts I Werner's Syndrome, *Gerontology* **29:**73-82.

Tomkins, G. A., E. J. Stanbridge, and L. Hayflick, 1974, Viral Probes of Ageing in the Human Diploid Cell Strain WI-38, *Proc. Soc. Exp. Biol. Med.* **146:**385-390.

23

Reprinted from *Proc. Natl. Acad. Sci. (U.S.A.)* **50**:390–395 (1963)

*ANEUPLOIDY IN THE DEGENERATIVE PHASE OF SERIAL CULTIVATION OF HUMAN CELL STRAINS**

By Eero Saksela† and Paul S. Moorhead

THE WISTAR INSTITUTE OF ANATOMY AND BIOLOGY, PHILADELPHIA

Communicated by Warren H. Lewis, June 26, 1963

The factors affecting chromosomal stability of mammalian cells in long-term culture are not well understood, aside from the effects of experimentally applied agents such as irradiation and viruses. Species differences[1-3] and perhaps cell types appear to be important with respect to the occurrence of spontaneous alteration or heteroploid transformation.[4, 5] Cells cultured from mouse tissues seem to be extremely labile with regard to chromosome changes and associated morphologic and growth alterations.[1, 6, 7] On the other hand, in metaphase studies to date, cultured human cells of the fibroblastic type remain diploid and show no tendency to undergo spontaneous transformation.[8-11]

Interest in spontaneous occurrences of *in vitro* transformation of mammalian cells has been based upon hope for its confirmation as a model of the *in vivo* process of malignant change. The two processes have many features in common,[5, 12] and early proposals that viruses present in the original tissue or serum might be causative agents have been strengthened by the numerous studies showing that certain viruses do produce transformation *in vitro*. Experiments with SV40 virus and human cells have provided another reproducible cell-virus system for the study of transformation throughout its course.[13, 14] Furthermore, the discovery of chromosome lesions in human cells in association with measles infection[15] has increased our interest in the nature and extent of chromosomal changes which occur *in vitro* in the absence of any known agent.

This report is concerned with a study of metaphase chromosomes in two human cell strains during their total period of *in vitro* cultivation. In a previous less extensive study no aberrations were seen in various human diploid cell strains in material from the 9th to the 40th subcultivation *in vitro*.[10] In the present study spontaneous chromosome changes were encountered, but only in those cultures which had been subcultivated for 40 or more times, i.e., during the degenerative phase of their characteristically limited *in vitro* life. Although this aneuploidy seems to be associated with the degenerative period in long-term cultivation, on no occasion did transformation in terms of altered morphology, growth rate, or capacity for indefinite cultivation occur among many parallel cultures of these two strains. In long-term cultures of human fibroblasts Sax and Passano [16] have previously shown an association between "age *in vitro*" and increase in spontaneous rate of anaphase aberrations.

Materials and Methods.—*Cell strains:* Cell strains WI-26 and WI-38, male and female, respectively, were derived from fetal lung tissue by L. Hayflick according to procedures previously described.[10] In essence, cultures were grown as monolayers of fibroblastic-like cells in Eagle's basal medium with 10% calf serum and 50 μg aureomycin per ml and subcultivated by trypsinization with 0.25% trypsin (Difco 250:1). Subcultivation from one to two milk-dilution bottles was performed twice weekly, i.e., when the monolayer had become confluent. Confluency was achieved every 3–4 days during the major period (4–5 months) of total *in vitro* cultivation. After approximately 35 total passages *in vitro*, cultures required longer periods (5–8 days) to achieve confluency;

eventually (50 ± 10 passages), all cultures of these diploid cell strains failed to replicate sufficiently to permit any further subcultivation. It is recognized that each subcultivation or "passage generation" is only an approximation to cell population doubling, and the total number of passages has no significance except as a crude measure of the *in vitro* stage any strain has reached. The terminal portion of the finite period of serial subcultivation is referred to as Phase III[10] or as the degenerative phase, and in the absence of heteroploid transformation this limitation on cultivation has long been recognized. Ampoules containing 2×10^6 cells were frozen and stored in liquid nitrogen by procedures described previously.[10] Substrain designations in Tables 1 and 2 refer to cultures reconstituted from such ampoules and carried independently.

Chromosome preparations: Chromosome studies were made from permanent mounts of Giemsa stained air-dried metaphase preparations.[17] Ordinarily cells were harvested with trypsin on the second day after subcultivation, following 3–5 hr of treatment with Colcemid (CIBA), 0.05 μg per ml of medium. The cell suspension was concentrated in $^1/_2$ ml of trypsin solution, and a 4-fold volume of distilled water was added to swell the metaphase cells hypotonically for 8–10 min. Fixation was made with 3:1 methanol:acetic acid. Spreading was done by ignition of a drop of fixative, containing cells in suspension, immediately after its application to the surface of a clean wet slide. Suitable metaphases were selected under low power (150×) observation, and those judged to be free from excessive spreading were then studied under oil immersion optics. In all cells so selected, the chromosomes were counted, and 25–40% of the metaphases of each sample were subjected to detailed karyotypic analysis, involving the identification of individual chromosome pairs or groups of the human karyotype according to the Denver convention: Nos. 1, 2, 3, 4–5, 6–X–12, 13–15, 16, 17–18, 19–20, 21–22–Y. Metaphase counts considered to be artifacts because of scattering of some chromosomes or because of accidental contamination of one metaphase with chromosomes of another were not excluded from the data on the various samples studied. For each determination of the level of tetraploidy existent in the dividing cell population 250–300 unselected metaphases were examined and roughly estimated as being diploid or tetraploid.

Observations: WI-26: Chromosome counts, numbers of cells analyzed, and other karyologic observations from seven different cultures of strain WI-26 are presented in Table 1.

Chromosome preparations from diploid cell strain WI-26 were first made at its 19th subcultivation passage. In a sample of 100 cells the exact chromsome number was determined, and 36 cells were analyzed in detail. It could be seen that most of the cells had a normal male chromosome complement, and in the hypodiploid cells no pattern concerning the missing chromosomes could be deduced. Of the dividing cell population 3.1% were tetraploid, and no abnormal chromosomes were found.

Between the 28th and 37th *in vitro* passages, 106 cells were examined, and of these 28 were analyzed in detail. A normal distribution of chromosome numbers, skewed toward hypodiploid counts, was observed. The analysis of each of the 9 hypodiploid cells of the sample did not reveal any consistency as to the chromosomes which were missing, and these counts are presumed to be artifacts. A somewhat higher value of tetraploidy was recorded at the 28th passage, and again at the 37th passage, 4.5 and 4.4%, respectively. In the sample from the 32nd passage, 1.6% of the dividing cells were tetraploid. An acentric fragment was observed in one of 45 cells examined at the 37th passage. It was found in a cell with 46 (excluding the fragment) chromosomes, which was lacking a No. 4–5 chromosome and contained an extra chromosome in the 6–12 group. It is possible that the fragment resulted from breakage in the long arm of a 4–5 chromosome; this deficient chromosome would then be indistinguishable from members of the 6–12 group. Thus, from a total of over 200 cells counted in cultures of the 19th, 28th, 32nd, and 37th passages, of which 64 were studied in detail, only one cell was found which had an abnormality.

Between the 41st and 54th *in vitro* passage, 58 cells of WI-26 were karyologically examined. Six of 25 metaphases at the 41st passage contained abnormal dicentric chromosomes, while 3 out of 20 did so at the 54th passage. A small sample of 13 cells from 43rd passage material revealed no abnormalities, and the extremely low frequency of mitosis prevented extension of this sample. At the 41st passage, 3.6% of the cells were tetraploid, and the very high value of 16% tetraploidy occurred in the 54th passage material.

Substrains of WI-26 (X, XI, XIII, XIX, and XXIII) could not be subcultivated more than 40–49 total passages; XXVI survived for 56 passages *in vitro*.

TABLE 1

CHROMOSOME COUNT DISTRIBUTION AND KARYOLOGIC DATA FROM HUMAN DIPLOID CELL STRAIN WI-26

In vitro passages	WI-26 substrain designation	Chromosome Counts										Tetra-ploidy %	Identity of "Missing" Chromosomes							Total cells	Analyzed cells	Types of Abnormal Chromosomes			
		<	43	44	45	46	47	49	72	76	91	92		1–5	6–x–12	13–15	16	17–18	19–20	21–22–Y			D	f	m	other
19th	X	1	1	2	5	88						2	3.1	1	2	1				2	100	36				
28th	XI	1	1	2	2	16							4.5		2			1		1	20	4				
32nd	XIX			1	1	38	1						1.6		2		1				41	12				
37th	XXIII			3	1	42							4.4	1		1					45	12				1
41st	XXIII	1		1	4	19							3.6		4	2			2		25	12	6			
43rd	XIII		1	1	1	10															13	5				
54th	XXVI	1		1	5*	12†		1		1	1	1	16.0		2			2	1	1	21	6	4			

* Cell presumed 13/21 type translocation.
† Pseudodiploid cell; monosomic and trisomic.
D = Dicentric; f = acentric fragment; m = minute.

TABLE 2

CHROMOSOME COUNT DISTRIBUTION AND KARYOLOGIC DATA FROM HUMAN DIPLOID CELL STRAIN WI-38

In vitro passages	WI-38 substrain designation	Chromosome Counts									Tetra-ploidy %	Identity of "Missing" Chromosomes							Total cells	Analyzed cells	Types of Abnormal Chromosomes			
		<	43	44	45	46	47	53	64	92		1–5	6–x–12	13–15	16	17–18	19–20	21–22			D	f	m	other
4th	P	2	1	3	7	183	3‡		1		1.0	1	8	1	1	3	4	3	200	84				
14th	II					19					0.8								20	9				
21st	P			1		21					1.0		1						22	10				
31st	IV			1	2	47					1.0	1					1		50	22				
33rd	V			2	3	46†					0.4	2	2	1		1			51	19				
35th	IV				1	28													30	10				
37th	IV			1	5	43					3.8	1	2	2				1	50	18				
41st	P	1		2	9*	87†	1			1	2.7	8	4	5	2	4	3	3	104	51	3	1	1	
41st	IV			1	6*	30†		1			5.1	5	2	5	1	1		4	40	18	2	1	1	
46th	II	1		1	7	11						4	2	3				3	21	10	6			2

* Cell with presumed 13/21 type translocation.
† Pseudodiploid cell; monosomic and trisomic.
‡ Artifact due to contaminating chromosome from another cell, two cases.
P = parental culture; D = dicentric; f = acentric fragment; m = minute.

WI-38: At the fourth subcultivation passage 200 WI-38 cells were examined and, of these, 84 were subjected to detailed karyotypic analysis. The distribution of the exact chromosome counts obtained is given in Table 2. Among 14 hypodiploid cells no consistent pattern as to the identity of the missing chromosomes could be determined. Three cells with 47 chromosomes were observed, but in two of these contamination from another metaphase was definitely indicated by a differential state of condensation in the extra chromosome. In all other cells examined, no deviation from the normal female karyotype was observed. At this early stage *in vitro*, the frequency of tetraploid cells in the dividing population was 1.0%.

Essentially similar findings were obtained concerning the chromosomal constitution of various WI-38 substrains from the 14th to the 37th passage; 223 cells were examined and 98 of these karyotypically analyzed. A single pseudodiploid cell was observed in the 33rd passage sample. The level of tetraploidy remained low, except for one slightly higher value at the 37th passage, 3.8%.

From WI-38 cultures between the 41st and 46th *in vitro* passages, the chromosomes of 165 cells were counted, and 79 of these were karyotypically analyzed. Aneuploid changes and a marked increase in frequency of hypodiploid cells are evident in all of these late passage samples. Although no obvious pattern as to the missing chromosomes can be observed (Table 2), the fact that 2% (41st passage) to 30% (46th passage) of the cells exhibited obviously abnormal chromosomes would imply that perhaps many of the hypodiploids were not artifacts of technique. The abnormalities observed were mainly dicentrics formed by translocations between two chromosomes of the complement, in some cases identifiable chromosomes. Less frequently (Table 2), fragments, minute chromosomes, and abnormal monocentric chromosomes were observed. In this late passage WI-38 material, the frequency of tetraploid cells (2.7–5.1%) was elevated above values seen in lower passage levels (Table 2). The original or parent strain of WI-38 was cultivated for a total of 48 passages, and none of the three substrains (II, IV, V) could be subcultivated more than 43 to 48 passages *in vitro*.

Discussion.—Aneuploid changes appeared in different substrains of each strain studied following a long period of apparent chromosome stability. The earliest chromosome aberrations were coincident, with the beginning of the decline of each strain during its serial subcultivation. Setting aside the question of criteria for chromosomal or karyotypic "normality," we must consider whether these observations are generally applicable to human fibroblast strains in long-term culture. Since strains WI-26 and WI-38 were selected for extended investigation for the sole reason that one is male and the other female, it seems warranted to regard them as representative of such serially propagated strains, at least for those of embryonic origin. Further, two other human fibroblast strains cursorily examined also showed some aneuploid cells at high passage levels of cultivation.

That our previous study[10] failed to reveal aneuploid changes with increased "culture age" is attributed to the fact that only two of the 16 samples studied from 13 strains were from cultures approaching or in the period of decline, i.e., above the level of the 35th passage. At its 40th passage, WI-12 revealed no abnormalities in a sample of 33 metaphases, and only 17 cells were available from the 39th passage of WI-1 because of the low growth rate in its declining phase.

No aneuploidy was observed in earlier studies on long-term cultivated human fibroblasts. Among cultures from individuals 1–41 years of age, Tjio and Puck[8] reported "no variation in chromosome number and morphology." Their cultures were studied over periods comparable to those reported in the present paper. In a later paper[9] these authors remarked upon the constancy of the human karyotype, finding no change in number or morphology over "20 successive harvests (transfers)... involving more than 40 generations." Makino and co-workers[11] reported the maintenance of a "normal complement of 46 chromosomes" in cells obtained from the 2nd to the 44th subculture; however, they also state, "In comparison with the results of chromosome counts in the primary cultures, cells with hypo- or hyper-diploid chromosome numbers occurred at a higher incidence in the subcultured specimens."

These authors did not present details concerning the proportion of their large total sample which was derived from the late-passage cultures mentioned. On the other hand, in this study we deliberately sought information during the period when the cultures' growth rates had begun to diminish. This fact may account for our observing aneuploidy, whereas they reported none. Our results, in respect to samples taken prior to about the 40th subculture, are in accord with the other studies. The constancy of the human karyotype in such material is impressive when compared to continuously cultivatable cell lines from mammalian tissue which are near-diploid, i.e., containing a large number of apparently euploid cells but with some pseudo-, hypo-, and hyper-diploid cells also detectable.[2]

For a priori reasons, criteria for the "normal" human karyotype should be based upon in vivo dividing populations (such as direct preparations from bone marrow) or upon primary or very early tissue cultures. This must be determined against a background of a certain amount of hypodiploid counts which, for technical reasons, are spurious. Within these limitations of technique and of the experience of the investigators, prior to the 40th passage level, both strains WI-26 and WI-38 may be regarded as normal or classic diploid. A minor reservation may be made with respect to strain WI-26, however, as the tetraploid percentages observed at the 28th and 37th passages were slightly above values usually observed. A number of studies[8, 10, 11] have shown that the tetraploidy level in presumably normal cultures of human fibroblast cells is seldom greater than 3 per cent. An increase in proportion of tetraploids in the metaphase population is a common feature of SV40 transformation of human fibroblasts.[18, 19] We have observed a similar association between tetraploidy increase and subsequent spontaneous heteroploid transformation in serially cultivated cells of the rhesus monkey.

Considerations as to the tissue's species of origin and the possible presence of inapparent viruses may explain the fact that adult rhesus monkey kidney tissue cultures ordinarily degenerate within a few weeks, surviving no more than 3 or 4 subcultivations. These monkey kidney subcultures may show an extremely high frequency of mitotic aberrations (16–40 per cent) even before the second week of in vitro cultivation.[20] That human cell types other than fibroblasts may show entirely different patterns with respect to their fate in vitro is well known. For example, extra-embryonic amnion can usually be cultivated for only a few subdivisions before it degenerates or, on occasion, transforms. It is possible that the inherently limited in vivo growth potential of amnion, a kind of "tissue age," may be expressed even in vitro.

The three WI-26 substrains (XIII, XXIII, XXVI) which revealed aneuploidy survived as serial cultures no longer than various parallel subcultures from this strain which were not examined for chromosome changes. The same applies to the particular substrains (Parental, II and IV) comprising the post-40th passage material in which aneuploid changes were observed in the WI-38 strain. Among the rearrangements and chromosome breaks observed, there was no consistency as to the particular chromosomes affected.

We cannot exclude the possibility that the two particular strains, WI-26 and

WI-38, were in some way predisposed to undergo aneuploid changes or that some unknown effect of the medium used has induced an instability. However, aneuploid changes were noted in different substrains, even though they were carried on medium from independent commercial sources. It therefore appears more likely that our findings may apply generally to such cells and confirm the study by Sax and Passano[16] in which anaphase bridges, lagging chromosomes, rod and dot deletions were found to increase approximately 3-fold during six months of serial cultivation of human fibroblast strains. Interphase nuclei with abnormal sizes and shapes are seen in daughter cells in later passage material[10] and are presumably the products of abnormal division.

The conditions in these human fibroblast cultures, after loss of their proliferative capacity, are superficially identical to those reported in cultures of mouse cells just prior to the usual reversal of the latter's declining growth rate.[1, 7] This increase in growth rate then leads to establishment of the mouse cell culture as a continuously propagated and usually heteroploid cell line. In spite of the degenerate state of late passage cultures of strains WI-26 and WI-38, with chromosome aberrations and reduced mitotic activity, none has undergone a spontaneous transformation, although numerous cultures have been observed 1–3 months after cessation of growth.

Note added in proof: It has come to our attention that M. C. Yoshida and S. Makino have reported quite similar findings in the *Japan. J. Human Genetics*, **5,** 39 (1963). These independent observations strengthen our conclusion that the presence of aberrations associated with *in vitro* decline of such cell strains is a general phenomenon.

* Supported in part by USPHS research grant CA-04534, contract PH-43-62-157, and career development award 5K3-CA-18,372 from the National Cancer Institute. The excellent technical assistance of Miss P. Mancinelli and Mrs. M. Lebowitz is gratefully acknowledged.

† Permanent address: University Department of Pathology, Maria Hospital, Helsinki.

[1] Rothfels, K. H., and R. C. Parker, *J. Exptl. Zool.,* **142,** 507 (1959).

[2] Yerganian, G., and M. J. Leonard, *Science,* **133,** 1600 (1961).

[3] Ruddle, F. H., *Cancer Res.,* **21,** 885 (1961).

[4] Levan, A., *Cancer,* **9,** 648 (1956).

[5] Hsu, T. C., *Intern. Rev. Cytol.,* **12,** 69 (1961).

[6] Levan, A., and J. J. Biesele, *Ann. N. Y. Acad. Sci.,* **71,** 1022 (1958).

[7] Todaro, G. J., and H. Green, *J. Cell Biol.,* **17,** 299 (1963).

[8] Tjio, J. H., and T. T. Puck, these PROCEEDINGS, **44,** 1229 (1958).

[9] Tjio, J. H., and T. T. Puck, *J. Exptl. Med.,* **108,** 259 (1958).

[10] Hayflick, L., and P. S. Moorhead, *Exptl. Cell Research,* **25,** 585 (1961).

[11] Makino, S., Y. Kikuchi, M. S. Sasaki, M. Sasaki, and M. Yoshida, *Chromosoma,* **13,** 148 (1962).

[12] Barski, G., *Rev. franç. études clin. et biol.,* **7,** 543 (1962).

[13] Koprowski, H., J. A. Pontén, F. Jensen, R. G. Ravdin, P. S. Moorhead, and E. Saksela, *J. Cell. Comp. Physiol.,* **59,** 281 (1962).

[14] Shein, H. M., and J. F. Enders, these PROCEEDINGS, **48,** 1164 (1962).

[15] Nichols, W. W., A. Levan, B. Hall, and G. Östergren, *Hereditas,* **48,** 367 (1962).

[16] Sax, H. J., and K. N. Passano, *Am. Naturalist,* **95,** 97 (1961).

[17] Moorhead, P. S., and P. C. Nowell, in *Methods in Medical Research,* (Chicago: Yearbook Publishers), in press.

[18] Yerganian, G., H. M. Shein, and J. F. Enders, *Cytogenetics,* **1,** 314 (1962).

[19] Moorhead, P. S., and E. Saksela, *J. Cell. Comp. Physiol.,* in press.

[20] Kleinfeld, R., and J. L. Melnick, *J. Exptl. Med.,* **107,** 599 (1958).

A Rapid Rise in Cell Variants during the Senescence of Populations of Human Fibroblasts

Stephen J. Fulder and Robin Holliday
National Institute for Medical Research
Mill Hill
London NW7 1AA, England

Introduction

When it was first discovered that human fibroblasts have a defined lifespan in culture, it was suggested that genetic damage might accumulate with time, until eventually the load of mutations or chromosome abnormalities became so great that growth was no longer possible (Hayflick, 1965). Alternatively, mutations could arise as secondary consequences of other events associated with ageing, such as protein errors leading to a loss of accuracy in replication or repair (Lewis and Holliday, 1970; Holliday and Tarrant, 1972; Orgel, 1973). The first hypothesis states that mutations should accumulate linearly with time, while the second predicts a much more rapid increase at the end of the lifespan.

Unfortunately, conventional procedures for measuring mutation frequency cannot be applied to these populations of cells. If, for instance, one is measuring the frequency of cells resistant to a toxic compound, such as 8–azaguanine, it is necessary to count occasional colonies arising from a large number of plated cells. This approach is ruled out for late passage senescent cells, since not only is their plating efficiency very low, but more important, their limited growth potential prevents colony formation.

A novel approach was therefore adopted in attempts to measure the frequency of mutation during the lifespan of MRC–5, a male fetal lung fibroblast strain. The experiments utilize a histological assay system to look for individual variant cells with altered glucose–6–phosphate dehydrogenase (G6PD) activity. This enzyme is X-linked and therefore hemizygous in the male human fibroblasts. A large number of different forms of the enzyme are known to exist in human populations. The most common forms utilize an analogue substrate, 2–deoxyglucose–6–phosphate (dG6P), with an activity less than 5% of the normal substrate, glucose–6–phosphate (G6P). However, there are rare forms, for example, G6PD Markham, which have up to a

50 fold increase in activity with the analogue substrate, and one or two others, for example, G6PD Hektoen, which have a supranormal activity with the usual substrate (Yoshida, Beutler, and Motulsky, 1971; Dern, McCurdy, and Yoshida, 1969). The enzyme can be assayed in fixed intact cells using a tetrazolium staining method developed by Wajntal and DeMars (1967). Sutton and Karp (1970) first suggested that it might be possible to pick out mutant cells within a population by substituting the analogue substrate in the assay.

The method has the obvious advantage that one can score the variant phenotype whether or not the cell is actually capable of further cell division. This is an important point, since the use of an analogous technique, in which the phenotype scored was due to enzyme activity, demonstrated that the frequency of radiation-induced genetic changes in the fungus Ustilago is dissimilar in viable and nonviable cells (Holliday, 1971). In other words, it cannot be assumed that measuring the mutation frequency in viable cells provides a true value for the whole population. In examining senescent populations, it is clearly essential to screen all the cells, rather than a continually declining fraction of viable ones.

Results

Stained Variant Cells

Initial experiments using dG6P at 6×10^{-3} M with early passage fibroblasts showed that there were rare heavily stained cells within the faint pink monolayer (Figure 1A). It was soon found that stained cells could also be detected using the normal substrate at 1.3% of its normal concentration. It therefore seems probable that most of these cells simply contain abnormal levels of G6PD rather than having an enzyme with altered substrate specificity. We realize that the scoring of these variants is to some extent subjective. In order to minimize possible error, we divided variants into a "very heavily stained" category, which stood out very clearly from the background, and a more frequent category of "heavily stained" cells, which were dark rather than pale pink.

It is clearly essential to obtain evidence that these variants are indeed due to underlying genetic events, either true mutations or possibly heritable "epigenetic" changes. One or two trivial possibilities can be ruled out. The variants are not tetraploid or giant cells, nor are they due to the overlapping of two or more cells with normal staining. Maltreatment of the culture (for instance, by serum starvation, treatment with toxic levels of 5–fluororacil, or high salt concentrations) did not result in an increased frequency of stained cells. On the other hand, exposure of cells to mutagens such as γ-rays or ethane methyl sulphonate (EMS) significantly increased the frequency of variants. Some of these results are shown in Table 1. It can be seen that 400 μg of EMS for 22 hr, which kills about 50% of

the cells, increased the number of highly stained cells about 3 fold. Subculturing the treated cells for two passages showed that the level of heavily stained cells was maintained, although there was some reduction in the proportion of the very heavily stained cells (Table 1). A γ-ray dose of 150 rads doubled the frequency of variants.

In addition to these experiments with mutagens, the screening of cells which were growing on coverslips demonstrated that a proportion of variants (∼ 20%) occurred as side-by-side pairs of cells. It is extremely probable that these arose from the divi-

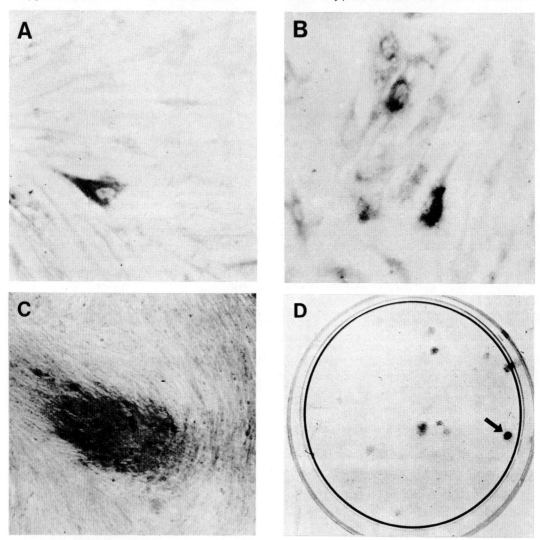

Figure 1(A) A Heavily Stained Cell in a Fibroblast Monolayer (× 43).
(B) Heavily Stained Cells in a Culture of Senescent Fibroblasts (× 43).
(C) A Heavily Stained Section within a Monolayer Colony of MRC–5 Fibroblasts Stained with dG6P.
(D) A Heavily Stained Colony of Filt-5.

sion of a single variant cell. Better evidence that the phenotype is heritable came from studies with colonies. We have had difficulty in obtaining high plating efficiency with MRC–5 cells, and it has not been possible to score more than a few thousand colonies. Among 4537 examined, none were fully stained, but 6 had very heavily stained sectors (Figure 1C). This is roughly the frequency we would expect from the scoring of individual very heavily stained cells. From microscopic examination it is quite clear that the sectors are monolayers like the rest of the colony, and the rather diffuse boundary is probably due to the mobility of fibroblasts. It should be emphasized that the scoring of colonies with colored patches or sectors has for many years been accepted as a valid criterion for gene mutation in bacteria or yeast (for example, Witkin, 1951; Nasim and Auerbach, 1967). Sectors corresponding to the heavily stained cells could not be readily distinguished by visual inspection from colony areas containing two or more layers of cells. Similar experiments were performed with the transformed hamster line Filt–5, because it is more suitable for cloning studies. Among 5569 colonies, one was very heavily stained (Figure 1D).

These observations do not alone prove that the phenotype is heritable, since it has not been possible to subculture clones and show that the high enzyme level persists. This cannot be done without a replica plating procedure, since the staining of the colonies obviously kills the cells, nor is there any method available for selecting cells with high G6PD.

Frequency of Variants with Age

Experiments were carried out to measure accurately the frequency of stained cells during aging. The cells were grown on coverslips, fixed, and stained, and coverslips of equivalent cell density sampled for variant frequency (see legend to Figure 2). In four experiments, the frequency of variants was counted every second passage throughout the lifespan of two consecutive cultures of MRC–5 fibroblasts, and representative results are shown in Figures 2a and 2b. [A full description of the results is given elsewhere (Fulder, 1975).] It is clear that there is a low frequency of spontaneous variants during much of the lifespan of the cells, followed by a dramatic rise during the senescent phase. The kinetics of the increase in variants with time are the same for the very heavily or heavily stained categories. In very old cultures over 3% of the cells are stained (Figure 1B). This photograph also demonstrates the increased heterogeneity in background staining which is always seen in senescent cultures. As shown in Figure 2b, the same frequency of variants was seen whether the dG6P or low concentrations of G6P were used. This strongly indicates that most of the variants simply have an increased enzyme level.

It is important to know whether the variants accumulate throughout the lifespan or only during the senescent phase. Accordingly, the results were subjected to regression analysis. An initial question which could be asked is whether the frequency of variants increases during the early stages in the lifespan. An arbitrary point was chosen, passage 48, after which the variant frequency clearly rises steeply. Regression analysis was carried out by a standard statistical computer program, and the results indicated that 3 of the 4 sets of data showed a highly significant ($P < 0.001$) increase in variant frequency during this part of the lifespan; the fourth set of data was marginally significant ($P = 0.01$–0.05). It is thus established that the frequency of variants increases during the early stages of the lifespan. Inspection of the data seems to show that the results of the entire lifespan might be described by an exponential relationship between variants and passage number. To test this a linear regression of the log of variant frequency on passage was fitted using a standard computer program. A highly significant regression was found with all 4 sets of data. The regression curve is shown for the data in Figure 2a.

The Rates of Origin of Variants per Generation

It is fortunate that a simple relation between the variant frequency and passage number has been obtained, because it then becomes possible to use the regression equations for calculation of the rate of increase. The rate per generation is not constant, but increases every generation, and it can be calculated in the following manner. The regression equation is in the form $\log m = a + bP$, where m is mutation frequency, P is passage number, and a and b are constants, or $m = e^a \times e^{bP}$. e^a is a constant for each set of data, so the relation between one passage (P) and the next ($P + 1$) is given by:

Table 1. Frequency of Variants after Mutagenesis with EMS

Passage Number	Phenotype	Frequency of Stained Cells ($\times 10^4$)	
		Control	EMS Treated
29	Heavy stain	26.4	73.1
	Very heavy stain	0.6	3.6
31	Heavy stain	38.4	83.4
	Very heavy stain	0.6	3.2
31*	Heavy stain		77.6*
	Very heavy stain		1.0

*Cells treated with EMS at passage 29, then grown for 2 passages and rescored.

251

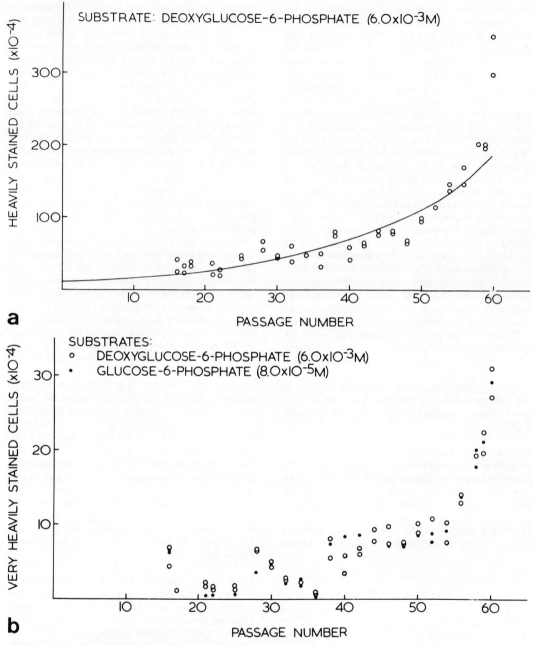

Figure 2. Frequency of Variants during the Lifespan of MRC–5 Fibroblasts.

Approximately 2 × 10⁵ cells were placed in a 35 mm plastic dish containing a sterilized coverslip. After incubation overnight, they were stained as in Experimental Procedures. An economic counting routine was developed and tested for accuracy and reliability. Coverslips were scanned stripwise at a magnification of × 100 to find and score heavily stained cells, following which the total number of cells were counted at higher magnification within squares along each strip. Two classes of stained cells were counted: (a) heavily stained which were noticeably darker than background, and (b) very heavily stained which were very much darker. The criteria of assignation of stained cells to either of these classes were inevitably somewhat subjective, but the ratio was consistent between experiments. Each point represents the average counts within three strips in each of two coverslips, which allowed 100–1000 heavily stained cells to be counted, depending upon age. Damaged and mitotic (rounded) cells were excluded. [Full details are given by Fulder (1975).]

$m_1 = \text{constant} \times e^{bP}$

$m_2 = \text{constant} \times e^{b(P+1)}$

from which

$m_2 = m_1 \times e^b.$

e^b is a constant with which the frequency at any one passage is related to the frequency at the next passage. The increment from one passage to the next is the number of variants occurring at that passage, that is, the rate per passage. The rate per passage can be converted to the rate per generation if one takes into account the fact that not all cells divide in every passage. From published data on the percentage of nondividing W1–38 cells, Good (1972) has calculated that the number of passages must be multiplied 3 fold to estimate the number of generations attained during the lifespan of the mass culture. Studies on the viability of MRC–5 at different ages (R. Cox, personal communication) indicate that the passage number should be multiplied 2 fold to obtain a reasonable approximation of the number of generations. Table 2 gives the calculation of the variant frequency and the rate at three representative points during the lifespan of the culture, that is, passages 0, 20, and 40. It can now be seen that the rather high frequency of variants observed gives a misleading impression of the actual rates of origin of the variants, which is quite low and more in line with what might be expected for mutation. These calculations assume that the variant cells divide at the same rate as the unstained ones, which seems to be the case for the heavily stained cells (Table 1). If the very heavily stained cells are selected again, as Table 1 indicates, then their rates of origin may be somewhat higher than the values in Table 2.

Table 2. Frequencies and Rates of Origin of Variants at Different Passage Levels

		Phenotype (Substrate: dG6P)	
	Passage	Heavy Stain	Very Heavy Stain
Frequency			
($\times 10^4$)	0*	12.1	0.59
	20	29.3	1.81
	40	71.0	5.52
Rate			
($\times 10^4$)	0*	0.28	0.017
	20	0.66	0.050
	40	1.60	0.158
% Increase at			
Any Passage		4.53	5.740

*Extrapolated from regression equations.

Discussion

We have attempted to measure the frequency of mutation during the senescence of human fibroblasts using a system which detects an altered phenotype in both viable and nonviable cells. The results indicate that at least some of the variants with altered G6PD staining are mutations. Not only is their frequency altered by mutagenesis, but the darkly stained phenotype has been shown to be inherited within clones. This does not demonstrate that all the stained cells actually observed and counted have heritable alterations. Further evidence on the genetic nature of these variants must await a technique to stain fibroblasts without killing them.

The heavily stained variants occur only in cells treated with both substrate and cofactor; if either is omitted there is no staining at all. The staining must therefore be due to the level of enzyme rather than to an independent reaction of the tetrazolium dye system. Moreover, the cells are entirely permeable when air-dried, so the variants cannot be connected with transport of metabolites into the cell. The variants are very similar in frequency and appearance both with the analogue substitute dG6P and low concentrations of the normal substrate. This shows that the variants have increased activity rather than reduced specificity of G6PD. This is confirmed by experiments (not reported here) in which monolayers were heat-inactivated before staining. This completely eliminates the background stain, but leaves a very few cells with detectable activity. These cells must have an increased level of normal enzyme, since enzyme with altered specificity would be expected to be more sensitive to heat, not less sensitive.

We know that the level of G6PD is regulated by its own structure, since the single amino acid substitution in the Hektoen variant results in a 4 fold increase in activity (Yoshida, 1970), but it would be surprising if more than a small fraction of the variants we see are due to mutation in the structural gene of the enzyme. It might be expected that the rate of passage of substrates along the pentose phosphate pathway would be under the genetic control of a number of loci and that G6PD synthesis might be sensitively governed because it is the rate controlling step of that pathway (Yoshida, 1973). Although we assigned variant cells to the heavily and very heavily stained categories, there appears to be a continuous spectrum of phenotypes. This indicates that there may be many direct or indirect effects of varying severity on the regulation of G6PD and the pentose phosphate pathway.

At the moment there is controversy about the nature of heritable variation in cultured mammalian cells: a number of authors maintain that the observed change in phenotype is due to "epigenetic"

changes rather than true gene mutation (for a critical review, see DeMars, 1974). We cannot distinguish between these alternatives, but the calculated
rate of origin of very heavily stained variants is not
unlike what might be expected for gene mutations
at one or a few loci (DeMars, 1974; Shapiro et al.,
1972).

The frequency of variants clearly increases
throughout the lifespan of MRC–5, but rises rapidly
during the senescent phase (phase III). Assuming
that a reasonable proportion of the variants are mutations, our results do not support the somatic mutation theory of fibroblasts aging, at least in its simplest form, since this predicts a linear increase
throughout. The observed incidence fits closely to
an exponential increase during most, if not all, of
the lifespan.

There have not been previous attempts to measure mutation frequency during clonal aging, with
the exception of experiments with the ciliate Paramecium. These results are in agreement with ours,
since the senescent phase is accompanied by a
rapid buildup of micronuclear mutations (Sonneborn and Schneller, 1960). There are, of course,
studies of chromosomes during the lifespan of fibroblasts. Again the results are consistent with
ours, since an increase in polyploidy, aneuploidy,
and chromosome abnormalities is seen only at the
end of the lifespan (Saksela and Moorhead, 1963;
Thompson and Holliday, 1975). Finally, Vogel
(1970) has reported that in the case of human germ
line mutations, the incidence of several dominants
increases more than linearly with age in males.

Orgel's (1963) theory of aging predicts that there
will be an exponential buildup in defects in proteins.
Some evidence has been obtained that fibroblast
aging is accompanied by the appearance of defective enzymes (Holliday and Tarrant, 1972; Lewis and
Tarrant, 1972; Holliday, Porterfield, and Gibbs,
1974; Goldstein and Moerman, 1975), but evidence
contrary to the theory has also been published (Holland, Kohne, and Doyle, 1973). It is now believed
that if protein errors do occur during ageing, there
will probably be a concomitant increase in mutations. Amino acid analogues have been shown to
be mutagenic (Lewis and Tarrant, 1971; Talmud
and Lewis, 1974), and the leu–5 strain of Neurospora, which contains a temperature-
sensitive leucyl tRNA synthetase, produces defective protein and is highly mutable at the restrictive temperature (Printz and Gross, 1967; Lewis and
Holliday, 1970). It is presumed that these effects on
mutation are due to the synthesis of faulty DNA
polymerase, since it is known that some genetically
altered forms of this enzyme generate a high frequency of mutants (Speyer, Karam, and Lenny,
1966; Bazill and Gross, 1973).

Our results are therefore in agreement with the
general error theory, which proposes that aging is
due to a breakdown in the fidelity of information
transfer between macromolecules (Medvedev,
1962; Holliday and Tarrant, 1972; Orgel, 1973). The
theory does not specify the relative importance of
protein errors or mutations in initiating a buildup
of errors, nor does it suggest which may be the
more important in finally killing the cells. With regard to DNA polymerase in fibroblasts, it has recently been shown that the rate of elongation of
DNA molecules (replicons) is reduced in senescent
cells (Petes et al., 1974), and the specific activity
of the main polymerase activity is greatly reduced
(S. Linn, unpublished results). Experiments are now
in progress to test the fidelity of in vitro replication
using DNA polymerase from cells of different age.

Experimental Procedures

Cell Culture
The cells are fibroblasts derived from lung tissue of a 14 week-old
male fetus. They have been characterized by Jacobs, Jones, and
Baillie (1970) and designated MRC–5. They divide at constant rate
until senescence, at which time growth slows down and they show
typical morphological degeneration. These cells normally live for
about 60 passages (Thompson and Holliday, 1973). The cells used
in these experiments were defrosted at passage 12 from stocks
stored in liquid nitrogen. They were grown at 37°C in 150 ml bow
bottles in Eagle's medium (GIBCO-BME Cat. No. G.13) containing
10% fetal calf serum (Tissue Culture Services Ltd.), 10% tryptose
phosphate broth (Difco-Bacto), 200 units of penicillin, and 200
μg/ml of streptomycin. The pH was adjusted with sodium bicarbonate to 7.2–7.3.

Cultures were trypsinized when confluent with Trypsin Versene
(trypsin at 1.25 g/l and versene at 0.1 g/l in phosphate buffered
saline) after washing with phosphate buffered saline (Dulbecco
PBS-A), and were split 1:4. The age of the culture was determined
by its passage number rather than by chronological time. Old cultures were split 1:2. Cultures were tested for mycoplasma contamination and found to be negative.

Filt–5 is a transformed cell line derived from a hamster fibrosarco-
ma. It was originally isolated by Dr. G. A. Girardi at the Wistar Institute, Philadelphia.

Mutagen Treatment
Cells were washed and cultured in medium containing 400 μg/ml
of EMS for 22 hr. Subsequently the cells were cultured in normal
medium for 7 days to allow expression of mutations.

Cloning
A sparsely confluent culture of young cells was carefully split and
the cell suspension examined under the microscope. Suspensions
containing pairs and clumps of cells were discarded. Cells were
distributed at a density of 40/ml in a cloning medium consisting
of F10 (Biocult) buffered with bicarbonate, 2 ml of nonessential
amino acids (Biocult) per 100 ml of medium, freshly added glutamine (292 μg/ml), and 17% of batch tested fetal calf serum (Biocult). Cells were placed in 50 mm Falcon plastic dishes with 4
ml medium in sandwich boxes in a humidified incubator gassed
with 4% CO_2. They were left undisturbed for 8 days. Plating efficiency was 20%. The colonies were then stained with deoxyglu-
cose-6–phosphate as described below.

Staining

Coverslips were washed in saline and left to drain at room temperature until dry. They were each placed in a 35 mm plastic dish (Falcon) together with a reaction mix containing freshly made up 2–deoxyglucose–6–phosphate or glucose–6–phosphate, phenazine methosulphate, and nitroblue tetrazolium (all from Sigma) in a Tris buffer (pH 8.0) containing triphosphopyridine nucleotide and magnesium chloride. The details of the method can be found in Wajntal and DeMars (1967). Dishes were incubated for 3 hr at 37°C with occasional shaking, and the coverslips dried and fixed. The reaction is completely substrate specific. Counting of stained cells is described in the legend to Figures 2 and 3.

Acknowledgments

We thank Mr. T.B.L. Kirkwood for his help in the quantitative analysis of the data.

Received October 18, 1974; revised June 19, 1975

References

Bazill, G. W., and Gross, J. D. (1973). Nature New Biol. *243*, 241.

DeMars, R. (1974). Mutation Res. *24*, 335.

Dern, R. J., McCurdy, P. R., and Yoshida, A. (1969). J. Lab. Clin. Med. *73*, 283.

Fulder, S. J. (1975). Ph.D. Thesis, Council for National Academic Awards.

Goldstein, S., and Moerman, E. J. (1975). Nature, in press.

Good, P. I. (1972). Tissue Cell Kinet. *5*, 319.

Hayflick, L. (1965). Exp. Cell Res. *37*, 614.

Holland, J. J., Kohne, D., and Doyle, M. V. (1973). Nature *245*, 316.

Holliday, R. (1971). Nature *232*, 233.

Holliday, R., and Tarrant, G. M. (1972). Nature *238*, 26.

Holliday, R., Porterfield, J. S., and Gibbs, D. D. (1974). Nature *248*, 762.

Jacobs, J. P., Jones, C. M., and Baillie, J. P. (1970). Nature *227*, 168.

Lewis, C. M., and Holliday, R. (1970). Nature *228*, 877.

Lewis, C. M., and Tarrant, G. M. (1971). Mutation Res. *12*, 349.

Lewis, C. M., and Tarrant, G. M. (1972). Nature *239*, 316

Medvedev, Zh.A. (1962). In Biological Aspects of Aging, N. W. Shock, ed. (New York: Columbia University Press), p. 255.

Nasim, A., and Auerbach, C. (1967). Mutation Res. *4*, 1.

Orgel, L. E. (1963). Proc. Nat. Acad. Sci. USA *49*, 517.

Orgel, L. E. (1973). Nature *243*, 441.

Petes, T. D., Farber, R. A., Tarrant, G. M., and Holliday, R. (1974). Nature *251*, 434.

Printz, D. B., and Gross, S. R. (1967). Genetics *55*, 451.

Saksela, E., and Moorhead, P. S. (1963). Proc. Nat. Acad. Sci. USA *50*, 390.

Shapiro, N. I., Khalizev, A. E., Luss, E. V., Marshak, M. I., Petrova, O. N., and Varshaver, N. B. (1972). Mutation Res. *15*, 203.

Sonneborn, T. M., and Schneller, M. (1960). In Biology of Aging, B. L. Strehler, ed. (Baltimore: Waverley Press), pp. 286–287.

Speyer, J. F., Karam, J. D., and Lenny, A. B. (1966). Cold Spring Harbor Symp. Quant. Biol. *31*, 693.

Sutton, H. G., and Karp, G. W. (1970). Genetics *64*, 63S.

Talmud, P. J., and Lewis, D. (1974). Genet. Res. *23*, 47.

Thompson, K. V. A., and Holliday, R. (1973). Exp. Cell Res. *80*, 354.

Thompson, K. V. A., and Holliday, R. (1975). Exp. Cell Res., in press.

Vogel, F. (1970). In Chemical Mutagenesis in Mammals and Man, F. Vogel and G. Röhrborn, eds. (Berlin-Heidelberg: Springer-Verlag), pp. 16–68.

Wajntal, A., and DeMars, R. (1967). Biochem. Genetics *1*, 61.

Witkin, E. M. (1951). Cold Spring Harbor Symp. Quant. Biol. *16*, 357.

Yoshida, A. (1970). J. Mol. Biol. *52*, 483.

Yoshida, A. (1973). Science. *179*, 532.

Yoshida, A., Beutler, E., and Motulsky, A. G. (1971). In Mendelian Inheritance in Man, V. A. McKusick, ed. (Baltimore: Johns Hopkins), pp. 565a-p.

25

Reprinted by permission from Nature **245**:316–319 (1973)

ANALYSIS OF VIRUS REPLICATION IN AGEING HUMAN FIBROBLAST CULTURES

John J. Holland, David Kohne, and Michael V. Doyle

University of California, San Diego

SINCE Swim and Parker[1] and Hayflick and Moorhead[2] documented the limited life span of diploid human fibroblasts in cell culture, many investigators have used these cells as possible models of cellular senescence. Most normal human fibroblasts exhibit marked cellular changes and eventually cease growth after fifty to seventy cell divisions. Hayflick[3] and Martin *et al.*[4] demonstrated an inverse relationship between the life span of cultured human diploid fibroblasts and the age of the donor. Furthermore, fibroblasts from children with premature ageing diseases have a markedly reduced potential for cell division[4]. Many investigators have studied human fibroblasts to determine whether their finite life span is related to ageing, and to determine the mechanism of 'senescence' in these cells[5]. There is no firm evidence, however, that fibroblast 'ageing' or limited fibroblast division potential is necessarily involved in the tissue and organ changes associated with ageing of individuals. Many types of cells which are obviously important in ageing of the whole organism undergo little or no division in adults (for example, neurones), and others which divide rapidly (for example, intestinal epithelium) have not been shown to behave as fibroblasts do, either in culture or in the organism.

Orgel[6] suggested that accumulating errors at the level of translation could eventually lead to an error catastrophe, and that such effects might be involved in ageing cells. Holliday and Tarrant[7] and Lewis and Tarrant[8] demonstrated that diploid human fibroblasts accumulate heat-labile enzymes during the final stages of their life span in culture. They showed that the RNA base analogue 5-fluorouracil induced premature senescence in these cells, and again altered enzyme was observed. Lewis and Tarrant[8] found that the ratio of active enzyme to cross-reacting material changed considerably as ageing cells approached death, and that the cells simultaneously lost the ability to discriminate between methionine and its analogue ethionine (the relative incorporation of ethionine increased markedly with senescence). These results are compatible with the error catastrophe theory, but as Lewis and Tarrant[8] point out, they do not conclusively support it since the accumulation of altered protein in ageing cells may not necessarily be a cause of ageing and death.

We have used virus infection to assess further the state of senescent human fibroblasts. The introduction of viral

Table 1 Virus Yields from Senescent and Control (Low Passage) WI38 Human Fibroblasts

Cell passage level	Cell No.	Total virus yield (PFU)	Virus yield per cell (PFU per cell)
Vesicular stomatis virus			
Passage 21 (control	1×10^6	1.6×10^{10}	16,000
Passage 59 (senescent)	1×10^6	2×10^{10}	20,000
Passage 24 (control)	2×10^5	3×10^9	15,000
Passage 51 (senescent)	2×10^5	2.5×10^9	12,500
Passage 32 (control)	1×10^6	2×10^{10}	20,000
Passage 60 (senescent)	1×10^6	3×10^{10}	30,000
Type 1 poliovirus			
Passage 23 (control)	1×10^6	1.8×10^9	1,800
Passage 61 (senescent)	1×10^6	2.6×10^9	2,600
Passage 22 (control)	1×10^6	3.9×10^9	2,000
Passage 59 (senescent)	1×10^6	2.4×10^9	2,400
Herpes simplex virus type 1			
Passage 23 (control)	1×10^7	5×10^{10}	5,000
Passage 60 (senescent)	1×10^7	5×10^{10}	5,000
Passage 32 (control)	1×10^6	2×10^9	2,000
Passage 61 (senescent)	1×10^6	1.6×10^9	1,600

High passage senescent cell cultures were from different stocks of the WI38 cell and some became senescent at later passage numbers than others. All senescent cells (above passage 50) used in these experiments were terminally senescent (cell division had ceased in nearly all cells). Control cells at low passages were rapidly replicating cultures. VSV plaque assays[13] were carried out on BHK_{21} hamster cells, poliovirus plaque assays on HeLa cells, and herpesvirus assays on a diploid human fibroblast cell strain established in this laboratory. In all of the above experiments but one, control and senescent cells were infected at a multiplicity of 50–100 PFU per cell. In the second VSV experiment (passages 24 and 51) a low multiplicity of infection was used (0.01 PFU per cell). In all cases virus yields were determined when cytopathology was maximal (all cells showed total cytopathology in control and senescent cultures by about 24 h following VSV and poliovirus infection and by 40 h following herpesvirus infection). Low multiplicity infection of senescent cells with herpesvirus gave very slowly progressing cytopathology and lower yields. The protein content per cell of control and senescent fibroblasts was approximately the same.

genes into ageing cells allows determination of their ability to replicate and translate viral genetic information independently of the mutational status of the cell DNA (except as cellular mutations may have affected the translational machinery). If the ageing fibroblasts are synthesizing a large percentage of abnormal proteins, then viral replication should be affected (due to altered viral replicases) and viral maturation should be defective since some altered proteins would be likely to interact in the assembly process involving hundreds or thousands of structural proteins per virion. Hayflick and Moorhead[2] showed that non-senescent diploid human fibroblasts are susceptible to various RNA and DNA viruses, and we report here the effect of senescence on susceptibility to three viruses.

Table 1 demonstrates that high passage senescent WI38 fibroblasts and low passage control cells were equally susceptible to poliovirus, to vesicular stomatitis virus (VSV) and to *Herpes simplex* virus type 1. Even though the senescent cells used were at the final possible passage before death, and all cells had ceased dividing, the virus yields per cell were equal to those in the non-ageing low passage control cell cultures. Viral cytopathology and death were complete in all cells in both high passage and low passage cultures. Table 2 shows that senescent cells produced virus of approximately the same specific infectivity as did low passage control cells. Senescent cells did not seem to be making an unusual number of inactive virus particles. This ability of ageing cells to support normal yields was surprising since these were terminal passage cultures in which all

Table 2 Specific Infectivity of Vesicular Stomatitis Virus Grown in Senescent Cells and in Control Low Passage Cells

Cells in which the virus pool was grown	Specific infectivity (PFU per mg protein)
Passage 23 (control WI38 cells)	2.1×10^{11}
Passage 61 (senescent WI38 cells)	1.9×10^{11}

Senescent and control cells were infected at a multiplicity of 1 and virus yields were collected and purified after 20 h at 37° C. Purification involved differential centrifugation, agarose gel chromatography and sucrose gradient rate zonal centrifugation. The amount of pure virus protein recovered was related to the total PFU determined from a small sample of the unpurified virus suspension, not from the purified virus which aggregates and loses considerable infectivity after the first concentration step. No correction was made for virus loss during purification, and it is assumed that purification losses were similar for senescent and control cells. Plaque assays were performed on BHK_{21} cells[13].

cells exhibited advanced signs of senescent degeneration (cytoplasmic granularity, highly branched cytoplasmic processes and so on).

Figure 1 compares the kinetics of production of VSV by senescent cells and low passage control cells. It can be seen that mature infectious virus was produced more rapidly by ageing fibroblasts than by rapidly growing control cultures. It is clear that the senescent cells are at least as efficient in replicating and assembling VSV as non-senescent controls.

It seemed possible that in spite of the efficiency with which infectious virus was produced, the virions from senescent cells might contain a significant proportion of abnormal structural proteins which would render them more thermolabile than virions produced by normal cells. Figure 2 demonstrates that this is not so. The infectivity of virions produced in senescent cells was thermally

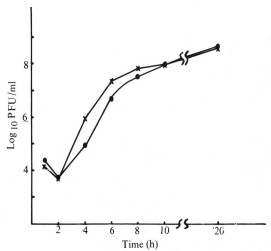

Fig. 1 Kinetics of production of infectious VSV by senescent cells and low passage control cells of WI38 strain diploid human fibroblasts. Senescent and control cells (2×10^5) were infected with Indiana serotype VSV at a multiplicity of 100, and 0.05 ml samples of the medium were withdrawn at indicated times following washing and incubation in 3 ml of fresh medium at 37° C. (All infectious virus was released into the medium since VSV matures by budding to the extracellular space from the plasma membrane.) ×, Senescent WI38 cells at terminal passage (passage 51); ●, control non-senescent WI38 cells at passage 24.

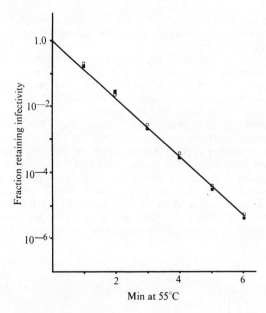

Fig. 2 Comparison of the kinetics of thermal inactivation at 55° C of VSV grown in senescent and low passage WI38 cells. Virus yields from both sources were collected 20 h after infection, adjusted to equal concentrations in Eagle's minimum essential medium containing 10% calf serum and heated in small samples for indicated times in this medium. □, Virus from terminal passage senescent (passage 51) WI38 cells; ■, virus from control non-senescent WI38 cells at passage 24.

inactivated at 55° C at a rate identical with the inactivation of virions produced by low passage cells. At 55° C, pseudo-first-order inactivation kinetics were obtained with virus from both non-ageing and senescent cells (Fig. 2). At 50° C, virus from low passage and from high passage cells again showed identical inactivation kinetics, but a complex multi-component curve was seen in both cases (data not presented). VSV is a complex membrane virus containing thousands of proteins some of which assemble into structures with helical and icosahedral or hexagonal symmetry[9]. If a significant proportion of abnormal viral structural proteins are synthesized in the senescent cells they must be excluded from the viral assembly process. Similar thermal stability was observed for poliovirus grown in senescent cells and control cells.

If generalised error catastrophe is involved in senescence of cells, errors in protein synthesis could lead to mutagenesis through synthesis of altered polymerases and the mutations could in turn lead to further errors in translation as outlined by Holliday and Tarrant. Therefore, we have compared the mutation rates of poliovirus grown in senescent cells and in low passage control cells. Poliovirus induces a replicase in infected cells[10] and, if the replicases synthesized in senescent cells include a significant percentage of abnormal molecules, the mutation rate might be altered. Growth of wild type poliovirus is inhibited by 10^{-3} M guanidine[11,12] and spontaneous mutation to guanidine resistance can be measured accurately.

Table 3 shows that the spontaneous rate of mutation to guanidine resistance is not significantly altered during replication in senescent cells. Virus yields from senescent cells exhibited the same low level of mutants for this character as were found in yields from low passage cells. It appears that highly mutagenic replicase molecules are not being synthesized at significant levels in infected senescent cells.

These results are not in accord with a situation in which generalised translation error is occurring, but they do not disprove error catastrophe in which translational errors are not generalised. For example, it is possible that errors might be restricted to certain unusual codons. Alternatively,

Table 3 Comparison of Mutation Rates (to Guanidine Resistance) of Poliovirus Grown in Senescent Cells and Control Low Passage Cells

Cells in which the polio-virus pool was grown *	Total virus yield (PFU)	Yield of guani-dine-resistant mutants (PFU)	Fraction of mutants in yield
Passage 23 WI38 (control)	1.8×10^9	8×10^4	4.4×10^{-5}
Passage 61 WI38 (terminal senescence)	2.6×10^9	7×10^4	2.7×10^{-5}

* Senescent or control WI38 fibroblast monolayer cultures were infected at a multiplicity of 50 by adsorption of type 1 poliovirus wild type to 1×10^6 cells for 30 min. Infected cells were washed thoroughly and incubated at 37° C in Eagle's minimum essential medium plus 10% calf serum. No guanidine was added. The virus yields were collected 18 h after infection, and the total yields from each culture were determined by plaque assay on HeLa cells under Eagle's medium containing 0.4% agarose. The yield of guanidine resistant mutants from each culture was determined by plaque assay on HeLa cells under Eagle's medium containing 0.4% agarose and 1 mM guanidine hydrochloride. Plaques were stained and counted after incubation at 37° C for 48 h (ref. 13).

viruses might have evolved so that they can replicate and mature efficiently in spite of translational error. Altered viral proteins might be relatively unstable or unable to compete with unaltered replicase molecules or with structural proteins in replication and assembly. The most obvious explanation for the difference between the present results and those of Holliday and Tarrant[7] and Lewis and Tarrant[8] is based on the possibility, which they pointed out, that the lowered specific activity of cellular enzymes might be due to accumulation of inactive enzyme in dead or dying cells. Since viral proteins are all newly synthesized in senescent cells, the presence of 'old' viral proteins was not a factor in the study reported here.

Further studies are needed before the error catastrophe theory can be proved or disproved. Obviously, careful quantitation of the synthesis, the specific activity and the turnover rates of a virus-induced enzyme could provide considerable insight into the synthetic capabilities of senescent fibroblasts. In all such investigations it must be remembered that fibroblast senescence in culture may or may not be relevant to the cellular changes undergone by fibroblasts and other cell types during ageing of the whole organism.

We thank Dr Leonard Hayflick for early and late passage WI38 strain diploid human lung fibroblasts, and Steven Tracy for some of his late passage cultures of the same cell strain. We also thank Drs Paul Price and Jonathan Gallant for helpful discussions. This work was supported by a grant from the National Cancer Institute.

[1] Swim, H. E., and Parker, R. F., *Am. J. Hyg.*, **66**, 235 (1957).
[2] Hayflick, L., and Moorhead, P. S., *Expl Cell Res.*, **25**, 585 (1961).
[3] Hayflick, L., *Expl Cell Res.*, **37**, 614 (1965).
[4] Martin, G., Sprague, C., and Epstein, C., *Lab. Invest.*, **23**, 86 (1970).
[5] Holečkova, E., and Cristofalo, V. J., eds., *Ageing in Cell and Tissue Culture* (Plenum, New York, 1970).
[6] Orgel, L. E., *Proc. natn. Acad. Sci. U.S.A.*, **49**, 517 (1963).
[7] Holliday, R., and Tarrant, G. M., *Nature*, **238**, 26 (1972).
[8] Lewis, C. M., and Tarrant, G. M., *Nature*, **239**, 316 (1972).
[9] Cartwright, B., Smale, C. J., Brown, F., and Hull, R., *J. Virol.*, **10**, 256 (1972).
[10] Baltimore, D., Eggars, H. J., Franklin, R. M., and Tamm, I., *Proc. natn. Acad. Sci. U.S.A.*, **49**, 843 (1963).
[11] Loddo, B., *Boll. Soc. ital. biol. Sper.*, **37**, 395 (1961).
[12] Crowther, D., and Melnick, J. L., *Virology*, **15**, 65 (1961).
[13] Holland, J. J., and McLaren, L. C., *J. Bact.*, **78**, 596 (1959).

26

THE LONGEVITY OF DIPLOID AND POLYPLOID
HUMAN FIBROBLASTS

Evidence Against the Somatic Mutation Theory of Cellular Ageing

K. V. A. THOMPSON and R. HOLLIDAY

SUMMARY

Diploid human foetal lung fibroblasts, strain MRC-5, were treated with colchicine for 3 or 6 h, and the surviving population was found to contain up to 60% polyploid (mainly tetraploid) cells. The lifespan of treated cultures was not significantly different from controls. Cultures of different passage level were treated with colchicine and the proportion of polyploids was measured at intervals throughout their subsequent serial subculture until senescence and death occurred. In all cases it was found that the proportion of polyploids remained constant until the end of the lifespan. This indicates that both the growth rate and the longevity of diploids and polyploids is very similar. The results are compatible with either the protein error or programme theory of ageing and provide evidence against any mutation theory which proposes that ageing is due to the accumulation of recessive genetic defects or mutations.

Since the discovery by Hayflick & Moorhead [5] that cultured human fibroblasts have a defined lifespan in vitro, many laboratories have used these cells as a model system for the study of the processes of ageing at the cellular level. One possibility, originally suggested by Hayflick [3], is that the cells eventually cease growth in culture because they accumulate genetic damage throughout their lifespan of 40–60 population doublings. If genetic defects are to accumulate over a long period of time, they must occur at fairly high frequency, more than one per cell division, and they must also be primarily recessive, since cells with dominant defects or mutations would be selected against. It follows that tetraploid cells should live longer than diploid ones, because they could accumulate more recessive defects before the symptoms of senescence are seen.

We have found that treatment of cultures of diploid fibroblast strain MRC-5 with colchicine gives rise to populations of viable cells with a high proportion of polyploids (mainly tetraploids), and we have used these populations to test the somatic mutation theory of ageing. Our results indicate that diploids and polyploids have the same, or similar, lifespans.

MATERIALS AND METHODS

Cells

The male fetal lung fibroblast strain MRC-5 was used in all experiments. The cells were obtained from J. P. Jacobs (National Institute for Biological Standards and Control, Hampstead, London), where the strain was originally isolated and characterized [10]. Two experi-

Table 1. *The induction of polyploidy with colchicine*

The culture of MRC-5 fibroblasts used was at passage 15

Initial split ratio	Colchicine treatment (hours)	Subsequent splits[a]	% Polyploid cells
1:4	3	1:4	44
1:4	6	1:4	46
1:4	3	3×1:4	51
1:4	6	3×1:4	58
1:16	3	1:1	55
1:16	6	1:1 and 1:4	63
1:32	3	M/C[b]	41
1:32	0 (control)	2×1:4	5

[a] For scoring the percentage polyploid cells, the survivors of the initial treatment must be subcultured (usually a 1:4 split) with a further colchicine arrest to accumulate cells in metaphase.

[b] The medium was changed after 8 days and the cells scored for polyploidy after a further 5 days without subculture.

ments were started, with cells at passage 15 and passage 23 respectively.

Growth conditions

Cells were grown as described in Thompson & Holliday [19] using Eagle's medium (Gibco-BME diploid [Earle's] cat. no. G-13) containing 10% fetal calf serum (FCS) (supplied by Gibco Biocult, Glasgow, Scotland), 10% tryptose-phosphate broth (Difco-Bacto) 200 U of penicillin and 200 μg/ml of streptomycin. The pH value was adjusted with sodium bicarbonate to 7.2–7.3. Cells were detached by trypsin/versene (1.25 mg/l, 0.1 g/l) and subcultured with a 1:2 or 1:4 split ratio. Medium containing colchicine (0.8 μg/ml of medium) was added 2 days after splitting for a 3 or 6 h period. Cells were then washed with phosphate-buffered saline (PBS) and incubation was continued with normal growth medium. Subculturing was continued in the usual way after cells reached confluency.

Cloning procedure

Ham's F12 Nutrient Mixture Medium (Gibco Biocult, Glasgow) was used, with the addition of 2% of a stock solution of non-essential amino acids (Gibco Biocult), 0.02% glutamine, 17% fetal calf serum (Gibco Biocult), 200 U of penicillin and 200 μg/ml of streptomycin. Cells were seeded for cloning in 10 cm Petri dishes at a density of 10^3 and 5×10^2/plate. Colonies were first isolated with a sterile glass ring (7 mm and 11 mm diameter) using sterile high vacuum silicon grease for firm adhesion to the plate. The cells were then detached by trypsin/versene and transferred to 30 mm plates. When confluent they were split at a 1:2 ratio into a 1 oz. glass bottle. Cells were transferred

similarly into a 2 oz. bottle, and eventually into 150 ml glass bow bottles. Lifespan studies on the cloned cells were continued in bow bottles, splitting at 1:2 or 1:4 ratios. Cell counts (Coulter Counter Model A) were made at each subculture.

Chromosome counts

Mid-logarithmic phase cells were arrested in division by a 3 h colchicine (0.8 μg/ml) treatment two days after the last splitting. Cells were fixed and stained according to the procedure by Moorhead & Nowell [16]. Chromosome preparations were made at roughly 10 passage intervals throughout the lifespan of the cultures. Slides were scanned under low power and numbers of diploids and polyploids were counted. Forty to 50 chromosomes were scored as diploid, over 50 as polyploids. The majority of these were in the tetraploid range (80–100). The frequency of metaphases with over 100 chromosomes was never higher than 2%. Threehundred to 700 metaphase cells were counted in each sample and the number of polyploids is expressed as a percentage of the total counted. The chromosome counts on cloned cultures were based on intact, not dispersed, metaphases.

RESULTS

Effect of colchicine treatment

In preliminary experiments cultures of MRC-5 lung fibroblasts at passage 15 were treated with colchicine. This treatment was given 2 days after the last splitting of the culture—before the cells became confluent. This is the time at which the culture can be expected to contain the maximum number of cells in division. After treatment and removal of colchicine, the non-dividing cells appeared shrunken and rounded off and there were small areas where the cells had detached from the glass surface. After 5–8 days incubation with normal growth medium, the cells had completely recovered and the culture bottles were confluent.

Slides for chromosome examination were then prepared using (*a*) cells immediately after they had reached confluence after treatment; and (*b*) cells which had undergone 1 or 2 further splittings after treatment. Bottles selected for chromosome preparations from (*a*) and (*b*) were split in a 1:2

Table 2. *Longevity of colchicine-treated cultures of MRC-5*

Colchicine treatment at passage	Longevity of parallel cultures (passages)	Average
26	66, 68, 70, 70	68.5
36	59, 61, 62, 65	61.7
46	64, 64, 66, 66	65
56	68, 70, 70, 70	69.5
65	66, 69, 69, 72	69
26 and 34	59, 61, 62, 65	61.7
26 and 44	64, 64, 66, 66	65
Control	71, 71, 72, 73	71.7

ratio and incubated for 48 h. Colchicine (0.8 μg/ml) was then added to arrest cells in metaphase and slides were prepared as described in Methods. Slides were also prepared from non-treated control cells. The results obtained are set out in table 1.

A high percentage of polyploidy was obtained in cells immediately after the colchicine treatment and this percentage remained constant even after 1–3 further cell doublings. This indicated that the cultures survived the colchicine treatment, that cells continued to divide and that subculturing had no effect on the frequency of polyploids. As there was no obvious difference in the percentage of polyploid cells ob-

served after 3 h or 6 h colchicine treatment, only the 3 h treatment was used in all subsequent experiments.

Lifespan of treated cultures

Experiments were carried out to determine (*a*) the longevity of cultures treated with colchicine in comparison with control cultures; (*b*) the effect of colchicine treatment at different passage levels; and (*c*) the percentage of polyploid cells throughout the remaining lifespan.

Starting with passage 23 cells, cultures were treated with colchicine for 3 h at passages 26, 36, 46, 56 or 65. Some of the cultures treated at passage 26 were also treated a second time at passage 34 or 44. Four parallel cultures were established for each treatment and their lifespans in passages measured. The results are shown in table 2. None of the treated cultures had a significantly reduced longevity; the overall slight decrease compared with the control is probably due to some cell loss after the colchicine treatments.

Slides for metaphase studies were prepared at intervals of about 10 passages from all treated cultures and the controls. The number of diploids and polyploids were

Table 3. *Induction of polyploidy in MRC-5 by colchicine*

Cells were treated at the indicated passage levels and then scored for polyploids at intervals until growth ceased

Control		Passage no. at time of treatment							
		26		36		46		56	
Passage	% Poly-ploids	Passage	% Poly-ploids	Passage	% Poly-ploids	Passage	% Poly-ploids	Passage	% Poly-ploids
34	3.4	26+ 4[a]	13.8	36+ 1	48.5	46+ 3	13.3	56+1	28.4
43	4.8	26+ 6	15.8	36+11	52.1	46+13	15.3		
51	7.1	26+15	17.8	36+19	51.9				
59	10.9	26+23	19.1						
		26+31	14.5						

[a] The overall age of the culture is given by the passage number at time of treatment plus the subsequent passages.

Table 4. *The numbers of chromosomes in populations (clones) derived from individual cells surviving an initial colchicine treatment*

Clone no.	Passage at time of cloning	Metaphases examined	Percentage of			Total lifespan[d]
			Diploids[a]	Subdiploid[b]	Polyploids[c]	
1	36	373	55.8	8.5	35.7	62
2	18	761	77.6	5.7	16.7	45
3	25	431	41.8	9.2	49.0	54
4	25	711	16.7	8.2	75.1	57
5	25	587	46.8	4.6	48.6	58
6	25	322	50.9	8.7	40.4	53

[a] Chromosome nos. 40–50.
[b] Chromosome nos. 15–40.
[c] Chromosome nos. 50–100, with <2% cells of higher ploidy.
[d] Population doublings calculated from the growth of one cell at the passage given to a confluent bow bottle plus the addition of subsequent subcultures.

counted using a ×40 objective. The results of the single colchicine treatments are shown in table 3. With the exception of the cells treated at passage 36, in these experiments the proportion of polyploids among the survivors of colchicine treatment was lower than that obtained in the experiments in table 1. Nevertheless, in every case the initial percentage of polyploidy was much higher than in the control culture. More important, it can be clearly seen that the proportion of polyploid cells remained constant throughout the subsequent lifespan. In the control the proportion of polyploid cells is 3–4% in mid-passage cultures but then rises to 10% in senescent cultures. This result is in agreement with that previously published [20].

Cloning

An attempt was made to clone some of the colchicine-treated cultures in order to obtain polyploid cell strains. Two cultures were used for cloning: (1) A culture was treated with colchicine at passage 26 and cloned at passage 36; and (2) a freshly defrosted young culture (passage 16) was treated at passage 17 and used for cloning

at passage 18 and passage 25. Out of 33 clones isolated, 6 could be subcultured by the 1:2 splitting procedure sufficiently long to enable us to make chromosome studies and also to calculate cell doublings of the individual clones. Clone 1 was isolated from culture (1) above, clone 2 from culture (2) at passage 18 and clones 3–6 from culture (2) at passage 25.

The results of the chromosome studies (table 4) show that none of the clones isolated contained only polyploids. This may indicate that the clones were not of single-cell origin. Fifty metaphases of each of the cloned cell strains were examined in detail under high power. Clone 2: Out of 50 metaphases analysed, 19 had endoreduplicated chromosomes. This was the only clone which showed this chromosome abnormality. Clone 4: This clone had the highest proportion of polyploids (75%) and also a fairly high proportion of subdiploids (8%). It is probable that this clone was tetraploid in origin, but that chromosome segregation resulted in the formation of diploid and subdiploid cells. Martin & Sprague [12] have shown that tetraploids of spontaneous origin can segregate cells with lower chromosome numbers, but we have no direct evidence for this. Clone 6: in 27 metaphases, breaks or other chromosome abnormalities were observed.

Population doublings of the clones are also given in table 4. The calculated cell divisions from a single cell to a confluent bow bottle are added to subsequent population doublings and the passage number at the time of cloning. The total lifespan of clones comes well within the range of normal MRC-5 diploid cultures [9, 19]. Cells of clone 2 had some of the morphological features of senescent cells, e.g. rather large cells with irregular growth pattern, and this may explain the somewhat shorter lifespan of this clone.

DISCUSSION

Our results show that blocking fibroblasts in metaphase of mitosis with colchicine leads to the formation of a high proportion of polyploids, which are capable of further proliferation. Most of these cells have chromosome numbers in the tetraploid range. Since the cell cycle of MRC-5 is about 18 h and the colchicine treatment was usually only 3 h, it would be expected that only about 16% of the cells would have entered metaphase. However, in many cases we recovered 50% of polyploids in the surviving population. This indicates that some cells which do not enter metaphase during treatment were sensitive to colchicine and did not survive. Our observations of treated cultures confirm that many cells do become detached from the glass surface following colchicine treatment.

Populations containing a high proportion of polyploid cells grew at the same rate as control cultures and the proportion of polyploids remained constant. These observations lead to two important conclusions. (1) Polyploid cells cannot be non-viable

since then the cultures containing 50% of such cells would grow at half the normal rate. (2) If the polyploid cells are growing they must be doing so at the same rate as diploid ones. If one or other type of cell grew faster, then the proportion of the two types would diverge during prolonged subculture, but this was not observed. We can therefore discount the possibility that colchicine treatment is inducing instability in diploid cells, which causes segregation of non-viable polyploids at high frequency over a long period of subculture. For this to be the case the degree of instability would have to be initially set at a given level and then inherited over many cell generations. Moreover, the rate of polyploid production would have to be exactly compensated by an increased growth rate of the parent diploids.

This is not to say that all the polyploid or diploid cells are completely stable. We know from previous studies that normal cultures of MRC-5 produce a few per cent polyploid cells [10, 20], and Martin & Sprague [12] have shown that tetraploid skin fibroblasts produce diploids or near diploids at a frequency of 4–17%. We hoped that our observations on clones would provide specific information about stability. However, groups of fibroblasts are more likely to grow than single cells and it is therefore possible that our "clones" were derived from more than one cell. Nevertheless, one clone contained 75% polyploid cells at least 20 population doublings after the colchicine-treated cells were plated at low density, which shows that these cells must retain a considerable degree of stability.

The most important result of our experiments is the demonstration that polyploid cells do not outgrow diploid ones during the senescent phase. Hoehn et al. [6] isolated tetraploid hybrids from human diploid skin fibroblasts and reported that these had longevities comparable to the parent cells. Both these observations provide strong evidence against the somatic mutation theory of cellular ageing. If mutations accumulate linearly with cumulative cell divisions, they clearly cannot be expressed during the long period of vigorous growth (phase II) which occurs before senescence. The mutations must therefore be mainly recessive. The theory rests on the supposition that the gradual increase in the genetic load of recessive defects leads eventually to the inactivation of at least one indispensable gene on both homologous chromosomes (for a full discussion, see Maynard-Smith [14]). Another version of the theory, proposed by Medvedev [15], is that some essential genes exist as multiple copies and that these are progressively inactivated until no functional copies remain. In either case, it would be predicted that tetraploid cells with twice the number of gene copies would be able to sustain much more damage than diploid ones and therefore have greater longevity. If deleterious mutations are dominant rather than recessive, they would be selected against and never accumulate over many divisions. They might conceivably have very slight deleterious effects and provided more than one such mutation occurred per cell division, they could accumulate until a final critical lethal level was reached. To explain the similar longevity of diploid and tetraploid cells, one would then have to suppose that the two-fold increase in mutation frequency per cell division of tetraploids over diploids was exactly compensated for by a two-fold decrease in the deleterious effect of each mutation, or the critical level, in the tetraploids. This seems an extremely unlikely possibility.

The observation that diploids and polyploids have much the same longevity is compatible either with the programme theory of fibroblast ageing [4, 7, 13], or with the protein error theory [8, 17]. The latter states that the initial events affect proteins in the cytoplasm, but later on as errors accumulate in the enzymic mechanisms of chromosome replication and repair, genetic damage would also be expected. Evidence that chromosomes, replicons, genes and DNA polymerase are all altered during the senescence of MRC-5 has previously been published [2, 11, 18, 20]. The error theory predicts that defects should accumulate exponentially and some results clearly indicate that this is the case [1]. A rapid build up of genetic damage at the end of the lifespan could be due in part to dominant mutations, and if this is so, then diploid and polyploid cells would be equally susceptible.

This research forms part of a collaborative programme with Professor K. Bayreuther, University of Hohenheim, Stuttgart, which is supported by the Fritz Thyssen Foundation, Cologne.

REFERENCES

1. Fulder, S J. Submitted for publication.
2. Fulder, S J & Holliday, R, Cell 6 (1975) 67.
3. Hayflick, L, Exp cell res 37 (1965) 614.
4. — Am j med sci 265 (1973) 433.
5. Hayflick, L & Moorhead, P S, Exp cell res 25 (1961) 585.
6. Hoehn, H, Bryant, E M, Johnston, P, Norwood, T H & Martin, G M, Nature 258 (1975) 609.
7. Holliday, R, Fed proc 34 (1975) 51.
8. Holliday, R & Tarrant, G M, Nature 238 (1972) 26.
9. Holliday, R, Huschtscha, L I, Tarrant, G M & Kirkwood, T B L, Science 198 (1977) 366.
10. Jacobs, J P, Jones, C M & Baillie, J P, Nature 227 (1970) 168.
11. Linn, S, Kairis, M & Holliday, R, Proc natl acad sci US 73 (1976) 2818.
12. Martin, G M & Sprague, C A, Science 166 (1969) 761.
13. Martin, G M, Sprague, C A, Norwood, T H & Pendergrass, W R, Am j path 74 (1974) 137.
14. Maynard-Smith, J, Proc roy soc Lond B 157 (1962) 115.

15. Medvedev, Z A, Exp gerontol 1 (1972) 227.
16. Moorhead, P S & Nowell, P C, Methods med res 10 (1964) 310.
17. Orgel, L E, Proc natl acad sci US 49 (1963) 517.
18. Petes, T D, Farber, R A, Tarrant, G M & Holliday, R, Nature 251 (1974) 434.
19. Thompson, K V A & Holliday, R, Exp cell res 80 (1973) 354.
20. — Ibid 96 (1975) 1.

Received July 7, 1977
Revised version received October 10, 1977
Accepted November 10, 1977

27

Reprinted from *J. Theor. Biol.* **93**:627-642 (1981)

Predictions of the Somatic Mutation and Mortalization Theories of Cellular Ageing are Contrary to Experimental Observations

R. HOLLIDAY

National Institute for Medical Research, The Ridgeway, Mill Hill, London NW7, U.K.

AND T. B. L. KIRKWOOD

National Institute for Biological Standards & Control, Holly Hill, Hampstead, London NW3, U.K.

(Received 23 March 1981, and in revised form 1 July 1981)

An accumulation of recessive lethal somatic mutations has often been proposed as a basis of cellular ageing. We have developed a mathematical model for the somatic mutation theory as applied to the finite *in vitro* lifespan of diploid fibroblast populations. Provided the mutation rate is sufficiently high, the model readily explains the cessation of proliferation of fibroblast cultures, but it predicts a much lower proportion of viable cells than is observed experimentally and also requires an unrealistically short cell division time. It is noted that the somatic mutation model is formally quite similar to the "mortalization" theory of Shall & Stein (1979), and that the mortalization theory is also incompatible with the same experimental data. We conclude that neither the somatic mutation theory nor the mortalization theory can explain the observed features of the growth of fibroblast populations *in vitro*. We discuss the possibility that deleterious mutations become important in the terminal stages of the lifespan, when they may accumulate as an indirect result of a general breakdown in information transfer between macromolecules.

1. Introduction

Human fibroblasts have a finite proliferative potential in culture. This was first fully documented by Hayflick (1965), who also suggested that the ageing of these cells might be due to the gradual accumulation of multiple mutations, or hits, in the genetic material. He did not develop this possibility quantitatively and although the somatic mutation theory is frequently referred to, so far as we are aware no adequate theoretical treatment with regard to fibroblast ageing has ever been published.

The accumulation of somatic mutations has, of course, frequently been proposed as the general cause of ageing of animals (Szilard, 1959; Curtis,

1966; Burnet, 1974). The theory gained support from the observation that ionizing radiation, a well known mutagen, is highly effective in shortening the lifespan of experimental animals. However, there are a number of observations which are hard to reconcile with the view that ageing is primarily due to the accumulation of deleterious mutations.

The first serious treatment of the theory was made by Szilard (1959). He considered ageing in diploid organisms and assumed the unit event in the ageing process to be a mutational "hit" on a genetic target rendering it incapable of performing its function. He also assumed hits were recessive, so that cells continued to function effectively unless both members of at least one pair of targets was hit, or unless one target was hit and the other already carried an inherited lethal fault. As an organism aged, the level of hits would increase, leading to progressive loss of cell viability, particularly among fixed post-mitotic cells like neurones, where there is no scope for cellular selection to replace sick cells by healthy ones. By making various assumptions, for example about the maximum proportion of non-viable cells that the organism could tolerate, Szilard was able to calculate the hit rate and mean level of inherited faults as functions of the number of targets in each cell, and to match theoretical against observed distributions of lifespan. For man, it turned out that the mean load of inherited faults would have to be impossibly high unless the number of targets was small, and Szilard was forced to assume that the targets were either whole chromosomes or large parts of them. This assumption, which may be reasonable in the context of radiation-induced damage, is rather extreme in the more general context of ageing.

We have constructed a modified version of Szilard's theory to see whether it is compatible with the observed ageing of fibroblast cultures [see Kirkwood & Holliday (1980) for a preliminary account]. We find that the theory predicts that fibroblast populations would die out at the observed population doubling level only if the mutation rate was at least two orders of magnitude higher than published estimates (see De Mars, 1974; Gupta, 1980; Cavalli-Sforza & Bodmer, 1971) and also that a high proportion of cells would be non-viable even quite early in culture lifespan. However, experimental observations show that the majority of fibroblasts are capable of cell division until quite late in culture lifespans.

Shall & Stein (1979) have developed a "mortalization" theory of fibroblast ageing. They suggest that during sequential cell division there is an increasing probability that non-viable (or non-dividing) cells will be produced, until growth ceases completely. Although their hypothesis is based on quite different assumptions to the somatic mutation theory, it also predicts that there should be a very high proportion of non-cycling cells long before the cultures finally die out. For this reason we think it very unlikely to provide a realistic description of the way these cultures age.

2. Somatic Mutation Model

To make quantitative predictions for cultures of human fibroblasts, we have formulated a mathematical model of the somatic mutation theory. We assume: (i) the targets for mutational hits are single genes, and there is a particular set of N genes, each of which is indispensible for the continued viability of the cell (we suppose N is in the range 10^3–10^4), (ii) 95% of these vital genes are on the autosomes, so they exist in duplicate copies, while the remaining 5% are on the X-chromosome and exist only as single active copies (in males there is only one X-chromosome, while in females one X-chromosome in each cell is totally inactivated), (iii) autosomal hits are recessive, (iv) the hit rate (h) per target per cell generation is constant in all cells in the culture and at all times, (v) all viable cells divide at the same rate. We can then calculate the probability that a cell is rendered non-viable by a lethal hit in any one cell generation.

A lethal hit can occur in two ways: either a vital gene on the X-chromosome is hit (with probability P_x), or a vital autosomal gene can be hit for a second time (P_d). The total probability of a cell being rendered non-viable is $P_x + (1 - P_x)P_d$, while the probability of a cell remaining viable is $(1 - P_x)(1 - P_d)$. P_x remains constant over time, while P_d increases steadily as the load of recessive single hits accumulates. Eventually, P_d reaches such a high level that the culture dies out.

To calculate P_d, let M be the average number of vital autosomal genes which have received a single hit after any given number of cell generations. The number of new hits on the entire genome during the next generation is $m = 1·95Nh$ where h is the hit rate per target per generation, since 95% of the genome is duplicated in the autosomes. Then, to a close approximation, $P_d = 1 - (1 - M/1·9N)^m$, since the probability of not getting a double hit $(1 - P_d)$ is the probability that none of the m fresh hits is on any of the M vulnerable second copies of the $0·95N$ vital autosomal genes. (The approximation arises since M increases slightly during the generation as new genes are hit for the first time: this effect is sufficiently small to be negligible for our present purpose.) P_x is simply $1 - (1·9/1·95)^m$. In the next generation, those cells which survive have M increased by an amount $m(1·9N - 2M)/(1·9N - M)$, the average number of fresh single hits; thus P_d increases towards its limiting value $1 - (0·5)^m$. As soon as P_d exceeds $(0·5 - P_x)/(1 - P_x)$, which must happen eventually for $m > 1$, the cell population ceases to grow, since less than half the progeny from any one cell division remain viable, and the culture becomes extinct.

TABLE 1

Predicted cell population lifespan (population doublings) for various values of m ($=1 \cdot 95Nh$) and N; h is the hit rate per target per generation, N is the number of vital genes, and m is the average number of new hits per generation. For these predictions, cells are assumed initially to be free of any mutations among the N indispensible genes (see text)

| | Population lifespan | |
m	$N = 10^3$	$N = 10^4$
2	150	1448
4	35	305
6	16	116
8	9	56
10	6	31

It is readily seen, therefore, that the model can explain the finite lifespan of fibroblast cultures, and Table 1 lists the predicted culture lifespan, in population doublings, for various values of *m*, the number of new hits per generation. [The lifespans in Table 1 were calculated for cell populations which were assumed to carry no initial faults among the *N* indispensible genes. Some recessive mutations are likely, however, to have been carried through the germ line, and these would shorten the lifespan. Also, in comparing theoretical lifespans with experimental ones some allowance must be made for the number of cell generations which will have taken place *in vivo* and in primary growth *in vitro* prior to setting up the cell cultures. We have discussed this elsewhere and estimated that for MRC-5 fibroblasts approximately 25 cell generations have occurred prior to passage 1 (Holliday *et al.*, 1977). This figure is based on an estimate of approximately 10 population doublings *in vitro* (10^3–10^4 cells from the tissue explant growing up to 2×10^6 at passage 1) and 15 population doublings *in vivo*. Considerable uncertainty is, of course, attached to the latter figure, especially if the number of divisions of fibroblast precursor cells is taken into account. However, the precise number of cell generations prior to passage 1 is not critical to our conclusions, and it does not matter greatly if this number were, say, between 10 and 40. We regard it as somewhat more likely that 25 cell generations is an underestimate than an over-estimate.]

To be compatible with the experimentally determined lifespan of human fibroblasts (50–60 population doublings for foetal cells), the value of h must be constrained to a narrow range. For $N = 10^4$, we require $h = 3$–4×10^{-4} hits per target per generation, and for $N = 10^3$ we require $h = 1$–2×10^{-3}. Either of these figures is considerably higher than seems reasonable on the basis of current experimental evidence. De Mars (1974) studied the incidence in human fibroblast cultures of variants resistant to 8-azaguanine and, among 29 out of 30 such cultures, observed frequencies in the range $0 \cdot 5$–37×10^{-6}. Fluctuation tests on four cultures gave estimates of the mutation rate as $0 \cdot 45$–$1 \cdot 8 \times 10^{-6}$ per cell generation. In other studies, Gupta (1980) estimated the frequency of mutants resistant to diphtheria toxin and thioguanine in young fibroblast cultures to be approximately 2×10^{-6}; the mutation rate would presumably be lower. [Estimates for germ-line mutation rates (Cavalli-Sforza & Bodmer, 1971; De Mars, 1974) are similarly low, or lower, especially if they are calculated as rates per cell generation. However, there may be differences between mutation rates in germ cells and those in somatic cells.] Even if we assume an improbably high estimate of $N = 10^5$ indispensible genes, we require a mutation rate of approximately 5×10^{-5} per gene per cell generation, which is still too large.

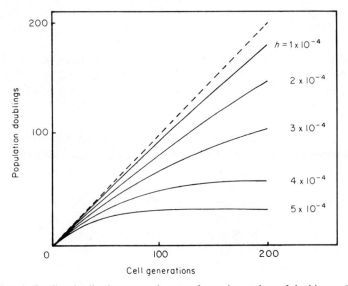

FIG. 1. Predicted cell culture growth curves for various values of the hit rate (h) per target per generation, assuming $N = 10^4$ vital genes. The broken line represents one population doubling per cell generation, as would be observed if all cells in the culture were viable.

The predicted growth curves of cell populations with $N = 10^4$ are shown in Fig. 1. An obvious feature of the curves for $h = 3 \times 10^{-4}$ and $h = 4 \times 10^{-4}$ is that, after the first few cell generations (which would have occurred *in vivo*), the growth rate per cell generation is quite low. This has two implications. Firstly, several cell generations must occur within one passage (growth of the cells through one population doubling). Secondly, a high proportion of cells must be non-viable. Even among an initial population of cells which carry no faults, the theory predicts a significant proportion of non-viable cells due to hits on the X-chromosome, and this proportion increases as the cultures grow and autosomal hits accumulate (see Fig. 2).

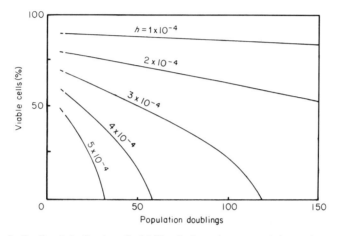

FIG. 2. Predicted decline in cell viability during culture growth for various values of the hit rate (h) per target per generation, assuming $N = 10^4$ vital genes. N.B. The curves do not extrapolate back to an initial viability of 100% because, even among cells which have accumulated no mutations, a fraction will be rendered non-viable by hits on the X-chromosome (see text).

Assuming, as described above, a total of about 25 cell generations *in vivo* and in primary growth *in vitro* before the cultures are established, the model predicts that for $N = 10^4$ and $h = 3-4 \times 10^{-4}$ cell viability at passage 1 will already be as low as 50–60%. (For 10 or 40 preliminary cell generations, this prediction is altered to 60–70% or 30–50% viability at passage 1.)

In deriving these specific predictions of the theory, it has been assumed that non-viable cells are passively transmitted during sub-culturing. This seems reasonable, since the majority of mutations which prevent a cell from undergoing further mitosis will not necessarily cause it to disintegrate immediately or prevent it from reattaching to the wall of the culture vessel.

The effect of altering this and other assumptions of our model will be discussed later.

3. Shall–Stein Mortalization Theory

A theory proposed by Shall & Stein (1979) to explain the finite proliferation of fibroblast populations makes predictions which are formally quite similar to those of the somatic mutation model. In Shall & Stein's terminology, each newborn cell in the culture has a certain probability, P_m, of "mortalization", or, in other words, of not dividing again. It is fairly obvious that if P_m increases in a time-dependent or generation-dependent manner from zero to unity, the population must eventually cease to proliferate. Thus, in its most general sense, the mortalization concept embraces a wide range of more specific models which postulate particular forms for the functional dependence of P_m on generation number or time. One special case is

$$P_m = P_x + (1 - P_x)P_d,$$

where P_x and P_d are as defined for the somatic mutation model in the previous section.

Shall & Stein proposed, however, that

$$P_m = \frac{t}{\gamma + t},$$

where t is the generation number and γ is a constant. Clearly, P_m defined in this way has the required property of increasing from zero to approach unity asymptotically as t becomes larger. For any particular value of γ, the number of population doublings that a culture may attain is strictly finite. Shall & Stein suggested a molecular mechanism for mortalization consistent with this formulation (based on control of initiation of DNA synthesis by "division" and "mortalization" proteins), but the molecular details need not concern us here.

The value of γ selected by Shall & Stein as appropriate for describing the growth of human fibroblast cultures was 160. The predicted growth curve and the decline in the percentage of viable cells for $\gamma = 160$ are shown in Fig. 3 and compared with those generated by the somatic mutation model for a culture with the same lifespan. As may be seen, the predictions of the two models are very similar; the main difference is that cell viability in the somatic mutation model is initially less than 100% because of hits on the X-chromosome.

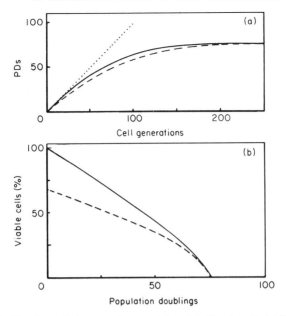

FIG. 3. Predicted population growth curve (a) and decline in cell viability (b) for Shall & Stein's (1979) mortalization theory (solid lines), and comparison with somatic mutation theory (broken lines). For the mortalization theory it is assumed that $\gamma = 160$, and for the somatic mutation theory $m = 7\cdot 1$ and $N = 10^4$ (see text for definitions of γ, m and N).

4. Experimental Observations

The standard procedure for measuring the proportion of cycling cells is to label nuclei with ^{3}H-thymidine for at least a 48 hr period of population growth, and then prepare autoradiographs. Any cell entering or completing S phase will incorporate label into DNA, whereas those that are blocked in cell division incorporate no label, or only a very small amount if there is any repair synthesis (unscheduled DNA synthesis). There have been several studies in which cells have been examined at intervals throughout their lifespan. The early experiments of Cristofalo & Sharf (1973), which are often cited, suggested that the proportion of labelled nuclei declined gradually throughout Hayflick's "phase II" of fibroblast growth, and then more rapidly as the cells entered the final senescent "phase III" of growth. This observation was challenged by Macieira-Coelho (1974, 1977), who pointed out that the reduction in the proportion of labelled nuclei could be due mainly to the slowing down of the division cycle and therefore an increase in the number of cells which do not enter S phase during the period of incubation in ^{3}H-thymidine. His claim that the majority of cells are dividing even very late in culture lifespans has been largely substantiated by several other studies (Holliday, Huschtscha & Kirkwood, 1981; Vincent & Huang, 1976; Mitsui & Schneider, 1976). In Table 2 we list the predicted frequencies of non-cycling cells for the two theories at four points during the lifespan and the observed frequencies from three separate studies.

TABLE 2

Comparison of expected and observed proportions of viable (cycling) cells at various points in culture lifespans. Expected viabilities are shown for the somatic mutation model and the mortalization theory. Observed viabilities are derived from three independent experiments using diploid human fibro-blasts and ^{3}H-thymidine labelling

Percentage of population doublings completed	Percentage of viable cells				
	Expected†		Observed‡		
	Somatic mutation	Mortalization	C & S	V & H	H, H & K
25	44	55	95	95	95
50	34	41	70	95	95
75	22	25	50	95	80
95	8	8	10	25	~55

† For calculation of expected proportions, it is assumed that 25 population doublings have occurred *in vivo* and in primary growth *in vitro* prior to the start of the culture lifespan (see text). Parameters for the models are as in Fig. 3, giving culture lifespans of approximately 60 population doublings.

‡ C & S Cristofalo & Sharf (1973), using W1-38 foetal lung fibroblasts; V & H: Vincent & Huang (1976), using W1-38 foetal lung fibroblasts; H, H & K: Holliday, Huschtscha & Kirkwood (1981), using MG4 foetal lung fibroblasts.

Another approach is to measure the ability of isolated cells of different age to divide once or more to form clones. This assumes that isolated cells attached to a surface are as likely to divide as those seeded at fairly high densities, which in many laboratories is certainly not the case. Nevertheless, Merz & Ross (1969) showed, by observation of WI-38 fibroblasts seeded at low cell density, that more than 95% of cells were capable of division early in the culture lifespan (passage 8–15) and that the proportion of non-dividers did not exceed 25% until about passage 35. In a major series of experiments, Smith and associates (Smith & Hayflick, 1974; Smith, Pereira-Smith & Good, 1977; Smith, Pereira-Smith & Schneider, 1978; Smith & Whitney, 1980) employed a technique which yielded a very high proportion of dividing cells, even at late passage. For example, Smith & Whitney (1980) found that 95% of cells divided at least once when 59%

of the culture lifespan was completed and 90% divided at least once when 89% of the lifespan was completed.

It can also be deduced from growth experiments that most of the cells in early or mid-passage populations must be capable of division. In the routine subculturing of these cells (usually by a $\frac{1}{4}$ or $\frac{1}{8}$ split), there is a lag after trypsinisation during which the cells attach to the surface of the container and spread out to form their characteristic fibroblast morphology. Most of these cells are in G_1, so they must also complete S phase and mitosis before the numbers increase. The lag before multiplication begins is about 20 hr for MRC-5. Thereafter, the cells grow logarithmically with a doubling time of about 18 hr. As they approach confluence, the rate of division slows down owing to "contact inhibition". If the cells are subcultured as soon as they become confluent the overall rate of population doubling is about 35 hr for MRC-5. This is illustrated in Fig. 4. The line C–D is the rate of increase during exponential growth and the line A–B is the rate of increase if the lag and confluence stages are also included. From the observation of single cells, mainly from time lapse photography, it is known that healthy foetal lung fibroblasts divide on average about every 18 hr (Absher, Absher & Barnes, 1974). Thus, both the slopes A–B

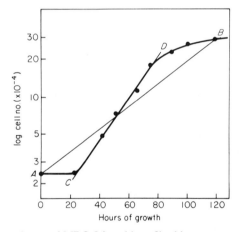

FIG. 4. The growth rate of MRC-5 foetal lung fibroblasts, passage 18. The doubling time during exponential growth, C–D, is 18·5 hr, and the doubling time from seeding to near confluency, A–B, is 34 hr. Cells were suspended in F15 medium, with 10% foetal calf serum, 100 units/ml penicillin, 100 μg/ml streptomycin and a glucose supplement of 1 mg/ml. 1 ml was added to replicate Linbro wells (1·75 cm^2) and cell counts were made as follows: the medium was removed, cells were rinsed with phosphate buffered saline and detached with 200μl typsin-versene (see Thompson & Holliday, 1978). A 100 μl Eppendorff pipettor was used to disperse the cells (20–30 strokes) and 100 μl of the suspension was counted with a Coulter counter. Each point is the average of three well counts.

and C–D are only explicable if almost all cells enter division, which is in complete agreement with the autoradiographic studies. If, for example, 50% of the cells were not cycling, then the remaining 50% would have to divide on average every 9 hr to account for the rates of population growth. It is impossible for fibroblasts to divide as quickly as this.

We conclude that the results from three distinct types of experiment make it very unlikely that either the somatic mutation or the mortalization theories of cellular ageing can be correct.

Other experiments also argue strongly against the somatic mutation theory of fibroblast ageing. Maynard-Smith (1962), in discussing the theory in relation to animal ageing, pointed out that increasing ploidy should have a drastic effect in increasing lifespan, since deleterious recessive mutations would be expressed much less frequently. The same argument applies to diploid and tetraploid fibroblasts, since the latter would retain at least one intact copy of each essential gene for a much longer period than the former. Two studies show that tetraploid or polyploid cells have the same longevity as diploid ones. Hoehn and associates (Hoehn et al., 1975; Hoehn, Bryant & Martin, 1978) fused diploids and recovered individual tetraploid cells. These were grown to the end of their lifespan. Their longevity was similar to the parent cells, or in the case of early × late passage hybrids was roughly intermediate between the parents. Thompson & Holliday (1978) discovered that up to 50% of diploid MRC-5 cells could be converted to a polyploid (mainly tetraploid) condition by colchicine treatment. The proportion of each type of cell remained constant during serial sub-culture until senescence, showing that their growth rates were the same and that the polyploid cells did not have an increased lifespan. Another approach is to treat fibroblasts with known mutagens. Gupta (1980) used either N-methyl-N′-nitro-nitrosoguanidine (MNNG) or ethyl methane sulphonate (EMS) and found that many sequential doses did not reduce lifespan; indeed, in one experiment with EMS, longevity was increased. All these observations provide further strong evidence against the somatic mutation theory of fibroblast ageing.

5. Mutations and Cell Death

In formulating the somatic mutation theory of fibroblast ageing, we assumed that mutations in essential genes are recessive. It is, of course, possible that they might be dominant or co-dominant. However, by definition, such deleterious mutations would have a direct effect on the phenotype, either slowing down cell division or preventing it entirely. In either case, the mutations would be continuously selected against in the population and mutation-free cells would always have the advantage. It is therefore very difficult to see how such mutations could accumulate to a point where the whole population died out. Fulder (1979) has suggested that fibroblast ageing might be due to a "mutation catastrophe". He supposes that individual recessive mutations would have no phenotypic effect, but as the numbers increased, there would be interactions between them leading to a progressively deleterious phenotype and finally cell death. This is a formal possibility, but there seems to be little, if any, evidence that such interactions between different recessive mutations are at all common in diploid organisms. The theory also requires a high mutation rate if multiple hits are to accumulate in every cell and therefore runs into the same difficulty as the somatic mutation theory we developed, namely, the mutations would be expressed on the X-chromosome and result in a high proportion of non-viable cells.

Two attempts have been made to measure the actual frequency of mutations during growth of fibroblasts in culture. The conventional approach is to screen populations for mutant cells which are resistant to and can form colonies in the presence of an inhibitor, such as 6-thioguanine (6-TG) or 8-azaguanine. Unfortunately, this measurement of mutation frequency cannot be applied to late passage cells, since these do not have sufficient growth potential to form colonies. Gupta (1980) measured the frequency of mutants resistant to 6-TG or diphtheria toxin and found it increased by 2–4 fold from 10 mean population doublings (MPD) to 75 MPD (total longevity was ~95 MPD). 6-TG resistance is expressed in diploids (since the structural gene is on the X-chromosome) and it is clear from his results, as well as from earlier data by De Mars (1974), that the mutation frequencies are in the range expected and very much less than the rates in Table 1. Fulder & Holliday (1975) developed a quite different method based on a procedure developed by Wajntal & De Mars (1967), in which the phenotype of individual cells is screened. This made it possible to use populations throughout the whole lifespan. The variants which were screened had an enhanced level of glucose-6-phosphate dehydrogenase (G6PD), which is detected histochemically and evidence was obtained that they were indeed the result of mutations, which probably affect the regulation of the enzyme. Again, there was a slow increase during early and mid passage cells, but in late passage there was a dramatic increase in variant cells. The results are consistent with an exponential increase in mutations, especially if a correction is made to convert cumulative population doublings into cell generations. The results are also completely consistent with those of Gupta (1980), if the last 20% of the *in vitro* lifespan is excluded, since it is only during this period of growth that mutations accumulate rapidly.

An exponential-like increase in mutations is predicted by the error theory of ageing (Orgel, 1973; Lewis & Holliday, 1971; Holliday & Tarrant, 1972). The theory proposes that cells are in a potentially unstable state, since they may propagate errors in protein synthesis exponentially, until a lethal level is reached [for a review of the theory of error propagation, see Kirkwood (1980)]. The formation of altered molecules of DNA polymerase, and other proteins required for accurate chromosome replication, would be expected to result in mutations. Evidence has been obtained that the fidelity of DNA polymerase α from senescent MRC-5 fibroblasts is less than that from young ones (Linn, Kairis & Holliday, 1976). The result was confirmed and further documented by Murray & Holliday (1981). They showed that the misincorporation of nucleotides by the enzyme, using several template primer systems, was closer to an exponential than a linear increase throughout the lifespan of MRC-5.

In fibroblast ageing it is reasonable to separate the intrinsic changes which must be occurring during the long period of normal growth, from the final degenerative changes which are eventually lethal. These degenerative changes could well be the result of mutations which are accumulating rapidly towards the very end of the *in vitro* lifespan. In this connection, it is interesting to note that Smith & Whitney (1980) detected a progressively larger population of cells during ageing which are capable of only very few divisions before the clone died. If mutation inactivates an essential gene and the protein gene product is relatively stable, there may be sufficient available for a few cell divisions before the clone ceases growth.

6. Conclusions

Several kinds of experimental evidence argue against somatic mutations being the primary cause of the cessation of growth of fibroblast populations. This adds to the evidence already put forward against the somatic mutation theory of ageing in whole organisms (Maynard-Smith, 1959, 1962; Clark & Rubin, 1961; Lamb & Maynard-Smith, 1964; Lamb, 1965). Nevertheless, there is some evidence linking increased somatic mutations with ageing, and it may well be that they are an important component of the general physiological deterioration and progressive loss of homeostasis that occurs in old age (see also Burnet, 1974). Such an effect is predicted by the error theory of ageing (Orgel, 1963, 1973) in which somatic mutations arise at increasing frequency as a secondary consequence of a feedback of errors in macromolecular information transfer.

A number of assumptions of our somatic mutation model may be questioned. For example, we have assumed from DNA values that 5% of vital genes are on the X-chromosome. Modifying this proportion will alter P_x and P_d correspondingly and will therefore introduce some changes in our specific quantitative predictions. Nevertheless, the general nature of our predictions will not be significantly affected, and the contradiction between theory and experiment cannot be resolved in this way. In fact, this difficulty applies quite generally to any mortalization-type theory in which P_m increases in a fairly gradual way. Brief consideration of the kinetics of growth of a hypothetical cell population shows that, if a non-proliferating state is to be reached in an acceptably short period of time, the proportion of non-viable cells must become appreciable relatively early in the culture lifespan. The only circumstance under which this difficulty may be avoided is if P_m is initially very small and increases only slowly for many cell generations, until finally it increases quickly and suddenly. This, however, approaches the supposition that each cell simply undergoes a set number of divisions and then ceases to divide. Such a hypothesis is at odds with much of what is known about the growth of human fibroblasts *in vitro* (see, for example, Holliday *et al.*, 1977).

The immediate fate of cells which are rendered non-viable, either by lethal mutation or by "mortalization" is, of course, unknown (see also Holliday *et al.*, 1981). In all of our calculations we have made the reasonable assumption that non-viable cells are transmitted passively through subsequent subcultures. (If this is so, it should be noted that the non-viable cells will in any case be carried through only a few subcultures before they are randomly discarded.) However, we should briefly note what happens if we adopt the other extreme and assume that non-viable cells are immediately lost. In this case, the observable proportion of viable cells will be increased, perhaps to as much as 100%. However, to assume that non-viable cells are very rapidly eliminated only increases the difficulty of reconciling cell cycle and the culture doubling time, since cell multiplication would need to be even faster to achieve the observed population growth rate.

Finally, we should point out that the only feature of fibroblast growth which is satisfactorily explained by the somatic mutation and mortalization theories is possession of a finite proliferative lifespan. However, there are several other important features of fibroblast populations *in vitro*, and to explain these we previously proposed the commitment theory of cellular ageing and subsequently tested its predictions with positive results (Kirkwood & Holliday, 1975; Holliday *et al.*, 1977, 1981). Although this theory may need modification to explain recent results with fibroblast clones (Smith & Whitney, 1980; Prothero & Gallant, 1981), its general form seems to offer a more satisfactory explanation of current data than either of those considered in this paper. Moreover, the commitment theory is fully compatible with the error theory of ageing (see Kirkwood, 1977), since it proposes that there are uncommitted cells, with the ability to grow indefinitely, which become committed with a given probability during division to finite growth. The theory of error propagation shows that cells can be in a steady state, with a given probability at every cell generation of moving into an unstable condition with an irreversible increase in errors (Orgel, 1973; Kirkwood, 1980). It may be that uncommitted cells have a steady state level of errors and that committed ones are on the downward path to a lethal error catastrophe.

REFERENCES

ABSHER, P. M., ABSHER, R. G. & BARNES, W. D. (1974). *Exp. Cell Res.* **88,** 95.

BURNET, F. M. (1974). *Intrinsic Mutagenesis: A Genetic Approach to Ageing.* New York: J. Wiley.

CAVALLI-SFORZA, L. L. & BODMER, W. F. (1971). *The Genetics of Human Populations.* San Francisco: W. H. Freeman.

CLARK, A. M. & RUBIN, M. A. (1961). *Radiat. Res.* **15,** 244.

CRISTOFALO, V. J. & SHARF, B. B. (1973). *Exp. Cell Res.* **76,** 419.

CURTIS, H. J. (1966). *Biological Mechanisms of Ageing.* Springfield, Illinois: C. C. Thomas.

DE MARS, R. (1974). *Mut. Res.* **24,** 335.

FULDER, S. J. (1979). *Mech. Ageing Dev.* **10,** 101.

FULDER, S. J. & HOLLIDAY, R. (1975). *Cell* **6,** 67.

GUPTA, R. S. (1980). *J. Cell Physiol.* **103,** 209.

HAYFLICK, L. (1965). *Exp. Cell Res.* **37,** 614.

HOEHN, H., BRYANT, E. M., JOHNSTON, P., NORWOOD, T. H. & MARTIN, G. M. (1975). *Nature, Lond.* **258,** 608.

HOEHN, H., BRYANT, E. M. & MARTIN, G. M. (1978). *Cytogenet. Cell Genet.* **21,** 282.

HOLLIDAY, R. & TARRANT, G. M. (1972). *Nature, (Lond.)* **238,** 26.

HOLLIDAY, R., HUSCHTSCHA, L. I., TARRANT, G. M. & KIRKWOOD, T. B. L. (1977). *Science, N.Y.* **198,** 366.

HOLLIDAY, R., HUSCHTSCHA, L. I. & KIRKWOOD, T. B. L. (1981). *Science, N.Y.* **213,** 1505.

KIRKWOOD, T. B. L. (1977). *Nature, Lond.* **270,** 301.

KIRKWOOD, T. B. L. (1980). *J. theor. Biol.* **82,** 363.

KIRKWOOD, T. B. L. & HOLLIDAY, R. (1975). *J. theor. Biol.* **53,** 481.

KIRKWOOD, T. B. L. & HOLLIDAY, R. (1980). *Structural Pathology of DNA and the Biology of Ageing* (L. Schoeller, ed.), p. 27. Deutsche Forschungsgemeinschaft.

LAMB, M. J. (1965). *Exp. Geront.* **1,** 181.

LAMB, M. J. & MAYNARD-SMITH, J. (1964). *Exp. Geront.* **1,** 11.

LEWIS, C. M. & HOLLIDAY, R., (1970). *Nature, Lond.* **228,** 877.

LINN, S., KAIRIS, M. & HOLLIDAY, R. (1976). *Proc. natn. Acad. Sci. U.S.A.* **73,** 2818.

MACIEIRA-COELHO, A. (1974). *Nature, Lond.* **248,** 421.

MACIEIRA-COELHO, A. (1977). *Mech. Ageing Dev.* **6,** 341.

MAYNARD-SMITH, J. (1959). *Nature, Lond.* **184,** 956.

MAYNARD-SMITH, J. (1962). *Proc. R. Soc.* B **157,** 115.

MERZ, G. S. & ROSS, J. D. (1969). *J. Cell Physiol.* **74,** 219.

MITSUI, Y. & SCHNEIDER, E. L. [cited in Schneider E. L. and Fowlkes, B. J. (1976). *Exp. Cell Res.* **98,** 298].

MURRAY, V. & HOLLIDAY, R. (1981). *J. mol. Biol.* **146,** 55.

ORGEL, L. E. (1963). *Proc. natn. Acad. Sci. U.S.A.* **49,** 517.

ORGEL, L. E. (1973). *Nature, Lond.* **243,** 441.

PROTHERO, J. & GALLANT, J. A. (1981). *Proc. natn. Acad. Sci. U.S.A.* **78,** 333.
SHALL, S. & STEIN, W. D. (1979). *J. theor. Biol.* **76,** 219.
SMITH, J. R. & HAYFLICK, L. (1974). *J. Cell Biol.* **62,** 48.
SMITH, J. R., PEREIRA-SMITH, O. M. & GOOD, P. I. (1977). *Mech. Ageing Dev.* **6,** 283.
SMITH, J. R., PEREIRA-SMITH, O. M. & SCHNEIDER, E. L. (1978). *Proc. natn. Acad. Sci. U.S.A.* **75,** 1353.
SMITH, J. R. & WHITNEY, R. G. (1980). *Science, N.Y.* **207,** 82.
SZILARD, L. (1959). *Proc. natn. Acad. Sci. U.S.A.* **45,** 30–45.
THOMPSON, K. V. A. & HOLLIDAY, R. (1978). *Exp. Cell Res.* **112,** 281.
VINCENT, R. A. & HUANG, P. C. (1976). *Exp. Cell Res.* **102,** 31.
WAJNTAL, A. & DE MARS, R. (1967). *Biochem. Genetics* **1,** 61.

28

Reprinted from Cytogenet. Cell Genet. **30**:92-107 (1981)

Cytogenetics of Werner's syndrome cultured skin fibroblasts: variegated translocation mosaicism

D. Salk,[1,2] K. Au,[1] H. Hoehn,[1,3] and G.M. Martin[1,3]

[1]Division of Genetic Pathology, [2]Department of Pediatrics, and [3]Center for Inherited Diseases, University of Washington School of Medicine, Seattle

Abstract. Skin fibroblast-like (FL) cells from patients with Werner's syndrome (adult progeria) regularly demonstrate frequent pseudodiploidy involving variable structural rearrangements that are clonal: variegated translocation mosaicism (VTM). Ninety-two percent of 1,538 metaphases from 29 independent strains derived from five patients with Werner's syndrome demonstrated this cytogenetic abnormality. In contrast, only eight (8.4%) of 95 non–Werner's syndrome FL cell cultures demonstrated VTM: seven with low-grade VTM (approximately 5% of 300 metaphases), and one with VTM affecting 90–100% of metaphases. Unlike the cytogenetic abnormalities observed in the terminal stages of normal FL cell cultures, VTM occurs throughout the entire lifespan of Werner's syndrome cultures. Ten of the identifiable break points in 1,005 banded metaphases accounted for 27% of all definable rearrangements. Baseline sister chromatid exchanges were not increased. Cocultivation of Werner's syndrome and normal strains did not induce VTM in the normal strain. The relationship between VTM and the reduced growth potential of Werner's syndrome FL cells is not yet understood, nor is the relationship between these in vitro abnormalities and the presumptive single gene defect that causes the progeroid clinical manifestations of Werner's syndrome.

Werner's syndrome is an autosomal recessive condition characterized by many features usually associated with aging (Epstein et al., 1966; Beadle et al., 1978; Goto et al., 1978; Martin, 1978; Nakao et al., 1978). Because some of these features differ quantitatively or qualitatively from those observed in the normal aging process, the syndrome has been referred to as a "caricature of aging" (Epstein et al., 1966) and as a "segmental progeroid syndrome" (Martin, 1978). The initial observation of reduced growth potential of cultured skin fibroblast-like (FL) cells from a patient with Werner's syndrome (Martin et al., 1965) has been confirmed in several laboratories, and detailed growth studies have been performed with 20 independently derived FL cell strains from three individuals with Wer-

Supported in part by NIH grants AG-00057, AG-01751, and GM-15253.

Request reprints from: Dr. Darrell Salk, Department of Pathology SM-30, University of Washington School of Medicine, Seattle, WA 98195 (USA).

ner's syndrome (SALK et al., 1981c). These studies demonstrate sharply reduced growth rates and replicative lifespans, compared with non–Werner's syndrome FL cells, but normal survival of postreplicative cells. Hybrid synkaryons have been formed between normal and Werner's syndrome FL cells with no apparent effect on the growth potential of the Werner's syndrome cells. This latter finding is potentially interesting, since Werner's syndrome is inherited as a recessive condition, and complementation of a defective enzymatic function might be expected.

In addition to the characteristically reduced growth potential, we have previously reported variable, pseudodiploid, clonal chromosome aberrations in cultured skin fibroblast-like (FL) cells from two patients with Werner's syndrome (HOEHN et al., 1975; NORWOOD et al., 1979). In the present report we confirm our initial cytogenetic observations and extend our studies to include 29 independently derived strains from five different patients. Our findings, and confirmation from two other laboratories, establish variegated translocation mosaicism in cultivated FL cells as a second characteristic feature of Werner's syndrome.

Materials and methods

Cell strains and cell culture

The Werner's syndrome cell strains used in these studies have been previously described and are summarized in table I. Patients 1 and 2, sisters of Japanese descent, were the subjects of earlier cytogenetic and cell culture studies (MARTIN et al., 1970; HOEHN et al., 1975; NORWOOD et al., 1979; SALK et al., 1981c), and patients 3, 4, and 5, unrelated Caucasian males, have been the subjects of cell culture studies (MARTIN et al., 1970; SALK et al., 1981c). Several sets of non–Werner's syndrome

cultures were examined. We studied 10 euploid and 22 constitutionally aneuploid strains from individuals ranging in age from newborn to 38 years. These strains were established and maintained in our laboratory during the same period and under comparable conditions as the strains from patients 2 and 3 (HOEHN et al., 1980). Sixty-three fibroblast-like cultures were established from various clinical specimens (biopsy, autopsy, and abortus) in the 14-month period from January, 1979, through February, 1980, using techniques and materials equivalent to those used for the Werner's syndrome strains. We also examined a selection of eight skin FL cell strains, cryogenically preserved in our laboratory, from individuals with a variety of progeroid disorders and diabetes (table II). Cultures were maintained and passaged as previously described (SALK et al., 1981c) and found to be free from mycoplasma contamination by periodic culture, ^{3}H-uracil uptake, and/or DNA fluorescence testing (KENNY, 1969, 1975; RUSSELL, 1975) (courtesy of Dr. GEORGE KENNY, Department of Pathobiology).

Cytogenetics

The principal techniques used have been previously described (HOEHN et al., 1975). Chromosome preparations were made on 1×3 inch glass slides using standard suspension or in situ (SCHMID, 1975) techniques, depending on \the number of cells available. Staining with pinocyanol (KLINGER and HAMMOND, 1971), G-banding (modification of KLINGER, 1972), or R-banding (SEHESTED, 1974) was performed, the slides were examined microscopically, and karyotypes were prepared from photographs according to the guidelines of the international cytogenetics nomenclature (ISCN, 1978).

Metaphases accepted for analysis were those in which chromosomal landmarks (as defined by ISCN, 1978) of the majority of chromosomes could be recognized. The number of metaphases analyzed from each passage varied due to the low mitotic index and/or poor chromosomal morphology obtained in some harvests. The details of the analyses of strains from patient 1 are given elsewhere (HOEHN et al., 1975). For patients 2 and 3, an average of eight R-banded karyotypes were prepared from each strain at each passage, and for patients 4 and 5, single chromosome preparations

287

Table I. Origin and cytogenetic evaluation of 29 fibroblast-like cell strains from five patients with Werner's syndrome

Patient	Sex (age[1])	Cell source[2]	Number of strains[3]	Metaphases analyzed	
				Number[4]	Abnormal (%)
1. H.McG.	F (57 yr)	A, skin	17	604[u+b]	87.3
2. M.I.	F (51 yr)	A, skin	8	740[b]	94.6
3. U.Va.	M (51 yr)	B, skin	1	165[b]	98.8
4. W.L.	M (37 yr)	B, skin	2	22[u+b]	100.0
5. R.N.	M (48 yr)	B, testes[5]	1	7[u]	100.0
Total			29	1538	92.3

[1] Age at time of tissue donation.
[2] A = autopsy; B = biopsy.
[3] Strains derived from independent tissue specimens. Patient 1: see table IV in HOEHN et al. (1975); patient 2: strains 2A-G (SALK et al., 1980a) and strain 2J (78–80 CL); patient 3: see SALK et al. (1980c); patients 4 and 5: see MARTIN et al. (1970).
[4] u = Unbanded; b = R- and/or G-banded.
[5] Tunica albuginea testis; bilateral orchiectomy performed as therapy for undifferentiated carcinoma of the prostate (MARTIN et al., 1970).

Table II. Origin of eight non–Werner's syndrome skin FL cell strains from individuals with progeroid syndromes and diabetes

Strain	Sex (age[1])	Diagnosis	Number of metaphases[2]
71-56	M (4 yr)	Progeria	26/7
70-22	M (17 yr)	Progeroid syndrome[3]	19/3
74-47	F (9 yr)	Cockayne's syndrome	31/8
73-63	M (10 yr)	Rothmund Thompson's syndrome	11/4
73-112	M (24 yr)	Progeroid syndrome[4]	31/7
72-87	M (58 yr)	Diabetes mellitus (DM)[5]	31/6
73-92	M (41 yr)	Juvenile onset DM	18/4
73-56	M (19 yr)	Juvenile onset DM	25/8
Total			192/47

[1] Age at time of tissue donation.
[2] Number counted/number karyotyped.
[3] Patient reported by MULVHILL and SMITH (1975).
[4] Patient reported by RUVALCABA et al. (1977).
[5] Alcoholic liver disease and colon cancer.

were made of each strain. The total numbers of metaphases examined from each Werner's syndrome patient are shown in table I, and for the non–Werner's progeroid patients in table II. At least 20 metaphases were counted for the 63 routine non–Werner's syndrome clinical specimens, and 30–60 metaphases were examined if there was evidence of nonconstitutional structural chromosome aberrations.

Analysis of sister chromatid exchange frequency was performed by administering 5 µg/ml bromodeoxyuridine (BrdU) in tissue culture medium for 90 h prior to harvest, staining air-dried slides with acridine orange, and examining 60 sequential metaphases by fluorescence microscopy (DUTRILLAUX, 1973).

Results

Werner's syndrome strains

Cultures from each of the Werner's syndrome patients demonstrated a high proportion of aberrant metaphases, ranging from 87 to 100% (table I). Only 118 of 1,538 total metaphases were classified as normal because they did not have demonstrable aberrations. Our experience suggests that some of these "normal" cells may have

subtle rearrangements that are difficult to appreciate. Figure 1 shows an exceptionally good quality karyotype of a metaphase with a complex rearrangement involving five chromosomes.

Table III summarizes selected strains from patients 1, 2, and 3. The majority (78%) of metaphases were pseudodiploid, and most aneuploid cells were the result of preparational artifact (overspreading) or hyperdiploidy that was associated with variable, clonal trisomies (+2; +7; +16; +18) occurring in metaphases that also displayed structural changes (strains 2A and 2E). Most Werner's syndrome strains demonstrated an increased frequency of polyploidy in the senescent phase, as do most non–Werner's syndrome cultures. There were only moderate amounts of unstable aberrations (rings, dicentrics, acentric fragments) during the proliferative phases of the cultures, although two cultures (2E and 3) demonstrated a marked increase in their terminal passages. There were only two or three chromatid exchange figures seen among 512 banded metaphases examined during the prolifer-

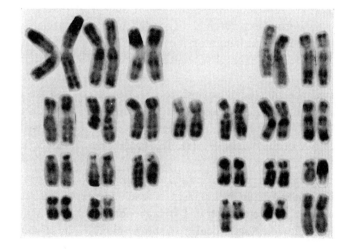

Fig. 1. An R-banded karyotype of a Werner's syndrome skin fibroblast-like cell showing pseudodiploidy with structural rearrangements involving five chromosomes: 46,XX,rea(4),t(6;11) (6pter→cen→11qter;11pter→ cen→6qter),t(7;21)(7pter→ 7q31::21pter→21qter::7q31→ 7qter),t(10;15)(10pter→ 10q21::15q14→15qter;15q14:: 10q21→10qter).

289

Table III. Frequency of numerical and structural changes in seven representative strains from Werner's syndrome patients 1, 2, and 3 analyzed by G- and/or R-banding

Strain	Number of metaphases	Chromosome counts			46,XX or 46,XY (normal)	Structural changes[1]					
							Clonal[2]				
		≤45	46	47		Unstable	a	b	c	d	Unique
1M[3]	20	8	11	1	5	1	6	3	2		0
2A	39	9	24	6	0	0	7	2	(20)	2	8
2B	49	5	42	2	1	2	(24)	23			1
2C	128	11	115	2	30	3	3	69	2		24
2D	49	5	42	2	1	5	39				9
2E	93	6	42	45	0	11	(40)	(41)			12
3	154	12	140	2	5	15	(114)	3	24		8
Total	532	56	416	60	42	37	424				62

1   A clone is defined as two or more cells containing identical structural rearrangements. A unique rearrangement was one observed in only a single cell.
2   Letters (a, b, c, d) indicate different rearrangements in each strain. Numbers in parentheses refer to a group of metaphases with one rearrangement in common, but with subsets of additional, distinct rearrangements.
3   Strain 73-18A (see table III in HOEHN et al., 1975).

ative phases of the strains from patients 2 and 3. No attempt was made to otherwise quantitate chromatid lesions, but the incidence in the Werner's syndrome cultures was not noticably different from that in non–Werner's syndrome FL cell cultures examined in our laboratory.

Structural changes observed in two or more metaphases were considered to be evidence of clonal expansion; 87% of the aberrant metaphases could be grouped into 18 cytogenetically distinct clones (table III). In five of the clones, a total of 11 additional distinct aberrations (clone variants) were observed that were classified as subclones because multiple metaphases of each were present. In certain strains it was possible to follow cytogenetically distinct clones throughout the in vitro life of one culture, as shown in table IV for the strain from patient 3, and as discussed in greater detail in SALK et al. (1981b). There are three clones in strain 3 (a, b, c) and there are four variants of one clone (a_1, a_2, a_3, a_4) that each qualifies as a subclone. There were two metaphases in this strain that had characteristic markers of more than one clone, an unusual finding observed in only one other strain (2B). Among all 29 strains that we analyzed, there were only three instances in which identical rearrangements were observed in independent cultures: (1) t(6;8)(p12;q14) in 2B and 2E, cultures initiated from widely separated autopsy sites, and (2) t(5;16)(q12;q22→23) and del(4)(p12) in 2F and 2G, strains that had been initiated

Table IV. Cytogenetic evaluation of 165 R-banded metaphases from patient 3

Passage	Chromosome counts				Unstable aberrations	Karyotype categories							
	45	46	47	Total		Normal (46,XY)	Clones 3−						Misc. unique[1]
							a_1	a_2	a_3	a_4	b	c	
2	1	5		6		6							
3	2	31		33	1	3	2			13	2		3
4	2	11		13	1		5	3		3			2
5	2	23		25	2	1	14		2	8			
6	3	29	2	34	1		27[3]			7			
7[2]													
8		7		7			5					2	
9	2	14		16			4				11	1	
10		12		12	1	1				1	9[4]	1	
11		8		8			2			2	1	2	1
14[5]				11	9								
Total	12	140	2	165	15	5	75	3	2	34	3	24	8

[1] Miscellaneous unique deletions and/or rearrangements.
[2] Harvest without analyzable metaphases.
[3] One metaphase with the characteristic markers of both clones a and b.
[4] One metaphase with the characteristic markers of both clones a and c.
[5] Screened for unstable aberrations; no adequate karyotype possible.

Clone 3-a_1: 46,XY,del(11)(q23)   Clone 3-b: 46,XY,t(15;17)(q21;p13)
  3-a_2: 46,XY,del(11)(q23),inv ins(B)   Clone 3-c: 46,?X,Y,t(2q;6q),del(2)(p16),10p−,−
  3-a_3: 46,XY,del(11)(q23),q13+   22,+mar
  3-a_4: 46,?X,Y,del(11)(q23)

from independent explants from the same tissue specimen. (The rearrangements observed in strains 2A through 2E are tabulated in SALK et al. [1981b], and those observed in strains 2F and 2G, referred to as E2 and R2, respectively, are tabulated in SALK et al. [1981a]).

Structural aberrations affected virtually every chromosome, but for each strain there was a unique pattern of involvement: strains from a single patient differed as much as strains from different patients (fig. 2). The number of aberrant chromosomes did not necessarily increase with longer time in cul-ture but was dependent on the particular rearrangements that were predominant during a given passage. Figure 3a shows an example of a culture with multiple, unique rearrangements involving many different chromosomes in passages 1–7, in which a predominant clone subsequently emerged that involved a simple deletion. Figure 3b illustrates the reverse situation: initial predominance of a clone with few rearrangements, followed by emergence of a different clone involving multiple aberrations.

Combination of the data from all five patients suggests a general correlation be-

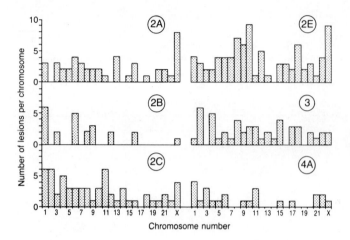

Fig. 2. Frequency distribution of structurally altered chromosomes in six skin fibroblast-like strains from three patients with Werner's syndrome. A unique pattern emerges for different strains from a single patient (2A, 2B, 2C, 2E), as well as for strains from different patients (2, 3, 4).

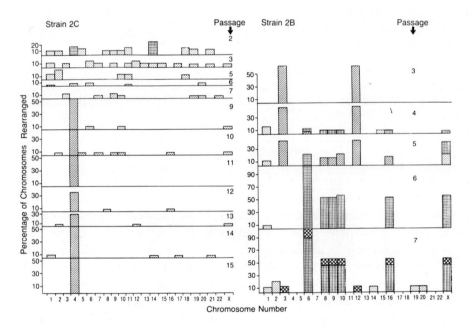

Fig. 3. Clonal succession during serial passage of two Werner's syndrome strains. The vertical scale indicates the frequency of structural rearrangements (as a percentage) for each chromosome. Strain 2C: Multiple, unique rearrangements involving many different chromosomes in early passages, followed by emergence of a predominant clone with a deletion of chromosome 4. Strain 2B: Predominant rearrangement in early passage involving two chromosomes, followed by a rearrangement in later passages involving multiple, different chromosomes.

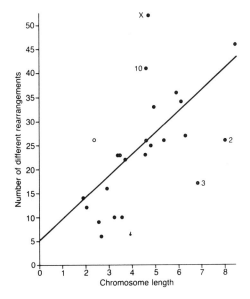

Fig. 4. Scattergram showing a correlation between chromosome length (ISCN, 1978) and the number of times the chromosome was involved in different rearrangements. Each clonal rearrangement was scored only once, and the length of the X chromosome has been adjusted to account for the number of male metaphases included. The regression line, calculated by the method of least squares, is $y = 4.19 x + 5.16$; the correlation coefficient, r, is 0.618, and the standard error of the estimate, $S_{y \cdot x}$, is 9.28. The empty circle represents the position for the X chromosome if each of the X homologs are considered separately (see discussion).

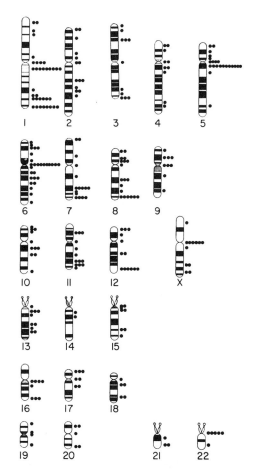

Fig. 5. Distribution of sites involved in rearrangements in 1,005 metaphases from 11 skin fibroblast-like cell strains from four patients with Werner's syndrome. Diagrammatic representation of chromosome bands is based on ISCN (1978).

tween chromosome size and the frequency of involvement in structural changes (fig. 4). A detailed analysis of 1,005 technically adequate, banded metaphases from four of the patients indicates, however, that there is a nonrandom distribution of breakpoints within certain chromosomes (fig. 5). Not all of the actual breakage sites are equally represented in this figure because many rearranged chromosomes could not be defined in detail. The X chromosome, for instance,

was involved in rearrangements to a much greater extent (fig. 4) than is apparent from the selected data in fig. 5. Nevertheless, it is clear that certain chromosomal sites have a greater frequency of involvement in structural rearrangements (table V). Four (3%) of 132 identifiable exchange sites accounted for 14% of the definable rearrangement

Table V. Summary of the frequency distribution of karyotypic locations of exchange sites in 1,005 metaphases from 11 skin fibroblast-like cell strains from four patients with Werner's syndrome (see fig. 5)

Number of events	Site of rearrangement	Total number of sites
1	See fig. 5	70
2	See fig. 5	31
3	See fig. 5	17
4	1cen, 5cen, 7q35, 16q11	4
5	7q31, 22cen	2
6	1q41, 8q24, 12q24, Xcen	4
7	None	0
8	1q44	1
9	1q12	1
10	6cen	1
11	5q12	1
Total		132

events. The 10 sites (8% of the total number) that were each involved with five or more events accounted for 27% of all definable rearrangements.

Preliminary co-cultivation experiments have been carried out to test if Werner's syndrome FL cells produce a substance that causes chromosomal aberrations or, conversely, if normal FL cells can complement the genetic defect in Werner's syndrome and correct the characteristic in vitro behavior of Werner's syndrome FL cells. We co-cultivated 5×10^5 cells from the fifth passage of Werner's syndrome strain 2B (female) with 1×10^5 cells from the ninth passage of a normal male FL cell culture (75-27). After 1 and 2 wk of co-cultivation the culture was examined cytogenetically. Forty cells were analyzed from each harvest, but no abnormalities were found in the normal male

strain, and the Werner's syndrome strain still consisted primarily of the characteristic clonal rearrangement 2B-b.

Sister chromatid exchange analysis of strain 2B at passage 6 revealed an average of 7.0 SCE per metaphase ($n = 60$). This is not different from values obtained with 3–5 μg/ml BrdU concentrations in non–Werner's syndrome. non–Bloom's syndrome strains in our laboratory.

Non–Werner's syndrome strains

Cytogenetic examinations were performed on several sets of non–Werner's syndrome strains, the first of which was 10 euploid and 22 aneuploid strains during their period of active growth. Aberrant metaphases were rare events in 30 of these strains during passages 8–16, and only two strains displayed more than 10 structurally altered metaphases at comparatively early stages of their in vitro lifespans. Both of these strains had been structurally normal prior to the 10th passage, whereas after the 15th passage each had high proportions of structurally aberrant metaphases. One of these two strains (trisomy 21 strain d in HOEHN et al., 1980) developed a pattern typical of VTM as seen in our Werner's syndrome strains: all metaphases displayed variable, stable, structural alterations, and there was a predominant clone that involved a t(11p;21q) insertional rearrangement. The second strain (euploid strain c in HOEHN et al., 1980) showed rearrangements that were not VTM: aneuploidy and unstable types of chromosome aberrations (dicentrics, rings, acentric fragments). There was a predominant pattern of dicentric chromosome formation via telomere to telomere rearrangements, a pattern unique among the many cultures maintained in our laboratory, and reminiscent of BENN'S

(1976) description of chromosomal changes in late-passage human embryonic cell cultures.

The second set of non–Werner's syndrome strains was 63 FL cell cultures established from clinical tissue specimens and maintained in our laboratory during a 14-month period (table VI). Seven (11%) of these cultures displayed a small number of pseudodiploid cells (approximately 5% of 300 total metaphases), but six of these seven cultures were established in two brief periods in October, 1979, and January, 1980. We have so far not been able to identify any reason for these clusters of cases; the cultures involved were not infected with mycoplasma. In summary, of the 95 non–Werner's syndrome FL cell cultures in these first two groups only 7 (7.4%) demonstrated low-grade VTM, one (1.1%) demonstrated VTM similar in degree

to our Werner's syndrome strains, and one culture demonstrated marked aneuploidy and structural rearrangements not defined as VTM.

In the third group of FL cell strains, from individuals with progeroid syndromes and diabetes (table II), 22 (11%) of 192 cells had ≤45 chromosomes, and the remainder were euploid. Only three cells (from three different patients) were observed that had structural aberrations among 47 cells that were karyotyped. This frequency of structural rearrangements is not unusual for normal FL-cell strains (table VI) and stands in sharp contrast to the frequency of rearrangements in Werner's syndrome strains.

Finally, selected non–Werner's syndrome cultures were also examined cytogenetically near the end of their period of active growth (entry into phase III: HAYFLICK and MOOR-

Table VI. Frequency of variable, nonconstitutional structural rearrangements in fibroblast-like cell cultures established in our laboratory during a 14-month period from clinical tissue specimens (biopsy, autopsy, and abortus)

Period	Number of cases	Number of cases with rearrangements
Jan.–Feb. 1979	4	0
Mar.–Apr. 1979	8	1
May–June 1979	14	0
July–Aug. 1979	8	0
Sept.–Oct. 1979	12	3[1]
Nov.–Dec. 1979	6	0
Jan.–Feb. 1980	11	3
Total	63	7

[1] One culture with variable rearrangements was initiated in another laboratory but was maintained in our laboratory after the first passage.

Table VII. Cytogenetic evaluation of nine representative non–Werner's syndrome strains at subterminal (∼0.25 population doublings per day, PD/day) and terminal (<0.1 PD/day) stages of their respective in vitro lifespans

Strain	Constitutional karyotype	Population doublings per day	Metaphases without rearrangements	Metaphases with rearrangements
71-95	46,XY	0.250	10	0
77-115	46,XY	0.260	20	0
71-33	46,XX	0.238	15	0
72-166	46,XY	0.095	0	1
75-160	47,XY,+18	0.095	0	6
77-144	46,XX	0.004	0	2
77-115	46,XY	<0	0	2
71-33	46,XX	<0	0	2
77-131	46,XY	<0	0	1

HEAD, 1961) and subsequently when the cultures were no longer doubling (table VII). R-banded analysis of 45 metaphases from three strains at subterminal passages revealed only structurally intact metaphases. In contrast, each of the very few available and analyzable metaphases from non-doubling cultures (<0.1 doublings/day) displayed structural chromosome changes, most of which were unstable (fragments or breakage, dicentrics, rings, and chromatid exchange figures).

Discussion

The results obtained with material from four additional patients are consistent with our initial observation in a single patient that Werner's syndrome FL cell cultures display pseudodiploidy with multiple, variable, predominantly stable chromosome aberrations that are clonal in nature. We have applied to this phenomenon the descriptive label "variegated translocation mosaicism" (VTM). In contrast, FL cell cultures derived postnatally in our laboratory from a variety of non–Werner's syndrome genotypes do not, as a rule, undergo structural chromosome changes during the peak of their proliferative activity, although cytogenetically marked clones may occasionally arise. We have observed approximately the same frequency of such aberrant clones in our non–Werner cultures as has been reported by others (7–14%) (LITTLEFIELD and MAILHES, 1975; HARNDEN et al., 1976). However, we have also noted a clustering of these events in time, which suggests that environmental factors may be involved in the development of such clones in normal cell strains.

Structural chromosome rearrangements were initially observed in FL cell cultures from Werner's syndrome patient 1 by Dr. WARREN NICHOLS and his colleagues (HOEHN et al., 1975). Two strains derived from different biopsy sites were established and grown in McCoy's medium, and an extra band was observed on either chromosome 1 or 22, or both in 88% of the 144 metaphases examined (W. NICHOLS, personal communication). Dr. M. FRACCARO has also recently observed multiple, variable structural rearrangements in skin FL cells from an unrelated patient with Werner's syndrome (personal communication). This combined experience indicates that VTM is a characteristic cytogenetic abnormality of Werner's syndrome cultured skin fibroblasts.

The bulk of our Werner's syndrome FL cell strains were derived from postmortem material from siblings (patients 1 and 2) after terminal illnesses (bronchopneumonia) requiring hospitalization, diagnostic X-ray exposure, and treatments with a variety of antibiotics. In describing patient 1 (HOEHN et al., 1975), we were careful to point out that epigenetic factors should be considered in explaining the high incidence of chromosomally aberrant cells. We here reiterate this concern as regards data of patients 1 and 2, but at the same time emphasize the apparent concordance between the autopsy (patients 1 and 2) and biopsy (patients 3, 4, and 5) results. Special significance must be attributed to the high ratio of aberrant metaphases from patient 3 (table I), since this culture was derived from a biopsy specimen and had never been frozen. VTM has been consistently observed in Werner's syndrome FL cell cultures initiated and maintained in three different laboratories, in at least four different media, with multiple lots of fetal calf serum, and in studies performed over a period of six years. Our cultures and those of Dr. NICHOLS were demonstrated to be free of mycoplasma, and the consistent observation of VTM in Werner's syndrome strains from a variety of sources, but not in many non–Werner's syndrome strains maintained in our laboratory during the last 10 years, argues against viral contamination acquired in vitro. In addition, unstable aberrations such as rings, dicentrics, breaks, and gaps known to be associated with viral and mycoplasma infection were not above the background level seen in our laboratory. It therefore seems unlikely that VTM in Werner's syndrome FL cells results solely from in vitro culture conditions.

The results of studies using peripheral blood lymphocytes are different from those using FL cells. Cytogenetic analysis of lymphocyte preparations from 19 patients with Werner's syndrome have been reported in the literature (McKUSICK, 1963; EPSTEIN et al., 1966; TAO et al., 1971; NORDENSON, 1977; BEADLE et al., 1978). All of these were described as normal (except NORDENSON's report, as discussed below), but banded preparations were not examined. In one published karyotype, there is a B-group chromosome with an unusual arm ratio (FRACCARO et al., 1962). A review of 100 metaphases from the nonbanded lymphocyte preparations from our patient 1 did not reveal obvious structural aberrations (HOEHN et al., 1975), although the original report of this patient describes a "typical Philadelphia chromosome" in 2 of 196 euploid metaphases (EPSTEIN et al., 1966). Our experience with Werner's syndrome FL cell preparations indicates that many chromosomal rearrangements would not be recognized without banding, and careful banded analysis

of peripheral blood lymphocyte cultures from patients with Werner's syndrome has yet to be performed. We have examined R-banded chromosomes from Epstein-Barr virus-transformed lymphoblastoid cell lines from two patients with Werner's syndrome (these were established and sent to us by Dr. W.T. BROWN). Both lines have a normal chromosome constitution, but occasionally demonstrate a high frequency of dicentric chromosomes in some harvests; studies of these lines are continuing.

In contrast to other reports, NORDENSON (1977) suggests that individuals with Werner's syndrome may have elevated levels of spontaneous chromosome and chromatid breakage in peripheral blood lymphocytes. However, she did not observe structural rearrangements (I. NORDENSON, personal communication) and if confirmed, these results could indicate that the chromosomal aberration that is demonstrable in FL cells as VTM is manifested differently in peripheral blood lymphocytes. Whether such different manifestations might be due to differences in tissue origin (dermal vs. hematopoietic) or to differences in culture conditions (long-term vs. short-term) is not known (SALK et al., 1981b).

The fact that we do not frequently observe chromatid aberrations, unstable rearrangements, or chromatid exchange figures in Werner's syndrome FL cells suggests that the causative event for VTM does not occur in the late-S, G_2, or M phases of the cell cycle. Sister chromatid exchanges would be expected to occur during this period, however, so it is perhaps not surprising that we did not observe increased spontaneous levels of this cytogenetic event. Consistent with our findings in FL cells, BARTRAM et al. (1976) previously reported normal levels of

297

sister chromatid exchange in phytohem-agglutinin (PHA)-stimulated lymphocytes from a patient with Werner's syndrome. However, others have suggested that there are increased levels of sister chromatid exchange, both baseline and induced by mitomycin-C, in lymphocytes from two patients with Werner's syndrome, and studies are in progress to explore this further (G. DARLINGTON and W.T. BROWN, personal communication). Once again, differences between FL cells and lymphocytes might represent tissue specificity or culture conditions.

VTM in Werner's syndrome FL cells is nonspecific at the chromosomal level in that there is no characteristic "Werner-chromosome" analogous to the Philadelphia-chromosome rearrangement associated with chronic myelogenous leukemia. Chromosomes are affected approximately in proportion to their length (fig. 4), but there is a nonrandom distribution of break points within chromosomes (fig. 5). With the exception of the 1qh heterochromatic region, the distal portion of 1q, and possibly the distal region of 7q (WELCH and LEE, 1975; HUTTNER and RUDDLE, 1976; ROWLEY, 1977; ATKIN and BAKER, 1978; MATTEI et al., 1979), these Werner's syndrome "hot spots" are different from those chromosomal regions frequently involved in other structural chromosome changes, such as fragile sites (SUTHERLAND, 1979), X-ray or chemical-induced aberrations (CASPERSSON et al., 1972; HOLMBERG and JONASSON, 1973; MEYNE et al., 1979), mitotic chiasmata in Bloom's syndrome (KORENBERG et al., 1978), spontaneous rearrangements in individuals with Fanconi's anemia (VON KOSKULL and AULA, 1973), or even sites of frequent constitutional rearrangements (YU et al., 1978; MATTEI et al., 1979).

It remains to be seen if the chromosomes in fig. 4 which appear to be underrepresented (2 and 3) or overrepresented (10 and X) are actually different from the other chromosomes or if this distribution is due to statistical variation within a small sample size. The X chromosome may be a special case, since (in females) this is the only pair of chromosomes that replicates asynchronously and that is differentially condensed during the G_1 phase of the cell cycle. If VTM results from some deficiency during replication, then each X chromosome might be at the same risk as each homologous *pair* of autosomes. If we consider each X chromosome to be a homologous pair one-half its size, the data point falls more nearly in line with a random distribution (open circle in fig. 4). If VTM results from an event during G_1, the relative overrepresentation of the X chromosome in fig. 3 might suggest that the inactivated X was at greater risk than other interphase chromosomes.

If VTM is a primary expression of the putative single gene defect in this autosomal recessive condition, then the defective gene product might be a protein involved in the maintenance of nucleoprotein integrity. Although there are reports of a retarded rate of DNA synthesis in Werner's syndrome FL cells, DNA repair functions appear to be normal (PETES et al., 1974; FUJIWARA et al., 1977; HIGASHIKAWA and FUJIWARA, 1978; NAKAO et al., 1978); thus, other components of the nucleoprotein complex may be implicated. Alternatively, VTM could represent a secondary phenomenon: accumulation of a precursor substance or metabolic by-product or excessive production of a normal protein involved in growth regulation might explain both the reduced growth potential and cytogenetic aberrations observed in Werner's syndrome FL cell cultures. Consistent with this hypothesis is the suggestion that inhibition of DNA replication occurs after fusion of Werner's syndrome FL cells with HeLa cells, in contradistinction to fusions between normal senescent cells and HeLa cells (NORWOOD et al., 1979). However, these observations have not been confirmed (TANAKA et al., 1979) and must be extended by further study. In the preliminary co-cultivation experiments reported here, there was no induction of numerical or structural chromosome changes in the normal indicator strain, nor was there any noticeable change in the pattern of VTM in the Werner's syndrome strain. Longer periods of co-cultivation will be necessary to establish if there is any effect on replicative lifespan or clonal evolution in either strain.

Variegated translocation mosaicism is a cytogenetic abnormality that appears to be characteristic of Werner's syndrome cultured skin fibroblasts: it is not observed in FL cells from other progeroid syndromes we have examined, nor does it occur in non–Werner's FL cells as they undergo in vitro senescence; the sporadic occurrence of VTM in proliferating non–Werner's syndrome cultures may be related to some as yet unidentified environmental influence. There is an increased incidence of neoplasia in Werner's syndrome, just as there is in ataxia telangiectasia, Fanconi's anemia, Bloom's syndrome, and xeroderma pigmentosum (GERMAN, 1972). However, in contrast to those other syndromes that display chromosomal instability, the cytogenetic aberration in Werner's syndrome is remarkable for its high degree of stability; the only cytogenetic manifestation is pseudodiploid, clonal rearrangements that may persist unchanged in culture for many weeks. In addition, the

aberration in Werner's syndrome VTM may not originate during G_2 or M, the phases of the cell cycle when the breakage and unstable rearrangements observed in the other "chromosome instability syndromes" occur. We are particularly intrigued that a single gene defect, such as in Werner's syndrome, can produce such a variety of progeroid clinical manifestations and at the same time be expressed cytogenetically at specific chromosomal sites. The nonrandom distribution of rearrangement sites in Werner's syndrome VTM reconfirms the idea that there is a heterogeneity of chromosome structure that has not yet been fully appreciated.

Acknowledgements

We thank BARBARA INGLIN for the cytogenetic analysis of clinical tissue specimens, SONJA MELJEW for assistance with illustrations, and VIRGINIA WEJAK, GINNY WALTERS, and CHRISTY COTA for manuscript preparation.

References

ATKIN, N.B. and BAKER, M.D.: Abnormal chromosomes and number 1 heterochromatin variants revealed in C-banded preparations from 13 bladder carcinomas. Cytobios *18:* 101–109 (1978).

BARTRAM, C.R.; KOSKE-WESTPHAL, T., and PASSARGE, E.: Chromatid exchanges in ataxia telangiectasia, Bloom syndrome, Werner syndrome, and xeroderma pigmentosum. Ann. hum. Genet. *40:* 79–86 (1976).

BEADLE, G.F.; MACKAY, I.R.; WHITTINGHAM, S.; TAGGERT, G.; HARRIS, A.W., and HARRISON, L.C.: Werner's syndrome, a model of premature aging? J. Med. *9:* 377–403 (1978).

BENN, P.A.: Specific chromosome aberrations in senescent fibroblast cell lines derived from human embryos. Am. J. hum. Genet. *28:* 465–473 (1976).

CASPERSSON, T.; HAGLUND, U.; LINDELL, B., and

ZECH, L.: Radiation-induced non-random chromosome breakage. Expl Cell Res. 75: 541–543 (1972).

DUTRILLAUX, B.; LAURENT, C.; CONTURIER, J., and LEJEUNE, J.: Coloration par l'acridine orange de chromosomes préalablement traités par le 5-bromodeoxyuridine (BrdU). C.R. Acad. Sci. 276: 3179–3181 (1973).

EPSTEIN, D.J.; MARTIN, G.M.; SCHULTZ, A.L., and MOTULSKY, A.G.: Werner's syndrome: a review of its symptomatology, natural history, pathologic features, genetics and relationship to the natural aging process. Medicine 45: 177–221 (1966).

FRACCARO, M.; BOTT, M.G., and CALVERT, H.T.: Chromosomes in Werner's syndrome. Lancet i: 536 (1962).

FUJIWARA, Y.; HIGASHIKAWA, T., and TATSUMI, M.: A retarded rate of DNA replication and normal level of DNA repair in Werner's syndrome fibroblasts in culture. J. Cell Physiol. 92: 365–374 (1977).

GERMAN, J.: Genes which increase chromosomal instability in somatic cells and predispose to cancer. In A.G. STEINBERG and A.G. BEARN, eds.: Progress in medical genetics, Vol. 8, pp. 62–101 (Grune and Stratton, 1972).

GOTO, M.; HORIUCHI, Y.; TANIMOTO, K.; ISHII, T., and NAKASHIMA, H.: Werner's syndrome: analysis of 15 cases with a review of the Japanese literature. J. Am. Geriat. Soc. 26(8): 341–347 (1978).

HARNDEN, D.G.; BENN, P.A.; OXFORD, J.M.; TAYLOR, A.M.R., and WEBB, T.P.: Cytogenetically marked clones in human fibroblasts cultured from normal subjects. Somat. Cell Genet. 2: 55–62 (1976).

HAYFLICK, L. and MOORHEAD, P.S.: The serial cultivation of human diploid cell strains. Expl Cell Res. 25: 585–621 (1961).

HIGASHIKAWA, T. and FUJIWARA, Y.: Normal level of unscheduled DNA synthesis in Werner's syndrome fibroblasts in culture. Expl Cell Res. 113: 438–442 (1978).

HOEHN, H.; BRYANT, E.M.; AU, K.; NORWOOD, T.H.; BOMAN, H., and MARTIN, G.M.: Variegated translocation mosaicism in human skin fibroblast cultures. Cytogenet. Cell Genet. 15: 282–298 (1975).

HOEHN, H.; SIMPSON, M.; BRYANT, E.M.; RABINO-

VITCH, P.S.; SALK, D., and MARTIN, G.M.: Effects of chromosome constitution on growth and longevity of human skin fibroblast cultures. Am. J. med. Genet. 7: 141–154 (1980).

HOLMBERG, M. and JONASSON, J.: Preferential location of X-ray induced chromosome breakage in the R-bands of human chromosomes. Hereditas 74: 57–68 (1973).

HUTTNER, K.M. and RUDDLE, F.H.: Study of mitomycin C-induced chromosomal exchange. Chromosoma 56: 1–13 (1976).

ISCN (1978): An international system for human cytogenetic nomenclature. Cytogenet. Cell Genet. 21: 309–404 (1978).

KENNY, G.E.: Serological comparison of ten glycolytic mycoplasma species. J. Bact. 98: 1044–1055 (1969).

KENNY, G.E.: Rapid-detection of mycoplasmata and non-culturable agents in animal cell cultures by uracil incorporation. In D. SCHLESSINGER, ed.: Microbiology 1975, pp. 32–36 (Am. Soc. for Microbiol., Washington, D.C. 1975).

KLINGER, H.P.: Rapid processing of primary embryonic tissues for chromosome banding pattern analysis. Cytogenetics 11: 424–435 (1972).

KLINGER, H.P. and HAMMOND, D.O.: Rapid chromosome and sex chromatin staining with pinacyanol. Stain Technol. 46: 43–47 (1971).

KORENBERG, J.R.; THERMAN, E., and DENNISTON, C.: Hot spots and functional organization of human chromosomes. Hum. Genet. 43: 13–22 (1978).

LITTLEFIELD, L.G. and MAILHES, J.B.: Observations of de novo clones of cytogenetically aberrant cells in primary fibroblast cell strains from phenotypically normal women. Am. J. hum. Genet. 27: 190–197 (1975).

MARTIN, G.M.; GARTLER, S.M.; EPSTEIN, C.J., and MOTULSKY, A.G.: Diminished life span of cultured cells in Werner's syndrome. Fed. Proc. 24: 678 (1965).

MARTIN, G.M.: Genetic syndromes in man with potential relevance to the pathobiology of aging. Birth Defects: Original Article Series, Vol. 14, No. 1, pp. 5–39 (The National Foundation, New York 1978).

MARTIN, G.M.; SPRAGUE, C.A., and EPSTEIN, C.J.: Replicative life-span of cultivated human cells: effects of donor's age, tissue and genotype. Lab. Invest. 23: 86–91 (1970).

MATTEI, M.G.; AYME, S.; MATTEI, F.J.; AURRAN, Y., and GIRAUD, F.: Distribution of spontaneous chromosome breaks in man. Cytogenet. Cell Genet. *23:* 95–102 (1979).

McKUSICK, V.A.: Medical genetics 1962 (paragraph 792). J. Chron. Dis. *16:* 599–603 (1963).

MEYNE, J.; LOCKHARD, L.H., and ARRIGHI, F.E.: Nonrandom distribution of chromosomal aberrations induced by three chemicals. Mutat. Res. *63:* 201–209 (1979).

MULVIHILL, J.J. and SMITH, D.W.: Another disorder with prenatal shortness of stature and premature aging. Birth Defects: Original Article Series, Vol. 11, No. 2, pp. 368–371 (The National Foundation, New York 1975).

NAKAO, Y.; KISHIHARA, M.; YOSHIMI, H.; INOUE, Y.; TANAKA, K.; SAKAMOTO, N.; MATSUKURA, S.; IMURA, H.; ICHIHASHI, M., and FUJIWARA, Y.: Werner's syndrome: in vivo and in vitro characteristics as a model of aging. Am. J. Med. *65:* 919–932 (1978).

NORDENSON, I.: Chromosome breaks in Werner's syndrome and their prevention in vitro by radical-scavenging enzymes. Hereditas *87:* 151–154 (1977).

NORWOOD, T.H.; HOEHN, H.; SALK, D., and MARTIN, G.M.: Cellular aging in Werner's syndrome: A unique phenotype? J. invest. Dermatol. *73:* 92–96 (1979).

PETES, T.D.; HARBER, R.A.; TARRANT, G.M., and HOLLIDAY, R.: Altered rate of DNA replication in aging human fibroblast cultures. Nature, Lond. *251:* 434–436 (1974).

ROWLEY, J.E.: Nonrandom chromosomal changes in human malignant cells. *In* R.S. SPARKES, D.E. COMMINGS, and C.F. FOX, eds.: Molecular human cytogenetics (Academic Press, New York 1977).

RUSSELL, W.C.; NEWMAN, C., and WILLIAMSON, D.H.: A simple cytochemical technique for demonstration of DNA in cells infected with mycoplasma and viruses. Nature, Lond. *253:* 461–462 (1975).

RUVALCABA, R.H.A.; CHURESIGAEW, S.; MYHRE, S.A.; KELLEY, V.C., and MARTIN, G.M.: Children who age rapidly—progeroid syndromes: case report of a new variant. Clin. Ped. *16:* 248–252 (1977).

SALK, D.; AU, K.; HOEHN, H., and MARTIN, G.M.: Effects of radical scavenging enzymes and reduced oxygen exposure on growth and chromosome abnormalities of Werner syndrome cultured skin fibroblasts. Hum. Genet. (1981*a*, in press).

SALK, D.; AU, K.; HOEHN, H.; STENCHEVER, M.R., and MARTIN, G.M.: Evidence of clonal attenuation, clonal succession, and clonal expansion in mass cultures of aging Werner's syndrome skin fibroblasts. Cytogenet. Cell Genet. *30:* 108–117 (1981*b*).

SALK, D.; BRYANT, E.; AU, K.; HOEHN, H., and MARTIN, G.M.: Werner syndrome cultured skin fibroblasts: reduced growth potential, isolation of hybrid synkaryons, and prolonged survival of post-replicative cells. Ms. submitted for publication (1981*c*).

SCHMID, W.: A technique for in situ karyotyping of primary amniotic fluid cell cultures. Humangenetik *30:* 325–330 (1975).

SEHESTED, J.: A simple method of R banding of human chromosomes, showing a pH-dependent connection between R and G bands. Humangenetik *21:* 55–58 (1974).

SUTHERLAND, G.R.: Heritable fragile sites on human chromosomes. II. Distribution, phenotypic effects, and cytogenetics. Am. J. hum. Genet. *31:* 136–148 (1979).

TANAKA, K.; NAKAZAWA, T.; OKADA, Y., and KUMAHARA, Y.: Increase in DNA synthesis in Werner's syndrome cells by hybridization with normal human diploid and HeLa cells. Expl Cell Res. *123:* 261–267 (1979).

TAO, L.C.; STECKER, E., and GARDNER, H.A.: Werner's syndrome and acute myeloid leukemia. Can. Med. Assoc. J. *105:* 952–954 (1971).

VON KOSKULL, H. and AULA, P.: Nonrandom distribution of chromosome breaks in Fanconi's anemia. Cytogenet. Cell Genet. *12:* 423–434 (1973).

WELCH, J.P. and LEE, C.L.Y.: Non-random occurrence of 7–14 translocations in human lymphocyte cultures. Nature, Lond. *255:* 241–242 (1975).

YU, C.W.; BORGAONKAR, D.S., and BOLLING, D.R.: Break points in human chromosomes. Hum. Hered. *28:* 210–255 (1978).

ERRATA

Page 96, line four from the bottom in the right-hand column should read:
"t(6;8)(p12;q24) . . . ."

AUTHORS' NOTE

Following are updated references for two of the papers in the reference list that were in press at the time of original publication.

Salk, D., Au, K., Hoehn, H., and Martin, G. M., 1981a, Effects of Radical Scavenging Enzymes and Reduced Oxygen Exposure on Growth and Chromosome Abnormalities of Werner Syndrome Cultured Skin Fibroblasts, *Hum. Genet.* **57:**269–275.
Salk, D., Bryant, E., Au, K. Hoehn, H., and Martin, G. M., 1981c, Systematic Growth Studies, Cocultivation, and Cell Hybridization Studies of Werner Syndrome Cultured Skin Fibroblasts, *Hum. Genet.* **58:**310–316.

29

Reprinted from *Mech. Age. Devel.* **19**:21–26 (1982)

HUMAN LYMPHOCYTES RESISTANT TO 6-THIOGUANINE INCREASE WITH AGE

A. A. MORLEY*, S. COX and R. HOLLIDAY

Genetics Division, National Institute for Medical Research, The Ridgeway, Mill Hill, London NW7 1AA (Great Britain) and Department of Haematology, Flinders Medical Centre, Bedford Park, South Australia 5042 (Australia)

(Received October 12, 1981)

SUMMARY

Using an autoradiographic technique we determined the number of circulating lymphocytes that were resistant to 6-thioguanine and which were presumably mutants at the hypoxanthine—guanine phosphoribosyl transferase locus. The number in normal individuals was found to increase exponentially with age. The data suggest a relationship between mutagenesis and ageing, perhaps by way of a decline with age in the fidelity of DNA replication or repair.

A number of theories have suggested that mutations might be important in the genesis of ageing. The simplest of these suggest that mutations accumulate as the result of random hits on single-hit or multiple-hit targets in chromosomes [1]. A more complex theory linking mutations and ageing is the error theory [2—4], which has stimulated a great deal of experimental work since its formulation. The error theory states that, owing to feedback between DNA, RNA and protein, any error produced in one class of macromolecules may lead to production of errors in the other classes; a circular propagation of errors may ensue and ultimately a rising level of errors may lead to death of the cell. With regard to mutations, the error theory predicts an exponential or quasi exponential increase in the number of mutations with increasing age.

The available data suggest that mutations do increase with age, although the quantitative nature of the increase is poorly defined. Measurement of mutations in Neurospora and Paramecium suggested an increase with ageing [3, 5]. Continuously maintained human fibroblasts have been suggested as a model for studying ageing but different workers using this system disagree on the presence and/or nature of any increase in mutation frequency during culture [6—8]. Direct studies on the fidelity of DNA polymerase extracted from cultured fibroblasts indicate that the number of errors introduced into syn-

---

*To whom correspondence should be addressed, at Flinders Medical Centre.

thetic DNA templates increases exponentially with successive passages [9, 10]. Chromosomal aberrations in liver cells of mice have been reported to increase, probably linearly, with age [11], and in human lymphocytes loss of the X chromosome in females and the Y chromosome in males increases in frequency, probably non-linearly, with age [12]. However, to our knowledge there are no data for higher animals on the relationship between somatic mutations and age.

Strauss and Albertini [13] recently described a system for directly measuring the *in vivo* mutation frequency of human lymphocytes at the hypoxanthine–guanine phosphoribosyl transferase (HGPRT) locus. Using a modification of this system we have studied individuals of various ages in order to search for any relationship between mutation and ageing.

MATERIALS AND METHODS

Individuals studied

Thirty-seven individuals ranging in age from 9 to 95 years were studied; 25 were aged less than 65 years. All were healthy and none were on any medication except for several female NIMR workers on the contraceptive pill. Twelve worked in various capacities in the laboratories at the National Institute for Medical Research. Twelve of the 37 individuals were aged over 65 and ten of them lived in an old people's home because they were unable to care for themselves for social reasons. All were ambulant and none of them had evident major disease, particularly heart disease or cancer, but several were taking either salicylates, an anti-inflammatory drug, or a diuretic.

Mutation assay

The assay is based on the principle that lymphocytes with a mutation at the HGPRT locus are resistant to 6-thioguanine (6TG) and, when grown in the presence of 6TG, are able to enter DNA synthesis, incorporate [3H] thymidine and become identifiable using autoradiography. Full details of our technique are described elsewhere [14] but, in brief, lymphocytes were separated from venous blood using Ficoll–Hypague and 10^6 cells/ml were cultured for 40 hours in modified McCoy's 5A medium containing 8% foetal calf serum, phytohaemagglutinin, and with or without 40 μg/ml 6TG. Preliminary studies in seven individuals 20–30 years of age showed that under our conditions the plateau of the 6TG dose–response curve developed at a concentration of approximately 10 μg/ml [14]. In order to check that the response to thioguanine was the same in old as in young individuals, full dose–response curves were performed in four of the 12 individuals aged over 65 and additional measurements at 20 μg/ml thioguanine were performed in a further three individuals.

After 24 hours of culture 2 μCi of [3H] thymidine (Amersham; specific activity 18–25 Ci/mmol) were added to each ml of culture. The cultures were terminated by adding Nonidet to lyse the cytoplasm of the cells and formalin to fix the nuclei. After fixation for 1 hour the nuclei were collected by filtration onto polycarbonate membranes

(Nucleopore) and washed three times with distilled water. For cultures without thio-guanine, nuclei from 1/100–1/1000 of a 1-ml culture of 1×10^6 cells were collected on one membrane; for cultures with thioguanine, nuclei from five tubes each containing 1×10^6 cells were collected on a single membrane. The membranes were affixed to slides using 2% gelatine, allowed to dry and dipped in nuclear emulsion (Ilford K2). They were exposed at 4 °C for 2–4 weeks and then developed. Labelled nuclei in cultures without and with 6TG were scored by counting the whole membrane and the frequency of 6TG-resistant cells calculated as a direct ratio.

RESULTS

The number of labelled nuclei enumerated on a membrane ranged from 59 to 3100 for cultures without 6TG and from 1 to 149 for cultures with 6TG. The logarithmic mean for cultures with 6TG was 8.9 and the range of logarithmic mean ±2 S.D. was 1.8–44.2. Studies using various concentrations of 6TG showed no difference in 6TG sensitivity be-tween lymphocytes from young and old individuals and the plateau of the dose–response curve appeared to begin at approximately 10 μg/ml in both groups.

The frequency of 6TG-resistant lymphocytes increased slowly with age up to about 70 years and then increased sharply. Within each decade the standard deviation was large compared with the mean and also increased with age. When frequency was plotted on a logarithmic scale against age (Fig. 1), the variation with age appeared reasonably uniform, suggesting that it was multiplicative rather than additive. Least-squares regression analysis of the logarithmically transformed data showed a highly significant linear increase with age ($n = 37, r = 0.67, p < 0.001$). The best-fitting straight line corresponded to the expo-nential relationship for untransformed data

frequency of TG-resistant cells = $0.31 \times (1.03)^{age}$

and this was equivalent to a 3% increase in mutation frequency per year.

Although the data showed no evidence of significant deviation from an exponential relationship, it is certainly possible, in view of the relatively wide scatter, that they could correspond to a quasiexponential rather than a truly exponential relationship. However, as indicated in Fig. 1, the exponential relationship did provide a better fit than the best-fitting linear relationship. The linear relationship (which appears curved owing to the logarithmic scale used) was obtained by regression analysis weighted to take account of the change of standard deviation with age.

The logarithmically transformed data showed a significant linear relationship both for males ($n = 16, r = 0.70, p < 0.01$) and for females ($n = 21, r = 0.66, p < 0.001$). There was a slight indication that laboratory subjects may have shown higher frequencies than non-laboratory subjects of similar age but this difference was not statistically significant. Seven of the laboratory subjects were above the regression line and five were below. When the laboratory subjects were excluded, the regression analysis on the logarithmically trans-formed data gave virtually. the same highly significant regression line ($n = 25, r = 0.71$, $p < 0.001$).

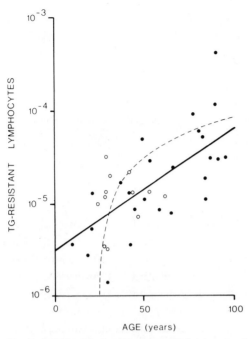

Fig. 1. Relationship between age and frequency of TG-resistant lymphocytes in laboratory (○) and non-laboratory (○) individuals. The regression lines of best fit for logarithmically transformed data (——) and untransformed data (-----) are shown. The latter line appears curved owing to the logarithmic scale used for the figure.

DISCUSSION

The evidence that 6TG-resistant cells are mutants is quite strong for systems where such cells are identified by their ability to proliferate and form colonies [15, 16]. The evidence that the cells detected by the present system are mutants is less compelling and is based on the presence of a plateau in the dose—response curve for thioguanine and on the observations that individuals treated with mutagens as part of cancer therapy show an increased frequency of resistant cells [13] and that X-irradiation of lymphocytes *in vitro* increases the frequency of 6TG-resistant cells [17, 18].

Although mutation is the most likely reason for the 6TG-resistant phenotype, other possibilities do exist. In particular, loss of the active X chromosome would give the same phenotype provided that lymphocytes without an active X were still able to synthesize DNA. Loss of an X chromosome in lymphocytes from females does increase with age, although it is not known whether the inactive or the active X is lost [12]. We are investigating this possibility further, but feel that it probably does not explain our findings, particularly since the 6TG-resistant phenotype also showed an age-related increase in frequency in males.

Strauss and Albertini [13] found a value of approximately 10^{-4} for the frequency of 6TG-resistant lymphocytes and were unable to detect an effect of age. By contrast we found a frequency of approximately 10^{-5}, *i.e.* at least an order of magnitude lower, and observed a highly significant age-related effect. A figure of 10^{-5} approaches the value observed for human fibroblasts cultured *in vitro* [16] although, since lymphocytes divide infrequently *in vivo* whereas fibroblasts divide rapidly *in vitro*, a close similarity between the values for lymphocytes and fibroblasts should not necessarily be expected. The discrepancy between our data and those of Strauss and Albertini [13] can probably be accounted for by the differences in the technique used. Our method was in substance the same as theirs but did contain some modifications. The duration of our assay was 40 hours whereas theirs was 30 hours. Cells were collected by filtration, which increased by a factor of 3–4 the number of scorable 6TG-resistant labelled cells and thus improved the counting statistics and the precision of the measurement. Repeated dose–response curves for 6TG showed that the plateau of the curve commenced at a concentration of approximately 10 μg/ml [14]. Thus the standard concentration of 40 μg/ml used for our assay provided very stringent conditions for selection of resistant cells and it is very unlikely that any wild-type cells were detected; in fact, it is quite possible that some mis-sense mutants, having decreased but not absent HGPRT activity, would not have been detected. The number and nature of thioguanine-resistant cells are critically dependent on the stringency of the selection procedure [14, 19], and we suggest that a difference in the stringency between our culture conditions and those of Strauss and Albertini could account for much of the discrepancy between our results and theirs. Evans and Vijayalaxmi [18] also used the autoradiographic technique but substituted azaguanine for thioguanine. They found a frequency of TG-resistant cells of 10^{-4}–10^{-3} although their data did suggest an age effect. Azaguanine is only a weak selective agent for HGPRT − minus cells and any difference between our results and theirs is again likely to be explicable by a difference in stringency of selection [14].

If the increase in TG-resistant cells with age does in fact represent an increase in mutations at the HGPRT locus and if events at the HGPRT locus are representative of events occurring elsewhere in the X chromosome or throughout the genome, then our observations have implications for the biology of ageing. An observed association between increased mutation frequency and increasing age suggests but does not prove a causal connection between mutation and ageing. However, if such a connection exists, then the exponential or quasiexponential relationship observed discriminates between several of the proposed theories linking mutations and ageing. The data argue against a process of random hits on a single-hit target since this would give a linear relationship. Random hits on a multiple-hit target would give a progressively rising relationship, but the data on the HGPRT gene, at least insofar as a single dose of X-irradiation [20, 21], ultraviolet radiation [22], or chemical mutagen [23] is concerned, point to a linear relationship and single-hit kinetics. The data are consistent with the error hypothesis since an accumulation of defects in the machinery for replicating and/or repairing DNA would result in an exponential increase in mutations with time. They are also consistent with any ageing process, including a programmed process, that leads to a progressive decline in the accuracy or activity of DNA repair mechanisms.

ACKNOWLEDGEMENTS

We thank Mrs. G. M. Tarrant for technical assistance and Mr. T. B. L. Kirkwood for help with the statistical analysis.

REFERENCES

1 L. Szilard, On the nature of the ageing process. *Proc. Natl. Acad. Sci. U.S.A., 45* (1959) 30–45.
2 L. E. Orgel, The maintenance of accuracy of protein synthesis and its relevance to aging. *Proc. Natl. Acad. Sci. U.S.A., 49* (1963) 517–521.
3 C. M. Lewis and R. Holliday, Mistranslation and ageing in Neurospora. *Nature, 228* (1970) 877–880.
4 L. E. Orgel, Ageing of clones of mammalian cells. *Nature, 243* (1973) 441–445.
5 F. M. Sonneborn and M. Schneller, Age-induced mutations in paramecium. In B. L. Strehler (ed.), *Biology of Aging,* Waverly Press, Baltimore, 1960, p. 286.
6 S. J. Fulder and R. Holliday, A rapid rise in cell variants during the senescence of populations of human fibroblasts. *Cells, 6* (1975) 67–73.
7 S. J. Fulder, Evidence for an increase in presumed somatic mutation during the ageing of human cells in culture. *Mech. Ageing Dev., 10* (1979) 101–115.
8 R. S. Gupta, Senescence of cultured human diploid fibroblasts. Are mutations responsible? *J. Cell. Physiol., 103* (1980) 209–216.
9 S. Linn, M. Kairis and R. Holliday, Decreased fidelity of DNA polymerase activity isolated from aging human fibroblasts. *Proc. Natl. Acad. Sci. U.S.A., 73* (1976) 2818–2822.
10 V. Murray and R. Holliday, Increased error frequency of DNA polymerases from senescent human fibroblasts. *J. Mol. Biol., 146* (1981) 55–76.
11 C. Crowley and H. J. Curtis, The development of somatic mutations in mice with age. *Proc. Natl. Acad. Sci. U.S.A., 49* (1963) 626–628.
12 P. A. Jacobs, M. Brunton, W. M. Court-Brown, R. Doll and H. Goldstein, Changes of human chromosome count distributions with age: evidence for a sex difference. *Nature, 197* (1963) 1080–1081.
13 G. H. Strauss and R. J. Albertini, Enumeration of 6-thioguanine-resistant peripheral blood lymphocytes in man as a potential test for somatic cell mutations arising *in vivo. Mut. Res., 61* (1979) 353–379.
14 A. Morley, S. Cox, D. Wigmore, R. Seshadri and J. L. Dempsey, Enumeration of thioguanine-resistant lymphocytes using autoradiography. *Mut. Res.,* in press.
15 R. De Mars, Resistance of cultured human fibroblasts and other cells to purine and pyrimidine analogues in relation to mutagenesis detection. *Mut. Res., 24* (1974) 335–364.
16 L. Siminovitch, On the nature of heritable variation in cultured somatic cells. *Cell, 7* (1976) 1–11.
17 A. Morley and J. Dempsey, unpublished observations.
18 H. J. Evans and Vijayalaxmi, Inducation of 8-azaguanine resistance and sister chromatid exchange in human lymphocytes exposed to mitomycin C and X-rays *in vitro. Nature, 292* (1981) 601–605.
19 M. Fox and M. Radacic, Adaptational origin of some purine-analogue resistant phenotypes in cultured mammalian cells. *Mut. Res., 49* (1978) 275–296.
20 A. A. Van Zeeland and J. W. Simons, Linear dose–response relationships after prolonged expression times in V-79 Chinese hamster cells. *Mut. Res., 35* (1976) 129–138.
21 R. Cox and W. K. Masson, X-ray induced mutation to 6-thioguanine resistance in cultured human diploid fibroblasts. *Mut. Res., 37* (1976) 125–136.
22 A. W. Hsie, P. A. Brimer, T. J. Mitchell and D. G. Gossiee, The dose–response relationship for ultraviolet-light-induced mutations at the hypoxanthine–guanine phosphoribosyl transferase locus in Chinese hamster ovary cells. *Somat. Cell Genet., 1* (1975) 383–389.
23 J. P. O'Neill and A. W. Hsie, Chemical mutagenesis of mammalian cells can be quantified. *Nature, 269* (1977) 815–817.

30

Reprinted from *Science* **220**:1055–1057 (1983)

DNA METHYLATION DECREASES IN AGEING BUT NOT IN IMMORTAL CELLS

Vincent L. Wilson and Peter A. Jones
*University of Southern California School of Medicine, and
Children's Hospital of Los Angeles*

Normal diploid cells do not replicate indefinitely in vitro unless they are altered by some heritable transformation (*1*) to produce immortal cell lines that may or may not be tumorigenic in nude mice (*2*). The mechanism underlying the spontaneous conversion of cells from senescence to immortality is unknown, but alterations in gene expression are probably involved in the process.

The methylation of DNA influences vertebrate gene expression and methylation patterns are tissue-specific (*3*). The patterns are inherited with a high but not absolute fidelity (*4, 5*), and alterations in them could cause changes in gene expression that might be involved in aging in culture. We have therefore determined the levels of cytosine methylation in senescing cells from mice, hamsters, and humans and compared them to cell lines derived from the same species.

The growth kinetics of normal populations of diploid cells from these three species were followed (Fig. 1). The number of population doublings (computed from the increases in cell number without correction for plating efficiency) that the cells could undergo before senescence was lowest for mouse embryo (C3HME) cells that showed a maximum of four to eight population doublings. Spontaneous regrowth of the culture occurred after 12 passages, which is characteristic of the well-known abilities of mouse cells to form immortal lines (*2*). Syrian hamster embryo (SHE) cells were maintained for 25 doublings before decreased plating efficiency and cell death caused a decrease in the computed growth curve. Human fibroblasts can replicate for 100 population doublings (*1*). Human fetal lung fibroblasts (IMR 90) were followed for more than 40 doublings without change in growth rate (Fig. 1).

The levels of DNA methylation in these cultures were measured as a function of population doubling (Fig. 2A). The percentage of cytosine residues modified to 5-methylcytosine in freshly explanted mouse fibroblasts was 3.8 percent; the ratio decreased dramatically with increased time in culture. This decrease was in marked contrast to the situation with two immortal mouse cell lines in which the level of methylation was stable, or increased slightly, over many passages representing a large number of cell divisions (Fig. 2B).

Hamster fibroblasts, which underwent approximately 25 population doublings (Fig. 1), showed a lower rate of loss of 5-methylcytosine as a function of time in culture and human fibroblasts showed the lowest rate of decrease (Fig. 2A). Thus, the cell populations that rapidly senesced in culture showed the greatest rate of loss of methyl groups from their DNA's.

The 5-methylcytosine content of other cell types available in the laboratory was also determined to assess the generality of these findings (Table 1). The rapidly dividing population of C3HC1V cells isolated in the experiment shown in Fig. 1 had a cytosine modification level of 2.95 percent at passage 5 which increased to 3.82 percent at passage 7 and stabilized at 3.65 percent between passages 11 and 23. This value was similar to those of the established mouse lines (Fig. 2B), but we do not know whether the increase was the result of de novo methylation or the selection of cells with high methylation levels from the aging population. A hamster cell line $A(T_1)C13$, which was derived by oncogenic transformation of hamster fibroblasts in vitro (6), showed a decreased 5-methylcytosine content relative to freshly explanted fibroblasts (Fig. 2A).

The decrease in DNA methylation with passage of the IMR 90 cells (Fig. 2A) was also found with a strain of human foreskin fibroblasts (T-1). Since normal human immortal lines are not available, we measured the 5-methylcytosine content of a human tumor cell line HT1080 (7), which showed a much decreased 5-methylcytosine content relative to diploid fibroblasts.

Our results clearly demonstrate that decreases in the levels of DNA methylation occur with time in culture in diploid mouse cells, in contrast to the situation with immortal cell lines. The increased levels of cytosine methylation of the newly derived mouse line relative to the senescing cultures was interesting and suggested that cells may require a certain level of methylation to survive in culture. However, tumorigenic hamster and human lines had lowered 5-methylcytosine levels compared to their diploid progenitors so that alterations in methyl-

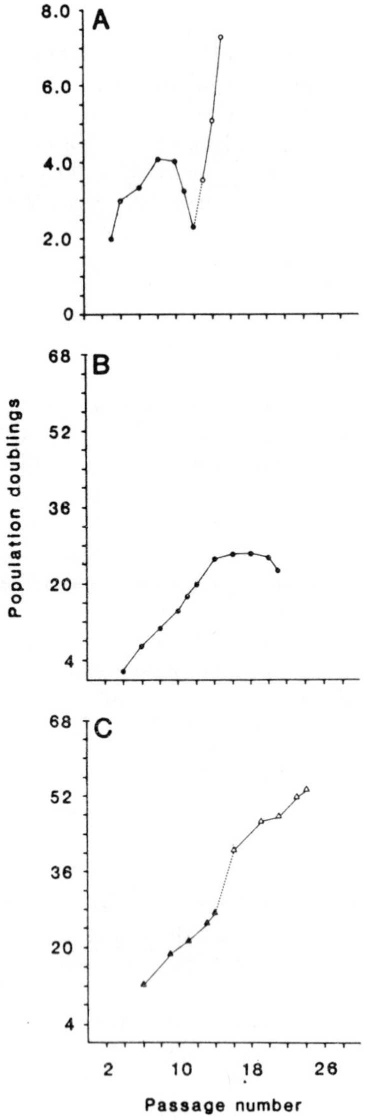

Fig. 1. Representative population growth curves. (A) Mouse. Freshly prepared C3HME cells (●) from C3H/HEN(MTV−) mice were passaged by trypsinization every 3 to 4 days; 3×10^5 cells were seeded per T-75 flask at each passage. The surviving 2.5×10^4 cells at passage 12 were seeded into a 60-mm dish and fed twice a week for 2 weeks, at which time a colony (C3HC1V) developed. C3HC1V cells (○) were then passed as above. (B) Hamster. Golden Syrian hamster embryo (SHE) cells were prepared from 10-day embryos; 2×10^5 cells were seeded per T-25 flask at each passage. (C) Human. Human fetal lung fibroblasts (IMR 90) were obtained from the Institute for Medical Research (Camden, New Jersey) at passages 6 (▲) and 16 (△) with known division histories. The cells were passaged by trypsinization every 7 days and 3×10^5 cells were seeded per T-75 flask.

Table 1. Methylcytosine content of cells.

Cell line	Description	Passage number	5-Methyl-cytosine* (%)	Number of samples
Mouse cells				
C3H C1V	Immortal	5	2.95 ± 0.19	3
		7	3.82 ± 0.31	3
		11	3.66	2
		23	3.64	2
Hamster cells				
A(T₁) C13	Oncogenic and immortal	Unknown	2.25 ± 0.24	5
Human cells				
Γ-1	Normal diploid	8	3.00	2
		22	2.88	2
		36	2.37	2
HT1080	Oncogenic and immortal	44	2.32	2

*The percentage of 5-methylcytosine (5mC) was calculated as follows: [(5mC)/(5mC + cytosine] × 100. Values represent the average of at least two samples, and standard deviations are provided where appropriate.

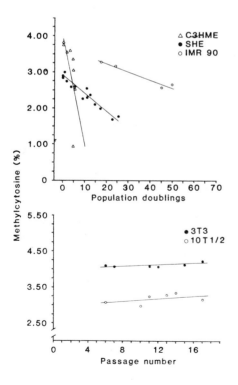

Fig. 2. The rate of DNA methylation in aging cells. Nonconfluent cells were grown for 24 hours in the presence of 30 to 100 μCi of [6-³H]uridine in 2 ml of medium. The cells were lysed, and the percentage of 5-methylcytosine was determined by two-dimensional thin-layer chromatography after hydrolysis of the DNA to bases (*15*). (A) Amounts of 5-methylcytosine were determined for normal diploid cells after various population doublings. Curves were drawn by linear regression analysis of the data for mouse (△) (slope = −0.136 ± 0.153), hamster (●) (slope = −0.043 ± 0.005), and human (○) (slope = −0.021 ± 0.004) cells (the ± is standard deviation of slope). The mouse and hamster data represent the compilation of data from two separate experiments and each point represents the average of two samples. (B) The amount of 5-methylcytosine in the DNA of BALB/3T3A31 Cl 1-13 cells (●) and C3H/10T½ Cl 8 cells (○) was measured at different passages. Each passage represents a minimum of three to ten doublings. Linear regression analysis provided positive slopes of 0.010 ± 0.003 for BALB/3T3 cells and 0.016 ± 0.014 for C3H/10T½ Cl 8 cells. Each point represents the average of at least two samples.

311

ation patterns may be more important than the final 5-methylcytosine content reached.

The present data do not establish whether the loss of methyl groups was related to the number of divisions or simply to time in culture. However, methylation patterns most probably change after the synthesis of DNA (3). Division may therefore lead to infidelity of maintenance methylation, as has been demonstrated at selective sites for other dividing cells in culture. Significant variability in methylation at specific sites in γ-globin gene sequences (8) and in X-chromosome sequences (9) of cloned and serially passaged human fibroblasts have been reported. The work of Wigler *et al.* (4) suggested that mammalian cells were capable of maintaining methylation of foreign DNA with 95 percent fidelity per site per generation.

The observed loss of 5-methylcytosine in aging cells might be partially accounted for by selective decreases of highly repetitive DNA sequences (10) known to contain the highest percentage of the compound (3). However, the loss of selective DNA sequences cannot completely account for the rapid decrease or for the total reduction in 5-methylcytosine content observed in aging C3HME cells or in SHE cells (Fig. 2A). Highly repetitive DNA constitutes only a small portion of mammalian DNA and contains less than 20 percent of the total 5-methylcytosine (11). The decreases observed here, which are more extensive than this, may therefore reflect the combined effects of removal of repetitive DNA sequences and infidelity in the maintenance of 5-methylcytosine patterns.

Some data are available on possible changes in DNA methylation during aging in vivo. Romanov and Vanyushin (12) reported decreases in 5-methylcytosine in the DNA of aging cattle and salmon, but Ehrlich *et al.* (13) failed to detect such decreases in tissues obtained from young and old humans. Ehrlich *et al.* (13) also found no significant changes

in the 5-methylcytosine content of human fibroblasts allowed to undergo approximately 10 to 15 doublings. However, from our data, this may not have represented a sufficiently long time span for measurable differences to be apparent in human cells.

The decreases in 5-methylcytosine that we have observed in cultured cells from three different species may account for aberrant gene expression in aging cultures (14). However, it remains to be seen whether this represents a response to the culture environment or has any significance in vivo.

References and Notes

1. L. Hayflick, *Exp. Cell Res.* **37**, 614 (1965); S. Goldstein, J. W. Littlefield, J. S. Soeldner, *Proc. Natl. Acad. Sci. U.S.A.* **64**, 155 (1969); J. R. Smith and L. Hayflick, *J. Cell Biol.* **62**, 48 (1974).
2. C. A. Reznikoff, D. W. Brankow, C. Heidelberger, *Cancer Res.* **33**, 3231 (1973); T. Kakunaga, *Int. J. Cancer* **12**, 463 (1973); K.-Y. Lo and T. Kakunaga, *Cancer Res.* **42**, 2644 (1982).
3. A. Razin and A. D. Riggs, *Science* **210**, 604 (1980); M. Ehrlich and R. Y.-H. Wang, *ibid.* **212**, 1350 (1981).
4. M. Wigler, D. Levy, M. Perucho, *Cell* **24**, 33 (1981).
5. R. M. Harland, *Proc. Natl. Acad. Sci. U.S.A.* **79**, 2323 (1982); Y. Pollack, R. Stein, A. Razin, H. Cedar, *Proc. Natl. Acad. Sci. U.S.A.* **77**, 6463 (1980).
6. W. F. Benedict, N. Rucker, C. Mark, R. E. Kouri, *J. Natl. Cancer Inst.* **54**, 157 (1975).
7. S. Rasheed, W. A. Nelson-Rees, E. M. Toth, P. Arnstein, M. B. Gardner, *Cancer* **33**, 1027 (1975).
8. R. J. Shmookler Reis and S. Goldstein, *Proc. Natl. Acad. Sci. U.S.A.* **79**, 3949 (1982).
9. S. F. Wolf and B. R. Migeon, *Nature (London)* **295**, 667 (1982).
10. R. J. Shmookler Reis and S. Goldstein, *Cell* **21**, 739 (1980).
11. If one assumes that the highly repetitive DNA amounts to 10 percent of genomic DNA and contains twice the level of 5-methylcytosine per cytosine residues when compared to the rest of the DNA [see (3); T. L. Kautiainen and P. A. Jones, unpublished results], the level of methylation in the remaining 90 percent of the DNA is thus 3.45, 2.73, and 2.95 percent 5-methylcytosine in C3HME, SHE, and IMR 90 cells, respectively.
12. G. A. Romanov and B. F. Vanyushin, *Biochim. Biophys. Acta* **653**, 204 (1981).
13. M. Ehrlich, M. A. Gama-Sosa, L-H. Huang, R. M. Midgett, K. C. Lo, R. A. McCune, C. Gehrke, *Nucl. Acids Res.* **10**, 2709 (1982).
14. Low levels of ectopic protein production by normal fibroblasts in culture have been reported by S. W. Rosen, B. D. Weintraub, and S. A. Aaronson [*J. Clin. Metab.* **50**, 834 (1980)].
15. V. L. Wilson and P. A. Jones, *Cell* **32**, 239 (1983).
16. We thank P. Billings for the C3H/HEN (MTV−) mice. Supported by grant R01-GM30892 from the National Institute of General Medical Sciences and Fellowship Training grant No. 1-T32-CA09320 from the National Cancer Institute.

EPILOGUE

In many quarters the science of gerontology is not regarded as a serious discipline. One reason for this is the diffuseness of the field, since the label "aging" has been applied to situations as diverse as the senescence of leaves at the end of summer, the brief adulthood of the mayfly, the rapid physical disintegration of the salmon after spawning, or the slow deterioration of dormant seeds. It is clear that these and many other biological situations that fall within the province of gerontology are quite separate phenomena; each provides a perfectly proper field of study in itself, but whether each contributes to an understanding of the aging of higher organisms such as ourselves is very doubtful. Another criticism of aging research is that much of the experimental work is strictly empirical: age-related changes in cells, tissues, organs, or organisms are documented at length in considerable literature, but for the most part this information cannot at present be easily interpreted. Therefore, research in the field tends not to be intellectually stimulating and does not attract experimental scientists, especially young scientists, in the same way that, for example, molecular biology or genetics can. A further problem is that some gerontologists, who should know better, and others on the fringe of the field often confidently assert that the primary aim of the research is to increase the maximum human lifespan and that this will be achieved. As I will explain, research in gerontology is important for quite different reasons.

Although stimulation and impetus to research can arise from formulation and testing of specific hypotheses using well-defined systems, these can be followed by frustration if the experimental procedures available turn out to be inadequate to make the appropriate test. This has certainly been the case with regard to the general error theory of aging. It has been clear for some years what information had to be acquired, but the experiments have not given clear-cut answers. Even so, it is commonly believed that the error theory has been disproved by experimental observations. This is far from the truth; until the accuracy of protein, RNA, and DNA synthesis is measured at

313

different times throughout a lifespan, no final judgment can be made about the validity of the theory. The development of many new powerful molecular techniques may make these measurements possible in the future. In contrast to the error theory, the program theory of aging is strongly supported by many gerontologists. The challenge to them is to formulate in molecular terms what the nature of the program might be, so that experimental testing of this theory can be carried out. To date, the concept of a clock or program is so vague that it is simply not possible to devise, let alone carry out, such tests.

It has often been stated that all biological processes can only be fully understood in terms of their evolutionary origin. To many people, the evolution of aging presents few problems, since they intuitively feel that it is either an inevitable result of the evolution of a highly complicated organism or just part and parcel of the whole evolutionary process. If evolution depends on sexual reproduction, the argument runs, then parents must sooner or later make way for their progeny, and aging is the obvious way of accomplishing this. The superficiality of this argument was first exposed by Medawar (1952) and Williams (1957), who pointed out that aging is rarely seen in animals in their natural environment, because most individuals die from disease, predation, or starvation. No reasonable explanation for the selective advantage of a program for aging has been put forward. In contrast, the error theory does provide a strong basis for the evolution of aging—namely, that a balance has to be struck between the investment of resources to maintain and to reproduce the adult organism (Kirkwood, 1977; Kirkwood and Holliday, 1979). Continual maintenance is expensive and wasteful if organisms in their natural environment have a high mortality from environmental hazards. Under these circumstances, the organisms' fitness will be increased by using more of the available metabolic resources for efficient reproduction during early adulthood and less to prevent aging. The theory makes specific predictions: first, that germ-line cells, which are potentially immortal, have a greater investment in proofreading, repair, and maintenance than somatic cells; second, that the somatic cells from long-lived species preserve by these means the integrity of their macromolecules more successfully, and for longer periods, than do those from short-lived species. The effective testing of these predictions lies in the future, although there is already evidence that long-lived species are more efficient in DNA repair (Hart and Setlow, 1974).

The significance of studies of cellular aging have recently been clearly highlighted by the discovery of important steps in the transformation of normal cells to malignancy. The immortality of certain cancer cells has been known for many years, but the finding that primary diploid cells can be immortalized by specific oncogenes is a major advance. An understanding of the genetic and biochemical control of mortalization and immortalization now seems within reach, and this must surely throw important light not only on the genesis of cancer cells but also on the process of aging of normal cells.

The aging of whole organisms like humans is baffling in its complexity. Almost all tissues and organ systems seem to follow their own age-related changes, and this has led to the view that there must be multiple causes of aging. If this is the case, however, how can a simple genetic defect such as the acquisition of a single, small chromosome in Down's syndrome or a single recessive mutation as in Werner's syndrome (Martin, 1978) have such profound consequences on the overall aging process? If, indeed, there is one or a small number of underlying causes of aging, then it is of great medical importance to discover what these may be. The reason is that vast amounts of laboratory resources are devoted to the study of age-related disease in humans. In general, each disease is studied in isolation. For example, research on heart and vascular disease is not only devoted to the development of more effective treatments but also to an understanding of the origin of the age-related changes that are seen, and the same is true for research on many other specific diseases that become prevalent in older people. Therefore, the overall strategy for biomedical research should include studies of the process of aging at the fundamental level (Holliday, 1984). This provides ample justification for future research. The ultimate aim would not be to stretch out the lifespan but to alleviate many of the debilitating, age-related diseases.

REFERENCES

Hart, R. W., and R. B. Setlow, 1974, Correlation between Deoxyribonucleic Acid Excision-Repair and Lifespan in a Number of Mammalian Species, *Proc. Natl. Acad. Sci. (U.S.A.)* **71:**2169–2173.

Holliday, R., 1984, The Ageing Process is a Key Problem in Biomedical Research, *The Lancet* **II:**1386–1387.

Kirkwood, T. B. L., 1977, Evolution of Ageing, *Nature* **270:**301–304.

Kirkwood, T. B. L., and R. Holliday, 1979, The Evolution of Ageing and Longevity, *Proc. R. Soc.* **205B:**531–546.

Kirkwood, T. B. L., and R. Holliday, 1986, Ageing as a Consequence of Natural Selection, in *The Biology of Human Ageing,* K. J. Collins and A. H. Bittles, eds., Cambridge University Press, New York.

Martin, G. M., 1978, Genetic Syndromes in Man with Potential Relevance to the Pathology of Ageing, in *Genetic Effects on Ageing,* D. Bergsma and D. E. Harrison, eds., Liss, New York, pp. 5–39.

Medawar, P. B., 1952, *An Unsolved Problem in Biology,* H. K. Lewis, London. (Reprinted in the *Uniqueness of the Individual,* 1957, Methuen, London.)

Williams, G. C., 1957, Pleiotropy, Natural Selection and the Evolution of Senescence, *Evolution* **11:**398–411.

AUTHOR CITATION INDEX

321

Yamada, M., 149, 217
Yamamoto, K., 83
Yang, W. -K., 195, 217, 295
Yanofskyu, C., 66
Yatscoff, R. W., 164
Ycas, M., 56, 74
Yerganian, G., 248
Yoshida, A., 169, 255
Yoshida, M. C., 105, 106, 113, 248
Yoshimi, H., 301

Young, I. T., 129, 146
Yu, C. W., 301
Yuan, P. M., 181, 190

Zaun, M., 181
Zavala, C., 83, 125, 150
Zech, L., 300
Zeelon, P., 31, 74, 226
Zimmerman, J. A., 164

SUBJECT INDEX

331

About the Editor

ROBIN HOLLIDAY is a 1955 graduate of the University of Cambridge where he also earned the Ph.D. degree in 1959. From 1958–1965 he was a staff member of the John Innes Institute, spending a year during that period in the Department of Genetics at the University of Washington, Seattle. His research was in microbial genetics, particularly the mechanism of genetic recombination.

In 1965 Dr. Holliday joined the Microbiology Division at the National Institute for Medical Research (NIMR), where he first developed an interest in cellular aging. Since 1970 he has been head of the Genetics Division at NIMR. Although Dr. Holliday's most recent work concentrated on the relationship between DNA methylation and gene expression in mammalian cells, he and his colleagues have continued to study microbial and molecular genetics, as well as aging and transformation.

Dr. Holliday is the author of numerous articles on genetic recombination, DNA repair, cellular aging, gene expression, and carcinogenesis. He also wrote *The Science of Human Progress* (1981, Oxford University Press) and coedited with J. Maynard Smith, *The Evolution of Adaptation by Natural Selection* (1979 Royal Society, London). Dr. Holliday was elected a Fellow of the Royal Society in 1976.